Learning by Doing

Edited by Peter Heering and Roland Wittje

T0139527

Learning by Doing

Experiments and Instruments in the History of Science Teaching

Edited by Peter Heering and Roland Wittje

Cover illustration:
The physics lecture room of the Anna-Gymnasium Augsburg, Germany, in the 1910s.
By courtesy of Karl-August Keil and the late Inge Keil

Bibliografische Information der Deutschen Nationalbibliothek:
Die Deutsche Nationalbibliothek verzeichnet diese Publikation in der Deutschen Nationalbibliografie; detaillierte bibliografische Daten sind im Internet über <http://dnb.d-nb.de> abrufbar.

Dieses Werk einschließlich aller seiner Teile ist urheberrechtlich geschützt.
Jede Verwertung außerhalb der engen Grenzen des Urheberrechtsgesetzes ist unzulässig und strafbar.
© 2011 Franz Steiner Verlag, Stuttgart
Druck: Laupp & Göbel, Nehren
Gedruckt auf säurefreiem, alterungsbeständigem Papier.
Printed in Germany.
ISBN 978-3-515-09842-7

CONTENTS

Introduction:
Neglected Uses of Instruments and Experiments in Science Education
Peter Heering and Roland Wittje 7

Tools for Investigation, Tools for Instruction: Potential Transformations of Instruments in the Transfer from Research to Teaching
Peter Heering 15

The Audience is Listening: Reading Writing about Learning by Doing
Pete Langman 31

The Role of Chemistry Textbooks and Teaching Institutions in France at the Beginning of the Nineteenth Century in the Controversy about Berthollet's Chemical Affinities
Pere Grapí 55

Instruments of Science and Citizenship: Science Education for Dutch Orphans during the Late Eighteenth Century
Lissa L. Roberts 71

The Scientific Culture in Eighteenth to Nineteenth Century Greek Speaking Communities: Experiments and Textbooks
Constantine Skordoulis, Gianna Katsiampoura, and Efthymios Nicolaidis 97

The Magic Lantern for Scientific Enlightenment and Entertainment
Willem Hackmann 113

The Establishment and Development of Physics and Chemistry Collections in Nineteenth-Century Spanish Secondary Education (1845–1861)
Mar Cuenca-Lorente and Josep Simon 141

The Death and Life of the Plant Specimen
Dawn Sanders 159

Learning in the Laboratory: The Introduction of "Practical" Science Teaching in Ontario's High Schools in the 1880s
Michelle Hoffman 177

Changing Images of the Inclined Plane, 1880–1920: A Case Study of a Revolution in American Science Education
Steven Turner 207

Reforming American Physics Pedagogy in the 1880s: Introducing 'Learning by Doing' via Student Laboratory Exercises
Richard L. Kremer 243

The Evolution of Teaching Instruments and their Use between 1800 and 1930
Paolo Brenni 281

"Simplex sigillum veri": Robert Pohl and Demonstration Experiments in Physics after the Great War
Roland Wittje 317

The Role of Instruments in Teaching Science: A Machian View
Hayo Siemsen 349

Notes on Contributors 361

INTRODUCTION: NEGLECTED USES OF INSTRUMENTS AND EXPERIMENTS IN SCIENCE EDUCATION

Peter Heering and Roland Wittje

The teaching of science and its best practice, whether in schools, colleges or universities, has been a major concern for a long time and an issue of debate for scientists, educators, and the public at large. In the public opinion, much is seen to be at stake if societies do not manage to educate citizens about their scientific and technological culture, as well as to train enough technicians, engineers and scientists sufficiently well in order to develop it further. No matter how diverse the positions, the debating parties share the conviction that personal experience in the form of experiments and observations either carried out by students, or performed as demonstrations by teachers, are essential to the pedagogy of science. Yet, this is not a timeless observation. Consequently, one should ask how this conviction has been put into practice.

This book focuses on the history of experiments and instruments within the history of teaching science. It brings together two themes which so far, according to our knowledge, have never been brought together in one edited volume. One theme is the history of experimental practice, scientific instrumentation and material cultures of science; the other is the history of teaching science. The history of experiments, scientific instruments and scientific practice came into focus in the history of science with the 'pragmatic turn' in the 1980s and 1990s.[1] More recently, material cultures have come into prominence with the rise of cultural studies of science on one hand, and the (re)discovery of material heritage and historical collections at universities and other scientific institutions on the other.[2] The origins and the development of science teaching, in contrast, is one of the topics that is just beginning to receive more attention within the larger scholarship on the history of science. Education still needs to be moved from the periphery of

1 Two important monographs in this respect—to which we refer in the title of our introduction—are Franklin (1986) and Gooding et al. (1989). Other influential examples from this period are Shapin and Schaffer (1985) and Galison (1987).

2 An example for the interest in scientific objects inspired by cultural studies is Daston (2004). In the last decade a number of national as well as international initiatives, networks, and conferences have been organised to protect scientific heritage at universities, and to mobilise it for teaching, research and for public exhibitions. See Lourenço (2005) about historical collections at universities in Europe.

science to its centre (Kayser, 2005).[3] This deficit in its recognition seems to be surprising as the importance of training in the formation of scientists has been stressed for some time by epistemologists as well as historians. Moreover, the standardisation of scientific education can be seen as a crucial step in the establishment of experimental science: "Laboratory training and access to scientific instruments are essential for the professionalisation of science in the middle of the last century" (Rabkin, 1992, p. 62).

Addressing this under-represented field in the history of science was one of our main motivations to organise the symposium 'Learning by Doing: Instruments and Experiments in the History of Science Teaching' that took place at the University of Regensburg on 4 and 5 April 2009. The other was to bring together historians of science and science teachers who are interested in the historical development of science teaching. As historians we firmly believe that, in order to understand the present state of science teaching as well as to envision potential future developments, one has to have some understanding of its history. This volume should therefore be of interest not only to historians of science and of education, but also to science educators. Most contributions have been presented at the Regensburg symposium and subsequently developed further. However, some others have been added.

Our motivation to initiate the symposium and to edit this volume was to address questions such as the following ones:

– How did demonstration experiments and experimental lectures emerge, and how have they further developed? When, why and how were student experiments and organised laboratory courses introduced in science teaching in different disciplines and different institutions? How have these changed?
– What are the relations between scientific research and teaching? How have scientific instruments in research and teaching developed? To what extent have the same instruments been used in both activities, and to what extent have separate teaching or research technologies evolved? How were research experiments transformed into teaching experiments?
– What were the differences between various types of educational institutions, like public schools, vocational schools and universities? How did regional, political and cultural differences matter? What were the differences in the various scientific subjects?

The papers presented in this volume offer some answers to these questions; at the same time, they raise some new questions. They present an overview of and an insight into the role of experiments in science teaching in different time periods

3 This observation has been made by Kayser (2005). Important exceptions are the studies by Kathryn M. Olesko (Olesko 1989, 1991, and 1993) and the Special Issue of the journal Science & Education: Textbooks in the Scientific *Periphery,* edited by José Ramón Bertomeu-sánchez, Antonio García-Belmar, Anders Lundgren and Manolis Patiniotis (2006). Studies in the history of science education include Foellmer (2007), Lind (1992), and Wickihalter (1984).

from the Enlightenment to the twentieth century as well as geographical contexts in Europe and North America. As an integral part of science as a social activity, the teaching of science is as old as science itself. Both are indisputably interlinked. So why would we want to drive a distinction between the two by singling out the teaching aspect? The answer is quite simple. Science teaching is still looked at rather as context or the boundary of scientific research in universities, both by historians as well as by scientists. Parallel to the *neglect of experiment* (Franklin, 1986), one could use the phrase of a *neglect of science teaching* in the history of science. As Ian Hacking (1983) has established for experimenting, science teaching also has a life of its own. In the first paper of this volume, Peter Heering discusses common aspects in the relation between research and teaching apparatus, denouncing the idea of a simple transfer from the former to the latter. Instead of reducing the history of the teaching of science to the context of research history, we have to treat it in its own right and examine the complex inter-relations between both. Heering identifies four strategies: simplification, downscaling, stabilisation and iconisation, which seem to be important in the transformation of many research instruments into teaching devices.

Conducting experiments in public lectures had a prominent place in the European Enlightenment. In many instances and contexts, public demonstrations or university lectures rather than written papers was the way of presenting new scientific discoveries.[4] But in the eighteenth century, public lectures and entertaining demonstrations were also presented to a public that was not part of the scientific societies, but rather a lay audience, as Pete Langman exemplifies for the case of Newtonian experimental philosophy in England. Many devices from the early remaining scientific collections were designed, along with popular texts, to illustrate scientific principles rather than to generate new experimental knowledge and competences.

It is generally assumed that textbooks of science do not take part in scientific controversies but convey generally accepted and therefore noncontroversial scientific knowledge to students and other newcomers in a field.[5] Pere Grapi challenges this view by arguing for an important role of chemistry textbooks and teaching institutions in the controversy about chemical affinities in early nineteenth century France.

School and university collections for science teaching started to emerge in the eighteenth century. Lissa Robert's paper on Dutch orphanages exemplifies the underlying peculiar political and economic, as well as pedagogical discourses. From her study, it becomes evident that it was not always the universities that had

4 One of the best known examples in this respect is the work of Joseph Black on latent heat. Black communicated his findings only in his lectures which appeared in print only after his death. However, while it was common knowledge throughout Europe then, it was also known that these findings were related to Black (McKie and Heathcote, 1975).

5 There are of course some exceptions; most notably see Frercks (2006) on the controversy on Lavoisier's chemistry.

the best equipped teaching collections. Likewise, her study illustrates that science teaching is not just influenced by scientific ideas and conceptions, but also by political beliefs that shape the educational systems. This aspect becomes even more explicit in the study by Constantine Skordoulis, Gianna Katsiampoura and Efthymios Nicolaidis of the role of experiments in science teaching in Greek speaking communities in the late eighteenth and early nineteenth centuries. Their paper demonstrates the political meaning of introducing natural philosophy experiments in the Greek science teaching. Adopting the experimental method was a way of connecting to the European Enlightenment. Advocates of the new Newtonian ideas were, however, confronted with representatives of a neo-Aristotelianism who claimed a specific Byzantine tradition and rejected the use of instruments.

An important function of many teaching devices is to show and enlarge visual representations of scientific phenomena. Willem Hackmann presents a very nice and almost timeless instrumental example in his discussion of the projections that could be made with a magic latern. This instrument shows several aspects that are relevant for the entire volume: it was used for entertaining and educational purposes at the same time (and one could ask when education was separated from entertainment). It is—according to Hackmann—the 'precursor of PowerPoint', a teaching technology which is very much state of the art in the early twenty-first century. And it is in some sense the ambivalence between an instrument that is considered to be part of natural philosophy on the one hand, and magic and illusions on the other. Whilst the latter association is fairly unique, the role of wonder and entertainment in science teaching and the related instruments is a topic that deserves further attention.

The nineteenth century has been described as a period of differentiation, institutionalisation and professionalisation both for science and for education. We are cautious to avoid a separation at the onset between science teaching at schools on one hand, and institutions of higher education, like universities, on the other. We rather wish to debate what connected teaching at different institutions and where the differences lay. Universities became a space for both, teaching and research, and to some extent both fields were intertwined as lecturing was one common way to publish new findings and theories. At the same time, universities were also the place to train teachers for secondary schools. Countries that view themselves in the periphery of the scientific and educational reforms are often found to copy the examples from the cultures which they perceive as centres and potential reference. Mar Cuenca-Lorente and Josep Simon show that the development of physics and chemistry collections in mid-nineteenth century Spanish secondary schools largely followed examples of France which were implemented through and combined with local developments.

Collections of an entirely different type are the focus of Dawn Sanders' contribution: turning from physics to biology, she presents a history of the changing use of plant specimens in botanical teaching in the nineteenth and twentieth centuries. In particular, the Royal Botanical Gardens in Kew had a strong agenda in teaching, offering teachers plant specimens for educational purposes.

These were not limited to biological courses but also geography was adressed whose teachers used such specimens to show the 'products of empire'. But her analysis is not limited to British institutions—she also draws explicitly on the example of high school biology in New York City between 1900 and 1925.

In the standard historiography of science, teaching laboratories started to emerge at universities from the middle of the nineteenth century, following the well-known example of Justus Liebig's chemical laboratory at the University of Giessen (see, for example, Holmes, 1989). Until then, it had mainly been the lecturer or demonstrator who performed the teaching experiments. With the model of laboratory teaching, students were supposed to perform their own experiments. This development has been studied by Michelle Hoffman for science teaching in Canadian high schools, and by Steven Turner, using the example of the inclined plane, for American science education. Chemistry and physics should be taught experimentally, and botany practically. The objectives of science education were both moral and intellectual. But the steady implementation of the new experimental method of teaching met, as Hoffman argues, bureaucratic hurdles of school administration and the challenges of everyday pedagogical practice. The amount and importance of student experiments constantly increased during the second half of the nineteenth century. Until the turn to the twentieth century, student experiments were also introduced in school teaching, as the importance of science teaching and experiments increased in the curriculum.

The growing school market was not always that interesting to the large and established makers and therefore accommodated local makers also, as the study of Steven Turner shows. Some of the teaching instruments were designed as didactical devices, whereas others were modelled after research instruments, but these were often not as precise as the research instruments and altered to the requirements of teaching. Richard Kremer challenges the claim made by other historians of science that American physics teaching in the late nineteenth century was largely influenced by, if not being a copy of, German teaching practices, methods and instruments.[6]

Paolo Brenni describes the evolution of teaching instruments in physics and gives an overview of the instrument market through the nineteenth and early twentieth century. The last decades of the nineteenth century saw a drastic demand for scientific education that led to an extreme growth in the market for didactic instruments.[7] After the Franco-Prussian War, German instrument companies such as Max Kohl and E. Leybold's Nachfolger took the leading position

6 It is remarkable that, at the beginning of the twentieth century, three US institutions served as a role model for technical education in Spain, namely, "the Worcester Polytechnic Institute, …; Sibley College, at Cornell University, …; and the Stevens Institute of Technology" (Roca-Rossell et al., 2006, p. 151).

7 One could ask whether this is the result of an increasing number of pupils at schools and students at universities, or whether this can also be taken as an indication of a growing importance of students' experiments in science education.

from the British and French makers to satisfy the growing mass market. For school teachers, the professionalisation of science left less and less space to carry out research and be an active part of the scientific community.[8] With the increasing transformation of research experimentation from table-top arrangements to complex, highly specialised laboratories and black-boxed instruments, teaching technology has distanced itself further from research technology. After the World War I, the character of demonstration experiments in experimental physics lectures changed. Roland Wittje presents us Robert Wichard Pohl as the most important renovator of lecture demonstrations in Germany during the inter-war period. After World War II, however, the importance of demonstrations in science teaching has decreased drastically.

Hayo Siemsen brings us to the present day science education, even though he is not advocating the 'standard model of science education'. In his paper, he develops a very particular perspective on the role of scientific instruments in educational processes, a perspective that is strongly inspired by the phenomenological positivism of the Austrian philosopher and physicist Ernst Mach. This paper clearly shows that the role and status of scientific instruments in science education even nowadays is not completely settled, but can be controversial. In this respect, Siemsen's paper forms an implicit demonstration of the necessity to have also a historical perspective on the use of instruments in science education.

Naturally, the papers represent only a partial, episodic and incomplete picture of this history of experiments and instruments in teaching science. Therefore, this collection is more to be understood as being a starting point in this field instead of a closure. We would have liked to include papers from other geographical regions in order to add some more facets to our picture, as well as case studies covering the post-war period. Case studies from teaching physics dominate, giving only little space to biology and chemistry teaching, and ignoring disciplines such as geology and medicine. We nevertheless think that this volume fulfils an important function in raising crucial issues concerning the role of experiments and instruments in the history of teaching science, and its relation to the history of science at large.

The title page shows the physics lecture room of the Anna-Gymnasium Augsburg; it is likely that the picture was taken in the 1910s.[9] It was made available to us by the late Inge Keil, whose generosity was only exceeded by her scholarship. We are grateful for financial support for this publication from the Regensburg University Foundation Hans Vielberth and the EWE Stiftung. We would also like to thank An Rettig for her editorial assistance.

8 There are of course exceptions such as Elster and Geitel who, being school teachers, managed to carry out substantial research in the early twentieth century (see http://www.elster-geitel.de/, accessed on 19 September 2010).

9 We owe this information to Karl-August Keil, who also granted us permission to reproduce the image.

REFERENCES

Bertomeu-Sánchez, J. R., A. García-Belmar, A. Lundgren and M. Patiniotis (eds). 2006. 'Textbooks in the Scientific Periphery', *Science & Education*, 15 (Special Issue), pp. 657–880.

Daston, L. (ed). 2004. *Things that Talk: Object Lessons from Art and Science*. New York: Zone Books.

Franklin, A. 1986. *The Neglect of Experiment*. Cambridge: Cambridge University Press.

Frercks, J. 2006. 'Die Lehre an der Universität Jena als Beitrag zur deutschen Debatte um Lavoisiers Chemie', *Gesnerus*, 63(3–4), pp. 209–39.

Foellmer, R. 2007. *Naturlehre: die Entwicklung des Physikunterrichts an den Volksschulen der Rheinprovinz (1815–1968)*. Köln: Böhlau.

Galison, P. 1987. *How Experiments End*. Chicago and London: Chicago University Press.

Gooding, D., T. Pinch and S. Schaffer (eds). 1989. *The Uses of Experiment*. Cambridge: Cambridge University Press.

Hacking, I. 1983. *Representing and Intervening: Introductory Topics in the Philosophy of Natural Sciences*. Cambridge: Cambridge University Press.

Holmes, F. L. 1989. 'The Complementarity of Teaching and Research in Liebig's Laboratory', *Osiris*, 5, pp. 121–64.

Kayser, D. (ed). 2005. *Pedagogy and the Practice of Science: Historical and Contemporary Perspectives*. Cambridge, Mass: MIT Press.

Lind, G. 1992. *Physik im Lehrbuch 1700 - 1850. Zur Geschichte der Physik und ihrer Didaktik in Deutschland*. Berlin: Springer.

Lourenço, M. C. 2005. *Between Two Worlds: The Distinct Nature and Contemporary Significance of University Museums and Collections in Europe*. PhD thesis, Conservatoire Nationale des Arts et Métiers, Paris.

McKie, D. and N. H. d. V. Heathcote. 1975. *The Discovery of Specific and Latent Heats*. New York: Arno Press.

Olesko, K. M. 1989. 'Physics Instruction in Secondary Schools before 1859', *Osiris* 5, pp. 92–118.

Olesko, K. M. 1991. *Physics as a Calling: Discipline and Practice in the Koenigsberg Seminar for Physics*. Ithaca, NY: Cornell University Press.

Olesko, K. M. 1993. 'Tacit Knowledge and School Formation', *Osiris* 8, pp. 16–29.

Rabkin, Y. 1992. 'Rediscovering the Instrument: Research, Industry and Education', in R. Bud and S. E. Cozzens (eds), *Invisible Connections: Instruments, Institutions and Science*. Bellingham, WA: SPIE Optical Engineering Press, pp. 57–72.

Roca-Rosell, A., G. Lusa-Monforte, F. Barca-Salom and C. Puig-Pla. 2006. 'Industrial Engineering in Spain in the First Half of the Twentieth Century: From Renewal to Crisis', *History of Technology*, 27, pp. 147–61.

Shapin, S. and S. Schaffer. 1985. *Leviathan and the Air-Pump*. Princeton: Princeton University Press.

Wickihalter, R. 1984. *Zur Geschichte des physikalischen Unterrichts: unter besonderer Berücksichtigung von Reformbestrebungen*. Thun: H. Deutsch.

TOOLS FOR INVESTIGATION, TOOLS FOR INSTRUCTION: POTENTIAL TRANSFORMATIONS OF INSTRUMENTS IN THE TRANSFER FROM RESEARCH TO TEACHING

Peter Heering

INTRODUCTION

Recent studies on scientific instruments have revealed that instruments can have a variety of purposes within the practice of science: they can serve to produce or investigate phenomena; check other instruments; determine material or natural constants; analyse aspects of technical artefacts; etc. However, instruments can also have very different purposes that are often overlooked: they can be used in science education to familiarise novices with the practices in a field or to train them to work with a particular device. On a more general level, they can produce data that shall enable learners to accept facts or improve their ability to memorise certain details; they can serve motivational purposes, etc. Both purposes seem to differ significantly even though there appears to be some overlap. On a general level, the former criteria can relate to research activities whilst the latter apply mostly to educational purposes.[1] As instruments are created purposefully, and are supposed to be adequate for the respective tasks, it should not be a matter of surprise that instruments that were designed for research purposes differ from those that are meant to be used in teaching situations. This poses a question that is central to this paper: some types of instruments appear in research as well as in teaching situations.[2] Moreover, some experiments from the history of physics

1 These characterisations are not completely distinguishable: researchers might take measures to familiarise themselves with new instruments and create strategies to develop the skills that are necessary to use these instruments. Consequently, the respective experiments correspond to some of the characteristics pointed out for the educational use of the instrument, even though they are carried out in a research context. Yet, as this paper forms a first attempt to discuss the relation between research and educational uses of instruments, some generalisations may occur where a more differentiated discussion would be useful. This, however, is beyond the scope of this paper.

2 There are also purely educational devices which have no predecessors nor relatives in research—modern examples are the so-called Dynamot which has become in the last decade very popular in German schools for teaching electricity and energy, or experiment kits (see Beek, 2009; Sueß, 2009). By the early eighteenth century, Willem Jacob 's Gravesande and Pieter van Musschenbroek developed devices for teaching Newtonian physics in the Leiden course on natural philosophy (see Clercq, 1997; 's Gravesande, 1737).

have become canonical in science education. To give but three simple examples: Jean Bernard Léon Foucault's determination of the speed of light with a rotating mirror, Henry Cavendish's gravitational balance, and August Kundt's tube to measure the speed of sound and demonstrate standing waves.

Yet, even though the experiments in education bear names that create a relation to the historical ones, it appears questionable whether the instruments were identical.[3] Moreover, even though some of the instruments (such as the Cavendish balance) still bear the name that relates them to the historical instrument, they are obviously not identical.

Even a brief look reveals some aspects that are similar (if not identical) and others that differ. The most obvious similarity lies within the scientific content: Foucault's speed of light experiment still aims at determining the speed of light, Cavendish's gravitational balance is still used to determine the mass of the earth, etc. However, even these examples show that the similarity within scientific content is only on the surface and is in some sense misleading: Cavendish's balance determines the attractive force between gravitating masses; however, his intention was not to determine the mass of the earth but actually to measure its density (Cavendish, 1798; Clotfelter, 1987). And in Foucault's case it was an important detail that he was able to determine the velocity of light in an optical dense medium, thus being able to support either the wave or the corpuscle theory of light.

An obvious difference can be seen in the materials—modern ones such as plastic and steel are found where brass, glass and wood were used; sealing wax is no longer the choice of fixing different materials. This is true mainly with respect to contemporary teaching instruments, which even as teaching devices have a long history and underwent several transformations. What about the devices that were used when the experiment was transferred from research to teaching at some point in history?

I am far from able to give a final answer to this question. Thus, the aim of this paper is to clarify potential connections and meanings of educational versions of research instruments and experiments. In this respect, I focus on four categories that are relevant with respect to a comparison of research and educational versions: simplification, downscaling, stabilisation and iconisation. By using these categories I am not implying that an individual characterisation gives a complete account of the particular relation, nor am I suggesting that only one transformation for a research device is possible. Yet, these four categories appear to be more than just keywords for individual cases. From my understanding, they are to be found in a variety of cases and periods. In this respect, these categories may serve as a first step towards a fuller understanding of how research instruments were transformed into teaching devices. Together with this, an understanding of the meanings that were ascribed to the instruments and experiments in the teaching process should be developed. From these insights, it should become possible to

3 Rieß (2007) gives examples where didactical experiments bear historical names but are something completely different – this is however not the topic of this paper.

discuss questions with respect to the influence of experimental training on later research strategies, experimental preferences, and finally even teaching priorities.

INSTRUMENTS FOR RESEARCH AND TEACHING

The first instrument I am going to discuss is (at least in some respects) exceptional; it is a device that was very ambivalent with respect to its use in research and education. One of the most popular instruments of the second half of the eighteenth century was the solar microscope, an instrument which can be characterised as a projection microscope that uses sunlight as a light source.[4] Most instrument makers had the device in their catalogues, and the number of still existing solar microscopes can be taken as an indication of its popularity. Even though it has been questioned whether it was actually a scientific device,[5] it can be argued that solar microscopes were fairly typical instruments for natural philosophers. They appear in most textbooks, they were listed by instrument makers as 'philosophical' instead of entertaining, and they were used in some experimental researches such as the ones by Albrecht Wilhelm Roth (1797). Moreover, modified solar microscopes were also used for different experimental purposes such as the researches by Thomas Young on ultraviolet light (1804) and by Jean Paul Marat (1779) on heat.

Yet, it has to be pointed out that there was not *the* solar microscope. The instrument underwent several modifications, even though in the 1780s some sort of a standard device evolved that was mainly sold by London instrument makers such as Benjamin Martin, the Dollond workshop, and Jesse Ramsden (Figure 1).

In 1791, a different type of solar microscope appeared in the market. It was advertised by the military chaplain Friedrich August Junker who sold a device that was—at least from the technological point of view—significantly less sophisticated than the state-of-the-art instruments. Instead of having a toothed wheel mechanism and a worm gear to readjust the mirror, Junker's instrument realised the working principle with a wooden disc and a cord. Technically speaking, the instrument regressed two steps in comparison with the instruments that were established. However, the instrument appears to have been a success anyway and several of them are still kept in museums (Figure 2).[6]

4 For a detailed discussion of the solar microscope, see Heering (2008).

5 The British microscopist Charles R. Goring pointed out in the nineteenth century that the solar microscope is a mere toy, "fit only to amuse women and children" (Goring in Bradbury, 1967, p. 159). This characterisation has been taken up by several historians of science. However, a closer analysis reveals that this criticism is not justified (Heering, 2008).

6 Two Junker instruments can be found at the Deutsches Museum Munich, two at the Utrecht Universiteitsmuseum, one in the Billings collection, Washington, D. C. (see Ey, 1974) and one in the Museum Boerhaave Leiden (see Fournier, 2003). For a technical discussion of the instruments of John and Peter Dollond and Junker as well as the resulting consequences for the practices with these devices, see Heering (2010).

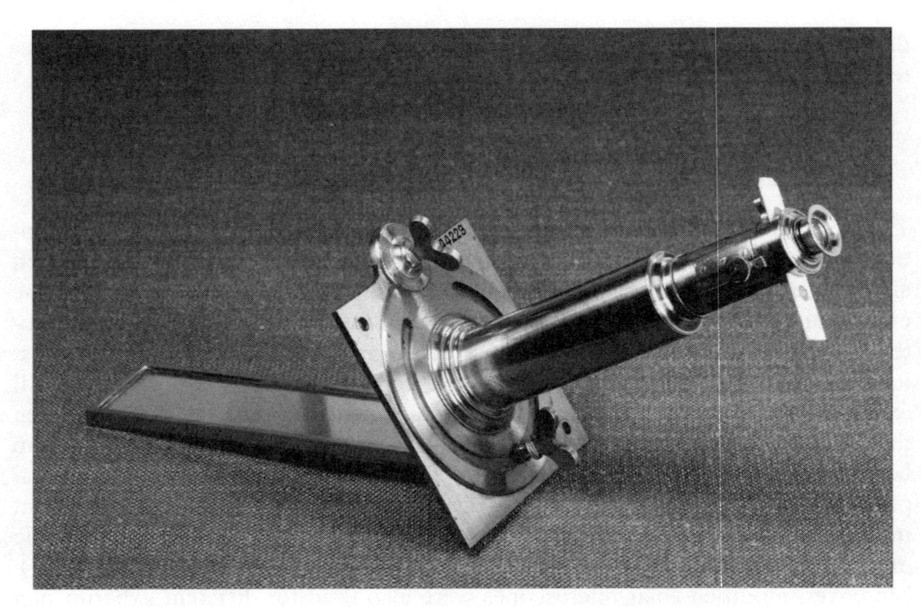

Fig. 1: Dollond's solar microscope. Photo: Deutsches Museum München.

This becomes comprehensible if one looks at the motivation Junker had to develop and market this instrument: in a leaflet he published to advertise his solar microscope (Junker, 1791), Junker stated explicitly that his instrument was intended to be used in schools. For this reason, he tried to make the instrument as cheap as possible so that as many schools as possible could afford the device. The consequence of its technical weakness—the instrument was not as user-friendly as the instruments by the London makers—appears to have been of little importance as the teachers who were supposed to use the instrument could develop the skills to handle it. In this respect, a school teacher had a different social status than a gentleman: consequently, the teacher could be expected to develop skills necessary to practice successfully with the solar microscope. This was not necessarily the case with the gentleman who would use the instrument to entertain himself and his guests with some demonstrations in natural philosophy.[7]

The instrument was also discussed in a research context: Johann Samuel Traugott Gehler mentioned the device (and its price) in his *Physikalisches Wörterbuch* (Gehler, 1795, p. 856). Moreover, a favourable review of the instrument was published in the 1791 issue of the *Magazin für das Neueste aus der Physik und Naturgeschichte* (Anonym, 1791). Both publications focused not so much on education but on research; consequently, the instrument seemed to have

7 For the role of gentlemen in eighteenth-century natural philosophy see Schaffer (1983).

been noteworthy also for researchers.[8] This example illustrates that instruments are not necessarily used either in a research context or in an educational context, but that specific instruments can be used in both. Even though research instruments are more likely to be specialised devices and thus serve only one purpose, there are certain examples where similar instruments can be found in the educational field as well.[9]

Fig. 2: Junker's solar microscope. Photo: P. Heering.

Though the demarcation between research and educational activities was not as clear as it is nowadays, it appears to be remarkable that even in this period economics played an important role: Junker stated explicitly that his instrument was about one-tenth the price of a London instrument. Moreover, he even offered a discount if a sovereign would order a larger quantity of instruments for the schools in his country. This policy is not unusual for instruments that are to be

8 The solar microscope is not unique with respect to this ambivalence; several of the eighteenth century electrical instruments can be found in teaching as well as research contexts. One may argue that this can be at least in part explained by the experimental approach in the Enlightenment age, which gave importance to collective experimenting and discourse on the experimental findings among lay persons also (Walters 1997). However, this issue requires a more thorough analysis which goes beyond the scope of this paper.

9 In the nineteenth century, gas discharge tubes were popular both among researchers and popular lecturers. This is not just a historical development—currently astronomers encourage amateurs to classify objects in images (a research task); at the same time, these images can be used for educational purposes. Likewise, the project Hands-on Universe encourages students to participate in research projects (http://www.handsonuniverse.org/, last accessed 15 October 2010). However, it should be understood that activities which combine research and education are exceptional in contemporary science.

used in public services.[10] However, it has some consequences for educational instruments. As already argued, the Junker solar microscope is technically simpler than the English instruments. Simplification appears to be a key to price reduction; consequently, this is an issue that can be found in many teaching instruments.

DOWNSCALING AND SIMPLIFYING THE RESEARCH INSTRUMENTS

A distinctly different approach can be identified when looking at the example of the experiments carried out by Johann Carl Friedrich Gauss and Wilhelm Eduard Weber in order to determine the earth's magnetic field in absolute measures. Gauss and Weber established an international network of experimenters who collaborated in order to map the magnetic field of the earth. In order to enable other researchers to carry out the observations and produce useful data, Gauss and Weber tried to give a very detailed account of their procedures while working with the magnetometer. In doing so, and in providing other researchers with an access to instruments that were technically identical to their own devices, they attempted to make these measurements comparable.

Yet, measurements of the magnetic field of the earth were relevant not just in this context. A few years after the initial measurements, the young student Gustav Kirchhoff complained in a letter about his training as a physicist which included "boring observations and even more boring calculations. Of the former, I had recently a little sample; I sat from 10pm until 2 am in the Albertinum behind a telescope, observing at a temperature of only 1° every 15 seconds a magnet, whose position I had to write down."[11]

From the description of the experiment it is evident that Kirchhoff took measurements that were very similar to the ones with the magnetometer. However, it seems questionable whether he really worked with a large instrument as had been used by Gauss and Weber (see Figure 3). Their experiment had been

10 Economic arguments can also be found with respect to eighteenth century medical electricity (Bertholon, 1788; Kühn, 1783). The stress on financial aspects in the fields of education and health is thus not a modern phenomenon.

11 Gustav Kirchhoff to Otto Kirchhoff, presumably 1843–44, quoted in Kärn (2002, p. 49). Kärn refers to "E. Warburg: Zur Erinnerung an Gustav Kirchhoff, Die Naturwissenschaften 13 (1925), Heft 11, S. 205–212" (ibid., p. 51). A remarkable aspect of this episode is the characterisation of the initiation that a student had to pass in order to become a physicist. Observations and calculations which appeared boring to the students were part of the syllabus, resulting in developing sufficient self-discipline. Consequently, one should ask what the purpose of the magnetic observations was—was the data still relevant for any sort of research, or was the procedure just a lesson in observations, where the data served merely as an indicator of the students' ability to make boring observations that were reliable? The educational concept which structured such practical tasks needs further attention in order to develop a better understanding of the role that was ascribed to experiments in the formation of scientists.

terminated, yet it became in a way as canonical as the ones with Charles Augustin Coulomb's torsion balance or the ice-calorimeter of Antoine Laurent Lavoisier and Pierre Simon Laplace. Consequently, textbooks such as the Müller-Pouillet (Pfaundler, 1905–14) still described the experiment many years later; however, the instruments look completely different.

Fig. 3: Moses Kärn and his reconstruction of the Gauss–Weber magnetometer. Photo: F. Müller.

Whilst the initial magnetometer was almost 3 metres high,[12] the educational version was of a size that enabled an individual to handle it and take measurements. Thus, the magnetometer probably fulfilled a completely different role: the actual measurement of the magnetic field of the earth was not the only relevant issue. At least another purpose was to familiarise students with the procedures required to become physicists. In this respect, the idea of formation is crucial. Students are to learn how to carry out an experiment adequately. The interesting part in this respect is the question, what can be accounted as being adequate—this is certainly no stable category, it may vary with time as well as with place and the level of education. Consequently, understanding what adequate means on different levels of education would enable a better understanding of the formation of researchers in a specific place, at a specific time. In doing so, a more thorough understanding of the similarities of researchers (which may result in their retrospective descrip-

12 For a discussion of the magnetometer of Gauß and Weber, see Kärn (2002).

tion as being part of a 'school') as well as specific limitations can be achieved. The latter appears to be as relevant for a historical analysis as the former, as this may help to understand why certain experimental approaches appeared to be inadequate in the view of several researchers. In terms of Ludwik Fleck's conception of style (Cohen and Schnelle, 1986), it can be argued that not only is a style of thought formed through the formal training of a student, but also the style of experimentation.

Yet, there are some aspects in the downscaling that are also relevant and interesting. First of all, downscaling is most likely related to simplification; yet, there are often difficulties involved in downscaling and miniaturisation. Therefore, a downscaled apparatus is not per se simpler or cheaper, even though this is probably the case in many examples from education.[13] The case of the Gauss and Weber magnetometer shows that there are advantages to downscaling: handling a 3 metre instrument is by no means easy; moreover, one needs sufficient space for handling such an apparatus. In this respect, the advantage of the educational version is that the instrument can be used in a standard laboratory. Thus, one aspect of simplification and downscaling can be related to the additional requirements. Requirements of special rooms or other special conditions (such as sunshine) make it more difficult to include formal training as a part of the syllabus. Thus, downscaling can be related (at least in some cases) to making a measurement less dependent on external resources or circumstances.[14]

Simplification is not always related to downscaling; in other words, simplification is not just a question of the dimensions of the research apparatus. An example can be James Prescott Joule's experiment on the mechanical equivalent of heat. Joule's set-up has been reconstructed at the Universität Oldenburg (Sibum, 1995); recently, Paolo Brenni demonstrated an educational version by the instrument maker Max Kohl that was part of the cabinet of the Istituto Tecnico Toscano in Florence.[15] On first look, both apparatuses are fairly similar, but a closer examination reveals that the two paddle-wheels are significantly different. One difference is the material—Joule's device was made of brass, the vessel being made of copper, whilst the Kohl device seems to be steel. More remarkable appears to be the design of the paddle-wheels: Joule's instrument consists of four levels of paddles with eight paddles on each level. Moreover, there are two differ-

13 Downscaling can also be a topic in research, particularly with respect to nano technologies. However, downscaling is not a new concern at least in technology.

14 Another example appears to be the educational versions of Cavendish's gravitational balance. The original masses in Cavendish's experiment were more than 150 kg each, certainly not what a teacher wants to set up once a year. Consequently, in the educational version, the masses are about 1 per cent of the original value (see Lauginie, 2007).

15 A video of this attempt is available at http://www.youtube.com/watch?v=PThq8fJpCLw&feature=related, last accessed 15 September 2010. The emails I quote from in the following discussion were posted on the mailing list "rete"; the archive of this mailing list is available at http://maillist.ox.ac.uk/ezmlm-cgi/2357, last accessed 15 September 2010.

ent designs of the paddles (see Figure 4) with four of them connected to the axis. Each of the other four paddles is attached to supports that connect two of the paddles which are fixed to the axis (see Figure 5). This differs from the design that is found in Joule's central publication (1850) where all paddles are directly connected to the axis.

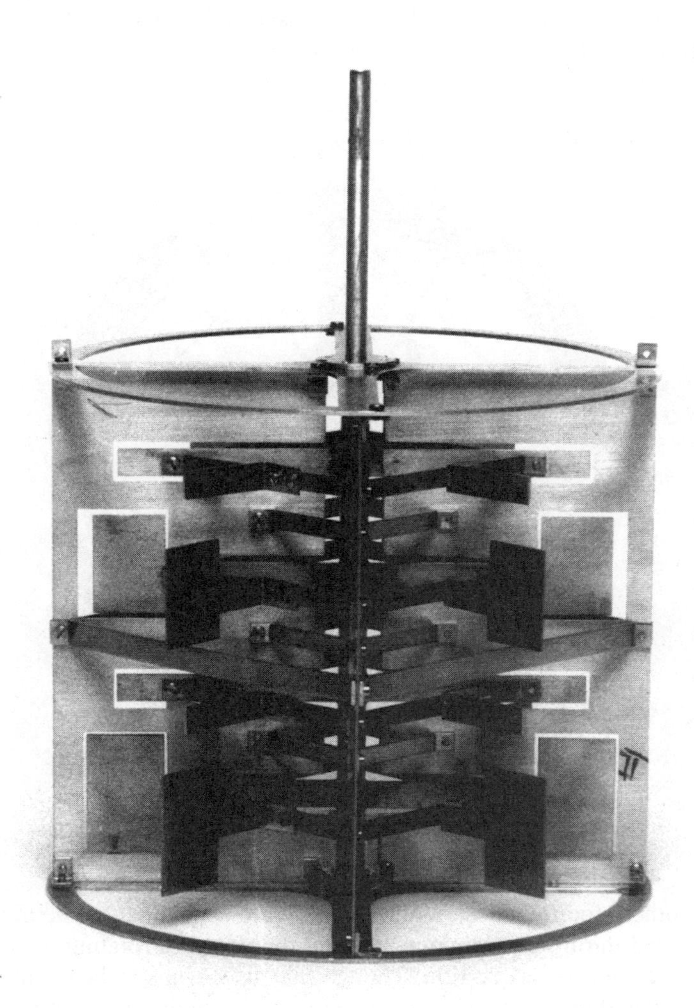

Fig. 4: Joule's paddle wheel, side view. Photo: W. Golletz.

In case of the educational device, there are only two levels of paddles, all of which are connected to the axis and have the same design. In the discussion that followed the publication of the video Paolo Brenni pointed out, "… I was extremely surprise [sic] of the speed of the paddle wheel!" (4 March 2010). Actually, the design of the paddles appears to be crucial for the terminal velocity, as experiments with different versions of the reconstructed paddle-wheel showed (Sibum, 1995; see also Heering, 1992). Consequently, this simplified version of the paddle-wheel should have an influence on the friction and thus on the terminal velocity of the weights.

Fig. 5: Joule's paddle wheel, bottom view. Photo: W. Golletz.

There are other modifications: Whilst in the Joule apparatus the thermometer has to be removed during the experiment when the wheel is rotating, this is not the case with the educational version. Though this also seems to be a simplification, there is one detail that deserves more attention as its modification appears to result in a more sophisticated design in the educational version. In Joule's apparatus, the roller is simply attached to the axis of the paddle-wheel after the weights had been wound up.[16] In the educational version, a mechanism connects and disconnects these parts; even though this is design is more elaborate, it is at the same time much simpler in performance. Thus this can be seen as an example where simplicity lies not in the apparatus itself but in the performance with the educational version of the apparatus—in this respect it can be interpreted as an inversion of the solar microscope example.

STABILISING AN EXPERIMENT

As the example of the Gauss–Weber experiment seems to indicate, simplification and downscaling is an issue in several educational versions of historical experiments. Yet, there is another aspect which is highlighted when Robert Andrews Millikan's experiment on the determination of the elementary charge is analysed. This experiment can be seen as being among the 'classical' experiments in science education; consequently, it is to be found in most textbooks and almost every company selling educational set-ups offers the instrument as part of their product range. However, taking a closer look at the printed materials reveals that in most examples, the experiment is de-contextualised (Niaz, 2000; Rodriguez and Niaz, 2004).[17] However, it is not only the description of the experiment that is changed and simplified (Parlow and Heering, 2009); in case of the oil drop experiment, it is also the experimental apparatus itself. This is not the same as saying that the apparatus is downscaled and simplified, which it is. One of the central ideas of school experiments (at least in the context of German science education) is that the experiment 'has to work'; in other words, the experimental outcome has to be consistent with theory and thus contribute to the content knowledge of the students. This has to be realised in an efficient manner, as teaching time is precious. Consequently, a criterion for the quality of the set-up is the reliability of data production, in particular with respect to the independence of the user of the instrument. To put it differently, an ideal educational version of Millikan's apparatus produces theory-consistent values even when operated by an unskilled experimenter. In order to make this possible, experimental set-ups and procedures are stabilised in a manner that makes the actual manipulation of the apparatus straightforward and controlled. Consequently, this is no longer an open experiment but just an experimental demonstration. This results in a modified meaning and in requirements that result in a simplified version of the apparatus, since not

16 The roller has to be disconnected from the axis as the weights are wound up; otherwise, the problem would be how to measure the energy transfer into the water during this procedure.

17 By de-contextualisation I am referring to the lack of context of the experiment; most descriptions of educational versions of historical experiments are limited to a 'scientific content' of the experiment, which is dissociated from its historical and other contexts.

all parameters need to be varied or controlled.[18] To give an example: Millikan placed his apparatus in a container full of oil in order to secure a constant temperature; in the educational version, the instrument simply remains at room temperature.

In some sense, these relations between the educational and the research version of an instrument are just the inverse in comparison with the solar microscope. In that example, the instrument maker Junker could assume that the user would take the trouble to develop the required skills in order to use the apparatus adequately. The gentlemen, however, would not necessarily need the skills as the device of Dollond would have already undergone a stabilisation process.

INSTRUMENTS AS ICONIC OBJECTS IN THE LECTURE ROOM

The eighteenth century was the period when natural philosophy experimenting was open to amateurs (in the positive sense of lay persons who are interested in participating in scientific knowledge). Things began to change at the end of the century. A professionalisation of experimental practice took place, which, among other aspects, also included the exclusion of lay persons from the production of scientific knowledge. Two instruments, which are in some sense related to this development, show that they can have a completely different meaning in science education compared to what we have discussed so far. The two instruments are Coulomb's torsion balance and the ice-calorimeter described by Lavoisier and Laplace. Both instruments became canonical devices during the nineteenth century; the plates which show the instruments are still reproduced in many modern textbooks on physics or chemistry.

Yet, when working with the devices, it becomes obvious that both instruments are not suited for demonstrative experiments in a classroom or lecture hall. Coulomb's torsion balance is extremely sensitive, and thus also very error sensitive (Heering, 1994). The measurement of an amount of heat with the ice-calorimeter takes several hours; during the measurement, a temperature between 4°C and 0°C is required; and there is nothing to be seen except an instrument whose lid is covered with ice (Heering, 2005). Yet, both instruments appear in the catalogues of instrument makers, and both instruments were seemingly used in lectures. The question that arises is what could be done with the instruments, as the study of the initial experiments seems to suggest that they cannot be demonstrated to an audience.[19] A clue to this question is provided by an instrument kept at the Norwegian

18 There might be an implicit control of some parameters due to the technical design of the apparatus. However, this means that the experimenter does not have to take any measures to control these parameters.

19 Due to the error-sensitivity, the torsion balance is hardly suited for lecture demonstrations. Therefore, it is not surprising that, in 1807, the Berlin lecturer Paul Louis Simon complained that the torsion balance is all too unreliable to be used in lectures. Consequently, he developed a new apparatus (Simon, 1807).

University of Science and Technology in Trondheim.[20] With this instrument the torsion micrometer turns out to be a dummy as it cannot move in the sleeve and is thus useless in the sense of being a measuring device. From this detail it is evident that the instrument was not intended to be really used for measurement—its quality lies on a different level. Apparently, what can be done with the instrument is to show it to an audience, to point at every detail in order to illustrate the working principle, and thus to familiarise the audience with the concept of the measurement. Even though no actual data can be taken with the instrument, it appears to be possible that Coulomb's data can be used in the lecture together with the device, thus familiarising students with certain aspect of this conceptual approach towards physics. A similar purpose can be ascribed to the educational versions of the ice-calorimeter—they appear to be too small to be used in a proper measurement.[21] However, they can be used as a demonstration device that is supposed to support the conceptual familiarisation of students with calorimetry as a quantitative approach. In this respect, the apparatus can be taken as the materialisation of a central device to the establishment of this concept—the ice-calorimeter is not to be perceived as a measuring device, but serves as a material representation of *the* ice-calorimeter by Lavoisier and Laplace.[22]

Yet, in both cases, one should ask why it had been so important for these particular instruments to be present in the lecture. If one examines the development of experimental practice in the late eighteenth and early nineteenth centuries, it is evident that both instruments had a particular meaning. They were staged as devices that represent a new style of experimentation which characterised (among other aspects) experiments by the claims of precision and sensitivity.[23] Thus, the instruments used in lectures were not only as representations, they were also established as icons of a new way of doing science.

20 I am indebted to R. Wittje who made me aware of this instrument and its peculiarities.

21 As already mentioned, the experiments with the calorimeter take at least several hours, and they need to be carried out in a temperature below 5°C. As the ice-calorimeters for educational purposes are substantially smaller than the one of Lavoisier and Laplace, it appears impossible to carry out proper experiments with them in a classroom or lecture theatre at room temperature.

22 Paolo Brenni comes to a similar conclusion with respect to the educational version of the Joule apparatus: "But I am sure that our Joule apparatus made by Max Kohl was not intended to do any measurement but simply to show how one could do it" (4 March 2010).

23 On the concept of style of experimentation (which is used as an expansion of Fleck's concept of style of thought and thought collective) see Heering (2007). The relevance of the torsion balance can be inferred (among other details) from its crucial role for French researchers to identify the Voltaic pile as an electrical device (see Brown, 1969; Sutton, 1981). On the relevance of the ice-calorimeter for establishing new standards of experimental practices see, in particular, Wise (1993) and Roberts (1991), but also the respective contributions in Beretta (2005).

CONCLUSION

These examples make evident that the relation between research and educational apparatus and experiments can be very diverse and in some cases rather complex. Yet, the examples also indicate that some criteria appear to be useful to describe the relation between research instruments and the respective teaching equivalent. Aspects such as simplification, downscaling and stabilisation appear to be identifiable in several examples. An open question remains with respect to the iconisation; here one has to ask whether this is still relevant nowadays or whether other methods of visualisation have replaced these devices.[24] Yet, one should be cautious about judging the aspect of simplification just from the technical point of view: both the example of the solar microscope and the one of the paddle-wheel make evident, in specific detail, that a more sophisticated apparatus can result in a much simpler performance. Consequently, strategies of simplification can also focus on aspects of performance with the instrument and not on the instrument itself.

It seems questionable at the moment whether these categories are sufficient to describe all transformations; one may also ask whether these categories are helpful for an analysis of the process of transformation. Yet, they appear to be at least an adequate tool for describing the relations between two set-ups that are related in a very specific manner.

Of course, there are certain aspects that are important with respect to such a relation and which are not discussed in this paper. To mention two of them: there has been no discussion of educational conceptions and how they influence the use of instruments in education (and one can also ask whether there are distinctions between school education and university courses as well as between demonstration experiments and students experiments). Moreover, the discussion is only one directional—from the research experiment to its educational match. It would certainly be relevant to discuss how the formation of a scientist with respect to experimental practice has an influence on her or his design of research experiments.

REFERENCES

Anonym. 1791. 'Nachricht von einem brauchbaren und wohlfeilen Sonnen-Mikroskop', *Magazin für das Neueste aus der Physik und Naturgeschichte*, 7(3), pp. 84–87.

Beek, Viola van. 2009. 'Experimental Spaces Outside the Laboratory: Experiment Kits and Instruction Manuals around 1900'. The Virtual Laboratory, http://vlp.mpiwg-berlin.mpg.de/references?id=art73

Beretta, Marco (ed.). 2005. *Lavoisier in Perspective*. München: Deutsches Museum.

24 It would probably be an oversimplification to argue that these material iconisations of some experimental set-ups are just a result of the lack of adequate methods of visualisation—magic lantern projections were already common before the nineteenth century.

Bertholon, Pierre. 1788. *Anwendung und Wirksamkeit der Elektrizität zur Erhaltung und Wiederherstellung der Gesundheit des menschlichen Körpers*. Translated and expanded by Carl Gottlob Kühn. Vol. 1, Weißenfels und Leipzig: Friedrich Severin.

Bradbury, Savile. 1967. *The Evolution of the Microscope*. Oxford: Pergamon Press.

Brown, Theodore M. 1969. 'The Electric Current in Early Nineteenth-Century French Physics', *Historical Studies in the Physical Sciences,* 1, pp. 61–103.

Cavendish, Henry. 1798. 'Experiments to Determine the Density of the Earth', *Philosophical Transactions of the Royal Society of London*, 88, pp. 469–526.

Clercq, Peter R. de. 1997. *The Leiden Cabinet of Physics: A Descriptive Catalogue*. Leiden: Museum Boerhaave.

Clotfelter, B.E. 1987. 'The Cavendish Experiment as Cavendish Knew it', *American Journal of Physics,* 55, pp. 210–13.

Cohen, R. S. and Thomas Schnelle (eds). 1986. *Cognition and Fact: Materials on Ludwik Fleck*. Dordrecht; Boston; Norwell, MA: D. Reidel.

Ey, John A. (ed.). 1974. *The Billings Microscope Collection of the Medical Museum Armed Forces Institute of Pathology*, second edition. Washington, D.C.: American Registry of Pathology.

Fournier, Marian. 2003. *Early Microscopes: A descriptive Catalogue*. Leiden: Boerhaave.

Gehler, Johann Samuel Traugott. 1795. *Physikalisches Wörterbuch*, Vol V. Leipzig: Schwickert.

Heering, Peter. 1992. 'On J.P. Joule's Determination of the Mechanical Equivalent of Heat', in Skip Hills (ed.), *The History and Philosophy of Science in Science Education*, Vol.1. Kingston, Ontario; The Faculty of Education Queen's University, pp. 495–505.

———. 1994. 'The Replication of the Torsion Balance Experiment: The Inverse Square Law and its Refutation by early 19th-Century German Physicists', in Christine Blondel and Matthias Dörries (eds), *Restaging Coulomb: Usages, Controverses et Réplications Autour de la Balance de Torsion*. Firenze: Leo S. Olschki, pp. 47–66.

———. 2005. 'Weighing the Heat: The Replication of the Experiments with the Ice-Calorimeter of Lavoisier and Laplace', in Marco Beretta (ed.), *Lavoisier in Perspective.* München: Deutsches Museum, pp. 27–41.

———. 2007. 'Das Konzept des Experimentierstils zur Beschreibung historischer Experimentalpraxis', in Bozena Choluj and Jan C. Joerden (eds), *Von der wissenschaftlichen Tatsache zur Wissensproduktion: Ludwik Fleck und seine Bedeutung für die Wissenschaft und Praxis*. Frankfurt/Main: Peter Lang, pp. 361–85.

———. 2008. 'The Enlightened Microscope: Re-enactment and Analysis of Projections with Eighteenth-Century Solar Microscopes', *British Journal for the History of Science,* 41, pp. 345–68.

———. 2010. 'Materialisiertes Handlungswissen – die Praxis mit dem Sonnenmikroskop', in Olaf Breidbach, Peter Heering, Matthias Müller and Heiko Weber (eds), *Experimentelle Wissenschaftsgeschichte*. München: Wilhelm Fink Verlag, pp.155–70.

Joule, James Prescott. 1850. 'On the Mechanical Equivalent of Heat', *Philosophical Transactions of the Royal Society of London*, 140, pp. 61–82.

Junker, Friedrich August. 1791. *Ueber das Sonnenmicroscop*. Magdeburg: Publisher unknown.

Kärn, Moses. 2002. 'Das erdmagnetische Observatorium in der Scheune: Messungen mit dem originalgetreuen Nachbau eines Magnetometers von Gauß und Weber', *Mitteilungen der Gauss-Gesellschaft Göttingen*, 39, pp. 23–52.

Kühn, Karl Gottlob. 1783. *Geschichte der medizinischen und physikalischen Elektricität und der neuesten Versuche, die in dieser nützlichen Wissenschaft gemacht worden sind*, Vol. 1. Leipzig: Weygandsche Buchhandlung.

Lauginie, Pierre. 2007. 'Weighing the Earth, Weighing the Worlds: From Cavendish to Modern Undergraduate Demonstrations', in Peter Heering and Daniel Osewold (eds), *Constructing Scientific Understanding through Contextual Teaching*. Berlin: Frank & Timme, pp. 119–48.

Marat, Jean Paul. 1779. *Découvertes sur le Feu, l'Électricité et la Lumière, constatées par une Suite d'Expériences nouvelles qui viennent d'etre vérifiées par MM. les Commissaires de l'Académie des Sciences*, Seconde Edition. Paris: Clousier.

Niaz, Mansour. 2000. 'The Oil Drop Experiment: A Rational Reconstruction of the Millikan–Ehrenhaft Controversy and its Implications for Chemistry Textbooks', *Journal of Research in Science Teaching*, 37, pp. 480–508

Parlow, Valentina and Peter Heering. 2009. 'Das Millikan Experiment und seine Behandlung in der Schule', in Peter Heering (ed), *Der Millikansche Öltröpfchenversuch zur Bestimmung der Elementarladung*. Oldenburg: Didaktisches Zentrum, pp. 35–46.

Pfaundler, Leopold (ed.) 1905–14. *Müller-Pouillets Lehrbuch der Physik und Meteorologie. In vier Bänden*, 10th edition. Braunschweig: F. Vieweg & Sohn.

Rieß, Falk. 2007. 'Short History of the Use of historical experiments in German physics lessons', in Peter Heering and Daniel Osewold (eds), *Constructing Scientific Understanding through Contextual Teaching*. Berlin: Frank & Timme, pp. 219–26.

Roberts, Lissa. 1991. 'A Word and the World: The Significance of Naming the Calorimeter', *ISIS*, 82, pp. 198–222.

Rodriguez, Maria A. and Mansoor Niaz. 2004. 'The Oil Drop Experiment: An Illustration of Scientific Research Methodology and its Implications for Physics Textbooks', *Instructional Science*, 32 (5), pp. 357–86.

Roth, A.W. 1797. *Bemerkungen über das Studium der cryptogamischen Wassergewächse*. Hannover: Gebrüder Hahn.

's Gravesande, Willem Jacob. 1737. *Mathematical Elements of Natural Philosophy, Confirm'd by Experiments, or, An introduction to Sir Isaac Newton's Philosophy*. Translated by J. T. Desaguliers. 2 volumes. London: Printed for J. Senex ..., W. Innys and R. Maney ..., and T. Longman.

Schaffer, Simon. 1983. 'Natural Philosophy and Public Spectacle in the Eighteenth Century', *History of Science* 21, pp. 1–43.

Sibum, Heinz Otto. 1995. 'Reworking the Mechanical Value of Heat: Instruments of Precision and Gestures of Accuracy in Early Victorian England', *Studies in the History and Philosophy of Science*, 26, pp. 73–106.

Simon, Paul Louis. 1807. 'Auszug aus einem Schreiben … an den Professor Gilbert in Halle', *Annalen der Physik*, 27, pp. 325–27.

Sueß, Michael. 2009. 'Historische physikalische Experimentierkästen: Untersuchung der im Schulmuseum Zetel vorhandenen Objekte und ihrer historischen Verwendung sowie die Entwicklung eines Ausstellungskonzepts'. Oldenburg: MA Thesis.

Sutton, Geoffrey. 1981. 'The Politics of Science in Early Napoleonic France: The Case of the Voltaic Pile', *Historical Studies in the Physical Sciences*, 11 (2), pp. 329–66.

Walters, Alice N. 1997. 'Conversation pieces: Science and Politeness in Eighteenth-Century England', *History of Science*, 35, pp. 123–154.

Wise, M. Norton. 1993. 'Mediations: Enlightenment Balancing Acts, or the Technologies of Rationalism', in Paul Horwich (ed.), *World Changes*. Cambridge, Mass. & London: MIT Press, pp. 207–56.

Young, Thomas. 1804. 'The Bakerian Lecture: Experiments and Calculations Relative to Physical Optics', *Philosophical Transactions of the Royal Society of London*, 94, pp. 1–16.

THE AUDIENCE IS LISTENING: READING WRITING ABOUT LEARNING BY DOING.[1]

Pete Langman

1745 saw the publication of the second edition of the late Jean Theophilus Desaguliers' *A Course of Experimental Philosophy*, a work which sought to explain Newtonian philosophy without recourse to Newton's complex, and often intractable, mathematics. Now a full two volumes, the book represented a considerable evolution from the somewhat slimmer *Physico-Mechanical Lectures* of 1717.[2] One of the leading exponents of the experimental lecture, an approach to disseminating natural philosophical knowledge pioneered by John Keill at Oxford, Desaguliers explained his rationale:

> The Thoughts of being oblig'd to understand Mathematicks have frighted a great many from the Newtonian Philosophy: [...] tho' its Truth is supported by Mathematicks, yet its Physical Discoveries may be communicated without. [...] since machines have been contriv'd to explain and prove experimentally what Sir Isaac Newton has demonstrated mathematically, and several of his own Experiments are shewn in publick Courses; a great many Persons get a considerable Knowledge of Natural Philosophy by Way of Amusement;[3]

Demonstration, therefore, was to replace the mathematics of Newton's *Principia mathematica* (1687) and *Opticks* (1704) for the layman.[4] The fashion for experimental philosophy reached the highest echelons of society over the course of the century, creating a thriving industry of instrument makers, demonstrators and lecturers. Performed in lecture theatres, coffee houses, and the lecturers' own

1 My thanks go to Matthew Dimmock, Annie Janowitz, Laura Miller, Tim Parnell and Cathy Relf for the provision of their proof-reading, draft-correcting, avenue-suggesting and sounding-board services, without which this essay would have been much the poorer.

2 J. T. Desaguliers, *Physico-Mechanical Lectures. Or, An Account of what is explain'd and demonstrated in the course of mechanical and experimental philosophy given by J. T. Desaguliers, etc*. (London: Printed for the Author, and Sold by him at his house in *Channel-Row, Westminster*, by *Richard Bridger*, 1717). For more on Desaguliers, see Larry Stewart, *The Rise of Public Science: Rhetoric, Technology and Natural Philosophy in Newtonian Britain, 1660–1750* (Cambridge: Cambridge University Press, 1992), pp. 119–41.

3 J. T. Desaguliers, *A Course of Experimental Philosophy*, 2 vols (London: Innys, Longman et al, 1745), II, A4r–v.

4 On reading the *Principia*, Gilbert Clarke, Cartesian natural philosopher and mathematician, wrote that 'I confess I doe not as yet well understand so much as your first three sections,' before noting that he practically 'despaire[d] of understanding' it (quoted in Mordechai Feingold, *The Newtonian Moment: Isaac Newton and the Making of Modern Culture* (New York and Oxford: New York Public Library, Oxford University Press, 2005), p. 32).

homes, the lectures were not cheap, with Hauksbee and Whiston charging two and a half guineas, and the companion texts encouraged further purchasing:

> All the above-mention'd Instruments, according to their Latest and Best Improvements, are made and Sold by Francis Hauksbee, (the Nephew of the late Mr. Hauksbee, deceas'd) in Crane-court, near Fetter-Lane in Fleetstreet, London.[5]

For Benjamin Martin, the companion text could prevent the lecture itself from being 'little more than a sublime and rational Amusement, for the instant Hours',[6] and he and Desaguliers seem to expect the text to be used independently of the lecture, as they were 'Designed for the Use of all such as have seen, or may see Courses of Experimental Philosophy'.[7] These books all contained plates of the apparatus, but the quantity of supporting material differed greatly. While Desaguliers' 1745 text contained detailed descriptions and mathematical formulae, suggesting that it was designed to allow the reader to carry out detailed study in the comfort of his own library, Hauksbee and Whiston provided only cursory explanatory material, suggesting that its primary use was alongside the machines, effectively as part of the demonstrator's apparatus.

With regards the attendance of women, Desaguliers and Martin perhaps protest too much, the former writing that 'I have with great Pleasure seen the Newtonian Philosophy so generally received among Persons of all Ranks and Professions, and even the Ladies, by the Help of Experiments', while the latter comments on his own work that

> In the last place, it was not the least part of my Design, to render this Course of Lectures plain and easy to be understood by all Capacities in general, and in particular by the Fair Sex: For why should not the Ladies understand and study the Sciences of Humanity, of which Philosophy holds the First Place? Knowledge is now become a fashionable Thing, and Philosophy is the Science a la Mode.[8]

The majority of books, however, were bought by men, and it seems likely that the experimental lectures themselves were a primarily masculine affair: one assumes there were no women present, for example, at the lecture given by Desaguliers when Peter Shaw requested that Thomas Topham, a strongman employed by Desaguliers to help demonstrate the advantages of mechanical apparatus, demonstrate naked to ensure that there was no trickery involved.[9] Desaguliers' use of

5 Francis Hauksbee and William Whiston, *A Course of Mechanical, Optical, Hydrostatical and Pneumatical Experiments* (London: 1715), p. [3], advertisement. This course ran over 26 days.

6 Benjamin Martin, *A Course of Lectures in Natural and Experimental Philosophy* (London: 1743), p. ii.

7 Desaguliers, *Physico-Mechanical Lectures*, LTP.

8 Martin, p. ii.

9 Stewart, p. 126. For example, George Adams Junior's 1794 *Lectures on Natural and Experimental Philosophy*, had 956 names on the subscription list, of whom a mere 4% were ladies, a number equalled by subscribers from the nobility, and well beaten by the clergy (who amounted to 10% of the total). A full 52.8% hailed from the London area (see John R. Millburn, 'An analysis of the subscribers to George Adams Junior's *Lectures on Natural and Experimental Philosophy* (1794)' (Aylesbury: for the author, 1985)).

experiments in the place of mathematics may have served to 'vastly increase the appeal of natural philosophy',[10] and Martin may have attempted to render Newtonian philosophy more attractive to the 'fair sex' by promoting its modishness, and describing the experiments as 'amusements', but the attention they pay to women in their works is more likely an indication of their absence than any significant attendance. Certainly neither women nor children generally possessed the independence of means or motion to attend such lectures, but it is the publishers who provide the real evidence in the form of titles such as *The Newtonian System of Philosophy Adapted to the Capacities of Young Gentlemen and Ladies* (London: John Newbery, 1761) and *Sir Isaac Newton's Philosophy Explain'd for the Use of the Ladies* (London: E. Cave, 1739). These popular and enduring books simply filled a gap in the market.

The books written by Desaguliers et al were designed to illustrate and explicate Newtonian Philosophy with or without the experimental demonstrations which underpinned the lecture texts themselves. Bursting with complex explanations, detailed illustrations and even mathematical equations, they assumed an audience comfortable with the medium of the technical manual. *The Newtonian System* (or to use its more popular name, *Tom Telescope*), and *Newton's Philosophy Explain'd for the Use of the Ladies* (or *Newton for the Ladies*) were written directly for an audience lacking the ability to comprehend the experiments on a technical level, and so they replaced the technical jargon with fictionalised versions of these same experiments.

REASON AND IMAGINATION

In 1748, the Earl of Chesterfield dispensed some fatherly advice to his son:

> Women, then, are only children of a larger growth; they have an entertaining tattle, and sometimes wit; but for solid, reasoning good sense, I never in my life knew one that had it, or who reasoned or acted consequentially for four-and-twenty hours together. Some passion or humour always breaks in upon their best resolutions.[11]

This view of women was anything but controversial, as the Renaissance commonplace that both women and children were unable to engage the faculty of reason endured, even while some few entered the realms of science. 'Children's Minds are narrow and weak [...]', wrote John Locke, '[the] natural Temper of Children disposes their Minds to wander. Novelty alone takes them; whatever that presents, they are presently eager to have a Taste of'.[12] Mary Wollstonecraft largely agreed:

> Women, in general, as well as the rich of both sexes, have acquired all the follies and vices of civilisation, and missed the useful fruit [...] their senses are inflamed, and their understand-

10 Stewart, p. 123.

11 *Letters Written by the Late Right Honourable Philip Dormer Stanhope, Earl of Chesterfield*, ed. by E. Stanhope, 4 vols (Dublin: E. Lynch et al, 1775), II, p. 21.

12 John Locke, *Some Thoughts Concerning Education*, thirteenth edition (London: A. Millar et al, 1764), pp. 244–45.

> ings neglected, consequently they become the prey of their senses, delicately termed sensibility, and are blown about by every momentary gust of feeling […] all their thoughts turn on things calculated to excite emotion; and feeling, when they should reason, their conduct is unstable, their opinions are wavering.[13]

Wollstonecraft considered this skittishness the result of faulty education rather than any innate deficiency, but the fact remained that women were prey to their feelings: reason took a back seat.[14]

Flighty and with roving imaginations, women and children might therefore invite reasonably similar, if not identical, methods of instruction, and yet this pair of texts seem to take markedly different approaches. In this essay, I shall argue that each text aims to recreate for its particular audience the *experience* of witnessing: *Tom Telescope* combines lecture with demonstration by presenting a fictional lecturer using familiar objects from childhood – it was subtitled *The Philosophy of Tops and Balls* – while *Newton for the Ladies* aims to harness the reader's imagination though the manipulation of poetic conceits. Both works write their audience *into* the experimental lecture, allowing them to *perceive* experiments they could not possibly *understand*: they were not designed for use *alongside* the apparatus, but to function *as* the apparatus.

The virtual witnessing these texts aim for is first created through the inclusion of an internal, literary audience (the characters within the book, who 'witness' the experiments in the first place), and then through the creation of images for the reader. In *Tom Telescope*, these images are embedded within the text through a combination of narrative and illustration, while in *Newton for the Ladies*, they are created by the narrator within the imagination of his fictional audience, and thus simultaneously within the readership.

13 Mary Wollstonecraft, *A Vindication of the Rights of Woman* (London: J. Johnson, 1792), K1r-v.

14 Education, Wollstonecraft argued, failed to cultivate women's understanding, as 'novels, music, poetry, and gallantry, all tend to make women the creatures of sensation, and their character is thus formed during the time they are acquiring accomplishments' (Wollstonecraft, K1v), a state of affairs she blamed on purposeful neglect: 'kind instructors! What were we created for? To remain, it may be said, innocent; they mean in a state of childhood' (Wollstonecraft, K2r). French educational theorist Francois Fenelon was espousing similar views a hundred years previously, though without the emphasis on intentional mis-education, writing 'Young Women without Instruction and Application, have always a roving Imagination. For want of solid Nourishment, their Curiosity violently turns them toward vain and dangerous Objects' (Francois Fenelon, *Instructions for the Education of a Daughter* (London: Jonah Bowyer, 1707), p. 11), while Locke ascribed the same malleability to children: 'I imagine the Minds of Children as easily turned this way or that, as Water itself' (Locke, A5v). See also Richard A. Barney, *Plots of Enlightenment: Education and the Novel in Eighteenth-Century England* (Stanford, California: Stanford University Press, 1999), p. 55.

THE PROBLEMS OF WITNESSING

The problem of how to report an experiment or phenomenon when observation is impossible has troubled scientists since the days of Francis Bacon. The Royal Society, in many ways Baconian in conception, followed the principle of 'show, don't tell' wherever possible, counting many experimental lecturers amongst its fellows, and employing Desaguliers as its official demonstrator.[15] But often such demonstration was impossible, and issues of authentication and authority reared their heads. Recent reconstructions of eighteenth century experiments, such as those by Peter Heering with the solar microscope, have noted similar problems:

> it is impossible to give an adequate written account of the impressions the images made on the viewer. Moreover, neither printed pictures nor even projections of these pictures can be used to communicate the experience of the projected images, since they provide only meagre and unsatisfactory reproductions of what could be seen in the darkened chamber.[16]

It is perhaps for this reason that these texts attempt to create within the reader, through identification, the feeling of witnessing. Bacon's approach was simplicity and clarity:

> in every new or slightly more subtle experiment, though (it seems to me) certain and proved, I shall subjoin a clear account of the way I performed it, so that after revealing its every detail, people can see if any latent fault clings to it, and push themselves to find (if any there be) more reliable and accurate proofs.[17]

Since Bacon's time, however, mathematics had emerged as the premium mode of demonstrating phenomena through virtual witnessing, but with neither mathematics nor physical demonstration to draw from, the writers and editors of *Newton for the Ladies* and *Tom Telescope* relied upon writing to recreate the sensation of 'being there' – the former harnessing of the power of the imagination and the latter, increasingly, through the imitation of the experimental lectures delivered by Desaguliers et al. Both texts evolve in such a way as to accentuate these properties, as a comparison between different editions of each text clearly shows.

15 See S. Shapin and S. Schaffer, *Leviathan and the Air-Pump: Hobbes, Boyle, and the Experimental Life* (Princeton: Princeton University Press, 1995); Steven Shapin, *A Social History of Truth* (Chicago: University of Chicago Press, 1995).

16 Peter Heering, 'The enlightened microscope: re-enactment and analysis of projections with eighteenth-century solar microscopes', *BJHS*, 41 (2008), 345–368 at 352. See also Elizabeth Calvecci, 'A Witness Account of solar microscope projections: collective acts integrating across personal and historical memory', in *BJHS*, 41 (2008), 369–84, in which she analyses her experience of witnessing Heering's Solar Microscope in operation relative to Schaffer and Shapin's account of collective witnessing.

17 Graham Rees, ed., *The Oxford Francis Bacon*, 15 vols (Oxford: Clarendon Press, 2004), XI, p. 41. See also Bacon on magisterial and initiative learning in Spedding, Ellis and Heath, eds, *The Works of Francis Bacon*, 7 vols (London: Longman et al, 1872), IV, p. 449.

PUBLICATION HISTORY

Both *Tom Telescope* and *Newton for the Ladies* were published in editions that might fairly be termed 'adaptations' – *Newton for the Ladies* three years after the first English translation, *Tom Telescope* thirty years after the first edition – in which the editorial changes seem to have been undertaken in order to reinforce and enhance the initial strategy of the recreation of the sensation of witnessing in the audience.

First published, and most likely authored, by John Newbery in 1761, *Tom Telescope* went through six editions in duodecimo between 1761 and 1798, eventually undergoing radical changes in 1820.[18] Newbery himself, however, died in 1767, his business consequently being fought over by his son and step-son on one side, and his nephew on the other. It took the retirement of Francis the son, the death of Francis the nephew, and finally the death of the step-son to unify the imprimatur in 1788 under the stewardship of Elizabeth Newbery, the nephew's wife – and she sold the business in 1802. With Elizabeth's stewardship began a series of editorial changes, with two editions published in the 1790s – in 1794 and 1798, under the imprimatur of Ogilvy et al[19] – which contained subtle but significant changes that increased their similarity to the 'grown-up' books of experimental lectures: *Tom Telescope* was edited 'according to their latest and best improvements'.[20]

First published in 1737 in handsome folio, *Il Newtonianesimo per le dame* was ostensibly an attempt by its Italian author, Francesco Algarotti, to do for Newton what Fontenelle's *On the plurality of worlds* had done for Descartes.[21] The first English edition, translated by Elizabeth Carter, was published in 1739 by Edward Cave.[22] Hot on the heels of Carter's much admired translation was another edition, however, one published by G. Hawkins 'at Milton's Head' in 1742,

18 See James A. Secord, 'Newton in the Nursery: Tom Telescope and the Philosophy of Tops and Balls, 1761–1838', *History of Science*, 23 (1985): 127–51. These changes are broad enough to suggest that it is less a new edition than a new work inspired by *Tom Telescope* – in a new century, with a new publisher, Tom literally grows up.

19 See S. Roscoe, *John Newbery and his Successors 1740–1814, A Bibliography* (Hertfordshire: Five Owls Press, 1973), pp. 2, 253.

20 Hauksbee and Whiston, advertisement; cf. Desaguliers, *A Course*, I, '*I haue endeavour'd to improue, by the Addition of new* Propositions *and* Experiments, *and by altering and changing my* Machines' (a1v-r).

21 Francesco Algarotti, *Newton Explain'd for the Use of the Ladies*, trans. by Elizabeth Carter (London: E. Cave, 1739), A5v. See Massimo Mazzotti, 'Newton for ladies: gentility, gender and radical culture', *BJHS*, 37 (2004): 119–146.

22 Carter was unsuccessfully courted by Thomas Birch, one-time secretary to the Royal Society, who wrote in the copy she presented to him 'Tho. Birch / Mais 31st. 1739 / *Ex dono interpretis / doctissime & elegan- / tissime* Elizab.Carter' (BL Shelfmark 535.b.10). *S.V.* DNB. Bound with this copy is the letter Carter sent to her publisher, which is itself interesting: Sr / I have sent you by the bearer / of this note 2 sets of Algarotti. One I desire you / to give to Mr. Birch with my compliments, the other / is for signore Algarotti himself. When I can get some / more bound I shall have a sett at Mr. Johnsons service.'

which Laura Miller describes as 'an embellished piracy of Carter's translation, inspired by imaginative literature as well as science', noting that in addition to stealing wholesale from Carter, it also included parts translated from a 1738 French edition by Adrien Duperron de Castera.[23] Neither text begins with Newton, instead imagining themselves into the environment of their audience, making the readership feel comfortable with their bibliographical instructors. *Newton for the Ladies* sees a chevalier and a marchioness whiling away the hours in a country house by indulging in a series of dialogues which shift from poetry to Newtonianism.[24] He then leads her on a whistle-stop tour of the various universal systems, leading her to become a fervent devotee of first Descartes, and then Malebranche, before she sees the error of her ways as they reach Newton, 'who brings you Light and Truth, who speaks with Candour.'[25]

While also being set in a great country house, *Tom Telescope* makes use of a far greater range of characters, taking an initially more theatrical, and ostensibly more moralistic, approach. When a group of children are deciding how to fill up their day, one suggests that they play at cards. Up jumps young Tom, and lambasts them, as 'Playing at cards for money, says he, is so nearly allied to covetousness and cheating that I abhor it [...] Parents, continued he, might almost as well teach their children to thieve as well as game; for they are kindred employments',[26] his views echo those of Locke:

> This has been that, which has given *Cards*, *Dice* and *Drinking*, so much Credit in the World: And a great many throw away their spare Hours in them through the Prevelancy of Custom, and Want of some better Employment to fill up the Vacancy of leisure, more than from any real Delight is to be found in them [...] I think the safest and best Way is never to learn any Play upon them.[27]

This moral priggishness resurfaces at several points within the book, but by then Tom's design, being effectively to save his companions 'from the insidious pastime of card playing' through his delivering a series of lectures on Newtonian Philosophy, has perhaps been forgotten.[28] The lectures are accompanied by demonstrations which, naturally for a work 'adapted to the Capacities of young

23 Catherine Talbot wrote to Miss Carter saying 'I must tell you of a Dutch Latin compliment made to you and your Algarotti, by Brucker in some new book, but I cannot transcribe it' (Talbot to Carter, Cuddesden, July 29, 1757 – in *A Series of Letters between Mrs Elizabeth Carter and Miss Catherine Talbot, 1741–1787*, 2 vols (J. Rivington: London, 1808), I, p. 416); Miller also points out that the 1742 edition, because of its French heritage, affords somewhat more space to Descartes than does Carter's translation (see Laura Miller, 'Publishers and Gendered Readership in the English Editions of *Il Newtonianismo per la Dame*', forthcoming).

24 Mazzotti, p. 120. The shift is occasioned by the chevalier's reading of poem, written and published by Algarotti in 1732, which celebrated Laura Bassi's graduation from the University of Bologna.

25 *Newton for the Ladies* (1742), II, B10r.

26 [John Newbery] *The Newtonian System of Philosophy* (London: John Newbery, 1761), A3v.

27 Locke, pp. 312–13.

28 Secord, p. 135.

Gentlemen and Ladies, and familiarized and made entertaining by Objects with which they are intimately acquainted', use 'familiar objects' such as cricket balls and spinning tops, perhaps following the teachings of Locke and Rousseau, that '*Learning* might be made a Play and Recreation to children; and that they might be brought to desire to be taught'.[29]

Through its use of familiar objects – it was subtitled the 'philosophy of tops and balls'[30] – *Tom Telescope* asks the reader to consider the necessity of complicated or specialised apparatus. Julia Douthwaite suggesting that 'it is the elaborate collection of paraphernalia that initially attracts Tom Telescope to the Marquis of Setstar's home, where he stages all his little lessons',[31] but in fact this movement is due to the Countess:

> The Countess was very desirous of knowing what sort of diversion could be made of Natural Philosophy; and finding her young visitors in the same disposition, she conducted them to the Marquis of *Setstar*'s, that they might have the use of proper instruments.[32]

For a text which seems to be wishing to recreate the sensation of witnessing it is perhaps inevitable that within it a debate should occur about with regards the efficacy of various instruments, and *Tom Telescope* is unclear, initially at least, where it stands on the subject. When, for example, Lady Caroline challenges Tom to prove that air has the property of 'spring', the Duke intervenes, saying 'That he cannot do, without the use of proper instruments'. Yet Tom points to a pop-gun on the window-sill: 'almost anything will do.'[33] While Tom and the Duke seem to disagree at this point, the text inexorably moves towards the position of the adults. It is almost as if the children grow up, intellectually, as the lectures progress. *Tom*

29 Locke, pp. 222–23; In *Emile*, Rousseau encourages the use of hand-made instruments and familiar objects to allow children to discover scientific principles (Julia V. Douthwaite, *The Wild Girl, Natural Man and the Monster* (Chicago: University of Chicago Press, 2002), p. 99). Interestingly, Tom lectures almost exclusively for boys, using very gendered toys for his demonstrations. Yet, of the seven copies at the British Library which bear inscriptions, the text definitely owned by a boy, and the one probably (shelfmarks c.119.15 and RB.23.A.21247), trumpet their status as gifts, whereas there are four copies owned by girls, and each is simply claimed by its owner. One, shelfmark CH.760, is actively claimed by her, as she writes 'Elizabeth Dickens | Her Book | March 12th 1805' on the first leaf, one (C.113.a.19), owned by Charlotte Mason, has 'This book belonged to dear mother' written on the LTP, while the other two, shelfmarks 12809.a.5 and 8704.a.7 have the names Maria Wilkinson and Elizabeth Kipping on their respective LTPs. Three of these copies have picture or floral boards. It may be of note also that the copy given to 'Mas. N. Belcher' in 1814 was an 1806 edition, while Elizabeth's copy is a 1761 edition, claimed in 1805.

30 *Tom Telescope* (1761), contents page (no sig.).

31 Douthwaite, p. 100. cf. Benjamin Martin's *Young Gentleman's and Lady's Philosophy*, in which Douthwaite notes the characters' 'infatuation' with 'proper' scientific instruments (Douthwaite, p. 100).

32 *Tom Telescope* (1798), B2r. Interestingly, the plate which faces the second lecture in 1761, namely Lecture II '*Of the* Universe, *and particularly of the* Solar System' (*Tom Telescope* (1761), C3r), namely the plate of the Marquis of Setstar's Observatory, is moved in 1798 to become the frontispiece.

33 *Tom Telescope* (1798), E5v.

Telescope seems to be asking whether there is any useful way in which it can imitate the experimental lectures of Desaguliers et al. The editorial changes of the 1790s enhance this feeling, as the texts become more realistic, dispensing with sensationalist language and accentuating philosophical and historical accuracy.

EDITING THE TEXT

The original dedication, 'The Substance of Six lectures read to the Lilliputian Society, By TOM TELESCOPE, A.M.', slowly vanishes. In 1794 the lectures are 'read to a select company of friends', while the 1798 edition removes the entire line. Both texts, however, trumpet their revised status on their respective letter-press title pages, with authorship now ascribed to William Magnet, F.L.S: 1794 is 'A NEW EDITION, | Revised and enriched by an Account of the late new | Philosophical Discoveries,'; 1798 reads 'A new improved Edition, | With many Alterations and Additions, to explain | the late new Philosophical Discoveries.'[34]

While it is edited in order to take account of new natural philosophical discoveries and historical landmarks, the most telling change to *Tom Telescope* is the gradual toning down of passages which sought to engage the reader emotionally. The 1790s editions also show a book more aware of its own status: in 1761 it thinks it is a lecture, but by the 1790s it knows it is a *report* of a lecture.[35] These changes all work to accentuate the lecture's 'reality': it is less a fictional account than a true report of the type of demonstration the reader might imagine being given around the country.

The 1790s also saw increased usage of specialised apparatus, such in the second lecture, '*Of the Universe, and particularly of the Solar System*', where Tom introduces the orrery, in order to 'illustrate our solar system':

> There are seven primary planets; and there are marked on the Orrery as follows: Mercury *b*, Venus *c*, the Earth *d*, Mars *e*, Jupiter *f*, Saturn *g*, and the Georgium Sidus (which being of such recent discovery, is not represented in this Orrery.)[36]

Tom here not only accentuates the use of modern, grown-up apparatus, but also subtly reinforces the state-of-the-art nature of his lecture, and therefore the book itself, as he shows how even the instrument makers have yet to catch up with this 'new' discovery (made by Herschal in 1781).[37]

34 cf. *Tom Telescope* (1761; 1794; 1798), LTP.

35 Tom is quoted as saying 'which I mentioned in my second Lecture (page 26), and 'which I described in my first Lecture (p. 11)' (*Tom Telescope* (1798), I1v, I4r). In 1761, Tom does mention 'my next course of lectures', three times, but does not give page numbers (*Tom Telescope* (1761), H4v, I1r, I1v), but these lines are omitted in the 1798 edition.

36 *Tom Telescope* (1798), C6v.

37 It isn't until the edition of 1794 that *Tom Telescope* notes this new discovery, however – he protests a little too much on this point.

Fig. 1: The Marquis of Setstar's Observatory (Tom Telescope, 1798, frontispiece).
Credit: author's own library.

(22)

Of the Solar Syſtem.

Now, by means of this Orrery, I will illuſtrate our Solar Syſtem; which contains the ſun (marked *a*) in the centre, and the planets and comets moving about it.

f d b a c e g

But how is it then, ſays Tom Wilſon, that we daily ſee the ſun riſe and ſet?

Your queſtion, replies Maſter Teleſcope, is very natural; for it was an opinion held by the ancients ſome thouſand years, that the earth was the centre of the Univerſe,

Fig. 2: The Orrery (Tom Telescope, 1798, p. 22). Credit: author's own library.

Having demonstrated using the orrery, and perhaps rather strangely pointing the reader and internal audience to the plain 2D map of the solar system, Tom goes on to demonstrate both solar and lunar eclipses. Here, he not only changes the demonstration, but, rather tellingly, the narrative point of view. In 1761, Tom explains the phenomenon using a cricket ball, an orange, and a spinning top:

> We will suppose this orange to be the sun, this cricket-ball the earth, and this top the moon; now if you place them in a strait line, with the ball in the middle, and then put your eye to the top, you'll find that the ball will entirely hide the orange from your view, and would prevent

> the rays of light (which always proceed in right lines) from falling upon it, whence would ensue a *total* eclipse.[38]

In 1798 Tom once again demonstrates via his home-grown methods, though now subtly changed:

> But I will endeavour to explain this to you more clearly, says our philosopher, taking an ivory ball suspended by a string, in his hand; we will suppose this ball to be the moon, the candle the sun, and my head the earth. When I place the ivory ball in a direct line betwixt my eye and the candle, it appears all dark, because the enlightened part is opposite the candle; but if I move the ball a little to the right, I perceive a streak of light, which is like the New Moon.[39]

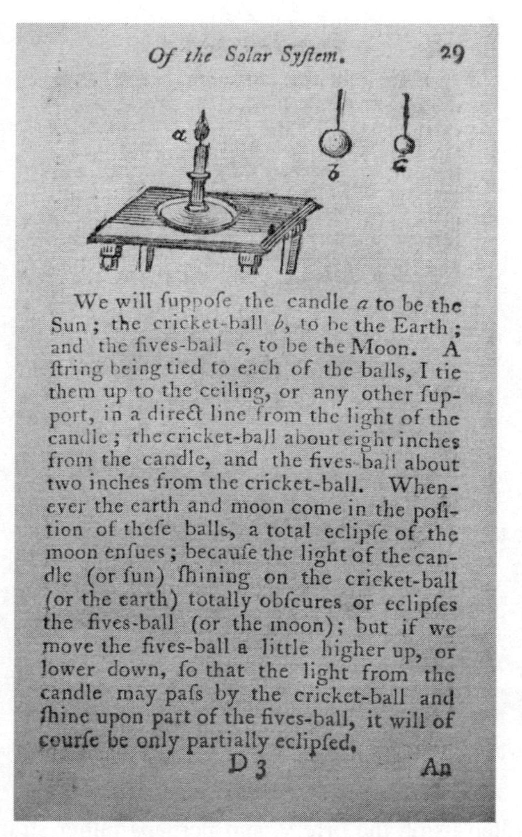

Of the Solar Syſtem. 29

We will ſuppoſe the candle *a* to be the Sun; the cricket-ball *b*, to be the Earth; and the fives-ball *c*, to be the Moon. A ſtring being tied to each of the balls, I tie them up to the ceiling, or any other ſupport, in a direct line from the light of the candle; the cricket-ball about eight inches from the candle, and the fives-ball about two inches from the cricket-ball. Whenever the earth and moon come in the poſition of theſe balls, a total eclipſe of the moon enſues; becauſe the light of the candle (or ſun) ſhining on the cricket-ball (or the earth) totally obſcures or eclipſes the fives-ball (or the moon); but if we move the fives-ball a little higher up, or lower down, ſo that the light from the candle may paſs by the cricket-ball and ſhine upon part of the fives-ball, it will of courſe be only partially eclipſed.

D 3 An

Fig. 3: Demonstration of the eclipse (Tom Telescope, 1798, p. 29).
Credit: author's own library.

Printed with the addition of two new illustrations, one of the orrery and the other of the re-imagined eclipse, this passage subtly privileges modern technologies

38 *Tom Telescope* (1761), D1v–D2r.

39 *Tom Telescope* (1798), D2r.

over the home-spun – though in this case the deficiencies in the audience lead Tom to resort to more home-spun methods – while accentuating Tom's activity as lecturer. More than this, however, the 1798 text has done something very interesting with the audience. In 1761, Tom writes 'if *you* place them in a strait line [...] *you'll* find that the ball will entirely hide the orange from *your* view',[40] while by 1798 he no longer talks of 'you' but 'I', and the experiment is now visible to the lecturer only. The audience in 1761 is part of the lecture, and Tom himself acknowledges his kinship with his audience. In 1798, however, Tom has taken a very different stance, and changes in his language show how he has assumed a much more authoritative position – he is now showing his audience, rather than sharing his knowledge with them. Tom is gradually growing up, gradually turning into a de facto experimental lecturer. This feeling is enhanced further by the back matter of the books. In the editions up to 1792, these pages are taken up with advertisements for children's books, prefaced by a publisher's note:

> Mention is made of the *Pocket Dictionary* (Page 95.) not because the Terms are better defined in that book than any other, but because it is sold only for *three Shillings* Bound.[41]

From the 1794 edition onwards, the reader finds that rather than trying to sell their books, the publisher is trying to sell experimental machines. From 1794, Tom Telescope is followed by a price list of experimental instruments, the bottom of which reads 'Orders sent to the Publishers of this Work, will be duly attended to'.[42]

This engagement with experimental culture is there from the first edition, but does change over time, with a good example being its management of one of the most popular and impressive pieces of experimental apparatus then available, the air pump immortalised by Joseph Wright of Derby. The 1790s texts once more take a far more technical approach than the fictive feel of 1761. Both text and illustration change, the former being increased by 50%, from 29 to 43 lines, while the latter became more technical, and linked by numbers directly to the text.

40 *Tom Telescope* (1761), D1v–D2r (my emphasis).

41 *Tom Telescope* (1761), M3r. In the 1761 edition the dictionary mentioned is, perhaps unsurprisingly, published by Newbery himself:
And now, Lady *Caroline*, prepare to hear a few hard words and I will finish this Lecture. Because I cannot deliver what I am going to say, Madam, without making use of the terms of art, says he, and those I must desire your Ladyship, and the rest of the good company, to learn from Mr. *Newbery's* pocket dictionary, or some other book of that kind. (*Tom Telescope* (1761), I6r.)
The 1798 text not only references another work but allows, once more, for a greater interaction between audience and lecturer:
And now, Lady *Caroline*, prepare to hear a few hard words and I will finish this Lecture. But why must it be finished in an unintelligible manner? Says the Lady. Because I cannot deliver what I am going to say, Madam, without making use of the terms of art; and those I would recommend your Ladyship, and the rest of the good company, to learn from *Jones's Pronouncing and Explanatory Dictionary*; which is a work no young reader should omit having in his library' (*Tom Telescope* (1798), K6r).

42 *Tom Telescope* (1798), N4r.

A Liſt of the Optical and Philoſophical Inſtruments mentioned in this Book; with the Prices at which they are made and ſold, by the Philoſophical Inſtrument Makers.

	£.	s.	d.
A Pocket Terreſtrial Globe, in a caſe	0	8	0
A Reflecting Teleſcope, one foot, in braſs	5	5	0
A Refracting ditto, from 10s. 6d. to	3	3	0
A Planetarium, according to the wheel-work, from 1l. 1s. to — —	10	10	0
An Accurate Map of the Moon —	0	1	0
An Armillary Sphere, on card paper	0	5	0
A Nine Inch ditto, all in braſs, completely and elegantly made — —	9	9	0
A Pair of Twelve Inch Globes, beſt ſort	5	5	0
A Triple Weather-glaſs [deſcribed in page 44.] — — — —	3	13	6
Air-Gun, for experiments only —	16	16	0
Air-Pump, with Receiver, from 4l 14s 6d to — — —	6	6	0
Apparatus to ditto, from 1l. 1s. to —	10	10	0
Electrical Machines, from 2l. 12s. 6d, to	8	8	0
Microſcope, with Apparatus — —	1	6	0
Glaſs Priſm — — — —	0	8	0

Orders ſent to the Publiſhers of this Work, will be duly attended to.

Fig. 4: Experimental price list (Tom Telescope, 1798). Credit: author's own library.

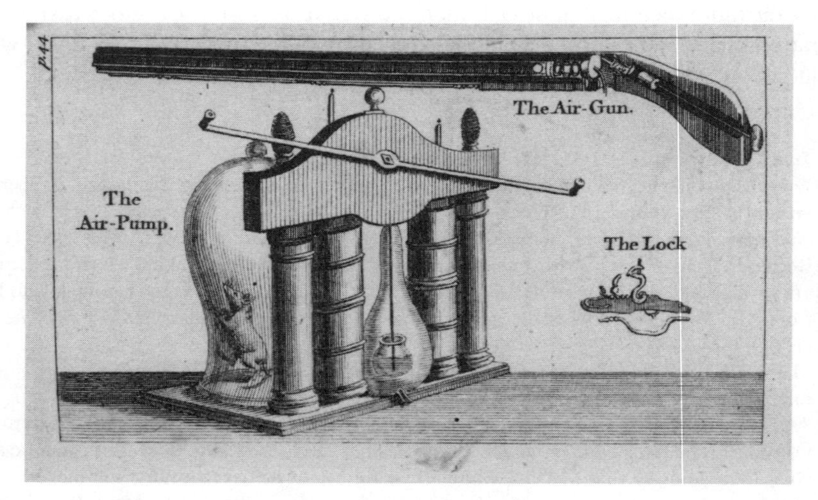

Fig. 5: The air pump (Tom Telescope, 1761, insert between p.44/45). Credit: British Library.

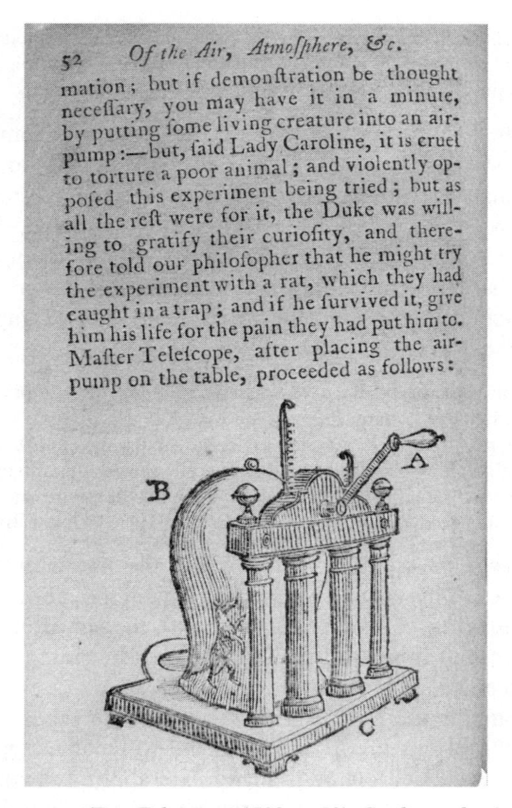

52 *Of the Air, Atmoſphere, &c.*

mation; but if demonſtration be thought neceſſary, you may have it in a minute, by putting ſome living creature into an air-pump:—but, ſaid Lady Caroline, it is cruel to torture a poor animal; and violently oppoſed this experiment being tried; but as all the reſt were for it, the Duke was willing to gratify their curioſity, and therefore told our philoſopher that he might try the experiment with a rat, which they had caught in a trap; and if he ſurvived it, give him his life for the pain they had put him to. Maſter Teleſcope, after placing the air-pump on the table, proceeded as follows:

Fig. 6: The air pump (Tom Telescope, 1798, p. 52). Credit: author's own library.

Both the 1761 and the 1790s texts begin with the same discussion of air, and the demonstration of the emotional instability of women provided by Lady Caroline's outburst that 'it is cruel to torture a poor animal',[43] but after the Duke's suggestion that they 'try the experiment with a rat', the texts diverge. The 1761 text dives straight into the rat's sufferings: 'when the air was partly exhausted, he appeared in great agony, and convulsed; and more air being pumped out, he fell on his side for dead; but fresh air being immediately admitted, it rushed into his lungs.'[44] In 1798, however, the text begins with technical information:

> Master Telescope, after placing the air-pump on the table, proceeded as follows:
>
> By the help of this machine, all that I have spoken concerning the weight and elasticity of the air, is demonstrated in the most simple and elegant manner. For by working the handle (marked A) all the air that is contained within the glass receiver (marked B) is pumped out; and if any living animal is put within the receiver, all the air in its body is pumped out like-

43 *Tom Telescope* (1798), F3r.

44 *Tom Telescope* (1761), D6v–E1r.

> wise: then, as mentioned before, air being the principle which preserves life, the animal dies, unless fresh air be immediately admitted, which may be done by turning a screw (at C). Our philosopher then put the rat into the receiver [45]

The two texts then synchronise, the later one accentuating Tom's active involvement in the demonstration, and maybe a hardening of attitude towards the rat, now called 'it' rather than 'he': the 1790s sees a more balanced, more technical and less emotional look at the principles and apparatus involved. This movement is accentuated by the editing of some of the book's more showy, flowery passages into a somewhat harder, more utilitarian style, further fostering the impression that the narrator actually witnessed the experiments. For example, after the children have been led to the Duke's observatory, 1761 elaborates:

> and the place itself is the best that can be conceived for enquiries of this kind, and for meditation. To see an extensive horizon thus shaded by the brow of night, and at intervals brightened up by the borrowed light of the moon dancing among the clouds, was to me inexpressibly pleasing. Nothing was heard but a gentle breeze whispering among the battlements; the dying murmurs of a distant cascade […] night, the nurse of nature, hushed all things else to silence—But the silence was soon broke by our philosopher, who thus began his Lecture.[46]

In 1798 this passage is replaced with 'When the company were seated, our Philosopher thus began his second Lecture'.[47] This removal of unashamedly literary language designed to convey atmosphere and the substitution of language of direct action is repeated in several parts of the work, lending the text a more direct, unmediated feel.[48]

As well as changing the book's attitude to its reader, the increasingly precise technical description brings it ever closer to the work of Desaguliers et al. Now called 'A new improved Edition, With many Alterations and additions, to explain the late new Philosophical Discoveries', the text was updated by the addition of new philosophical discoveries as well as circumstantial information, making current what was, by the 1790s, somewhat dated. From 1761 to as late as 1784, the text stated that 'the noble city of Lisbon was lately destroyed', where finally, in the 1790s, these events are 'some years ago'.[49] Tom changes names, too, replac-

45 *Tom Telescope* (1798), F3r.

46 *Tom Telescope* (1761), C3v.

47 *Tom Telescope* (1798), C2r. cf. the debate on slavery, where the lines 'When the ambassador had sat down with a sigh, and lady *Caroline* had wiped the precious pearly drops from her cheeks, our philosopher arose and thus pursued his lecture' (*Tom Telescope* (1761), K5v) is replaced by 'When silence was restored, our Philosopher arose and thus pursued his lecture.' (*Tom Telescope* (1798), M4r).

48 Tom himself is also made more pleasant and authoritative, less hectoring: in 1761, Tom 'commanded silence'; in 1798 he 'requested' it (*Tom Telescope* (1761), A5v; (1798), B3v). His audience, too, interact more: where Tom once talked of 'Sam Jones's knife', in 1798 we read 'my knife, said Sam Jones' (*Tom Telescope* (1761), B2r; (1798), B6r).

49 cf. *Tom Telescope* (1761), E6r; (1794), F3r.

ing Lord Anson with Captain Cook as the circumnavigator of choice, and perhaps makes a political statement by mentioning 'English miles'.[50]

The editing of *Tom Telescope*'s prose style and updating of its factual information, along with the subtly introduced self-awareness, work to produce a text more like a report than a literary representation of a lecture. It seems as if the editions of the 1790s are designed to *tell* rather than *show*. In his final lecture, on the 'Five Senses of Man, and of his Understanding', Tom's audience is augmented by 'a great deal of good company [...] on which account I could observe that Master Telescope took less pains to be understood by the young gentlemen and ladies; and addressed himself more particularly to those of greater abilities'.[51]

Tom is no longer discussing things that can be demonstrated, moving onto psychology, and perception, suggesting that 'all our ideas [...] are obtained either by sensation or reflection, that is, by means of our five senses [...] or by the operations of our mind'.[52] Lady Caroline asks that Tom explain what is meant 'by the term Idea', and Tom replies 'That, I apprehend, is sufficiently explained by what was said about the looking glass [...] but if your Ladyship requires another definition, you shall have it'.[53] He continues:

> By an Idea, then, I mean that image or picture, Madam, which is formed in the mind of anything which we have seen, or even heard talk of [...] Now when an image is formed in the mind from a view of the object itself, it may be called an adequate or real idea; but when it is conceived in the mind without seeing the object, it is an inadequate or imaginary idea.[54]

Tom's way of looking at the imagination is, it seems, at odds with that of Algarotti. Where *Tom Telescope* is continually updated in order to make the text more realistic, more like a de facto experimental lecture text, Algarotti sets out on an entirely different path. At first, however, he seems to criticise female intellectual weakness in terms that echo Lord Chesterfield,

> I have endeavoured to set Truth, accompanied with all that is necessary to demonstrate it, in a pleasing Light, and to render it agreeable to that Sex, which had rather *perceive* than *understand*.[55]

While, as we have seen, *Tom Telescope* is gently edited until it is a text which tells rather than shows, Algarotti is set on demonstration. He does this not through detailed technical descriptions of experimental apparatus, nor through illustrations, but in ways which Percy Bysshe Shelley would have understood. Shelley, in

50 See *Tom Telescope* (1761), B5r; (1798), B5r. See also *Tom Telescope* (1761), E6v; (1798), F5r, where Tom Wilson mentions Blanchard and Jefferies' crossing of the channel by balloon in 1785; *Tom Telescope* (1761), K3v; (1798), L3v, where 'nabob' changes to 'Tippo sahib'. The balloon interjection is also of interest because it does not follow the usual jeering course of Tom Wilson's questions, but is quite reasonable; Tom Wilson too, is growing up.

51 *Tom Telescope* (1798), L1v.

52 *Tom Telescope* (1798), L3v.

53 *Tom Telescope* (1798), L4v.

54 *Tom Telescope* (1798), L4v.

55 *Newton for the Ladies* (1739), I, pp. iv–v. This preface is cut from the 1742 edition.

his *Defence of Poetry*, writing some sixty years after Algarotti, theorises what Algarotti enacts:

> All things exist as they are perceived: at least in relation to the percipient. "The mind is its own place, and of itself can make a heaven of hell, a hell of heaven." But poetry defeats the curse which binds us to be subjected to the accident of surrounding impressions. And whether it spreads its own figured curtain, or withdraws life's dark veil from before the scene of things, it equally creates for us a being within our being.[56]

Shelley had begun his *Defence* considering the difference between reason and imagination, asserting that the imagination was the greater of the two faculties:

> According to one mode of regarding those two classes of mental action which are called reason and imagination, the former may be considered as mind contemplating the relations borne by one thought to another, however produced; and the latter, as mind acting upon those thoughts so as to colour them with its own light, and composing from them, as from elements, other thoughts, each containing within itself the principle of its own integrity. The one is the *to poiein*, or the principle of synthesis, and has for its objects those forms which are common to universal nature and existence itself; the other is the *to logozein*, or principle of analysis, and its action regards the relations of things simply as relations [...] reason respects the differences, imagination the similitudes of things.[57]

It was from the imagination that poetry sprung: 'Poetry', he wrote, 'may be defined as "the expression of the imagination"'.[58] Shelley's ideas are similar to those of Coleridge, whose *Biographia literata* of 1817 contained a more direct assertion of the power of the imagination:

> The imagination the I consider either as primary or secondary. The primary imagination I hold to be the living power and prime agent of all human perception, and as a repetition in the finite mind of the eternal act of creation in the infinite I AM.[59]

Where *Tom Telescope* uses an increasingly utilitarian style to reproduce the sensation of witnessing an experiment, relying on the reason of the reader for its effect, Algarotti specifically targets the imaginative faculty. As he has already pointed out, the imagination is stronger in women than reason, and it is because of this that the marchioness, in a crucial passage of *Newton For Ladies*, wants to *see* a crucial experiment, not to *hear* about one.

Just as Shelley equates poetry with truth, so Algarotti is demonstrating truth through the use of the imagination. According to Massimo Mazzotti 'the character of the marchioness functioned as a rhetorical device to naturalise the representation of certain attitudes and inclinations as specifically feminine', and that 'against this background of feminine deficiencies, the cognitive and moral virtues of the Newtonian philosopher became all the more apparent.[60] Mazzotti only tells half of the story, however, as Algarotti accommodates this 'feminine deficiency', the

56 Percy Bysshe Shelley, *A Defence of Poetry* (1820) in *Romanticism: An Anthology*, ed. by Duncan Wu, 3rd edn (Malden: Blackwell, 2006), p. 1197.

57 *A Defence of Poetry*, p. 1184.

58 *A Defence of Poetry*, p. 1185.

59 Samuel Tayler Coleridge, *Biographia literata* (1817), in *Romanticism*, p. 1691.

60 Mazzotti, intro, p. 7.

chevalier manipulating the marchioness's superior imaginative faculty to re-create in her, and thus in the imaginations of the female readers of his work, the sensation of actually witnessing an experimental demonstration. The chevalier criticises merely the over-active use of the imagination, citing Descartes as an example,[61] suggesting that he follows Fenelon in believing that it can be used to elicit positive results, as 'everything that reioyces or enlivens the Imagination, facilitates Study'.[62] While the marchioness is keen to move from 'consequences and Verses' to 'Evidence and Explications',[63] this evidence is witnessed within her imagination, a feature of the text accentuated by the changes made to the 1742 'edition'.

Instead of presenting an educator who explicitly performs experiments for his textual audience in the manner of *Tom Telescope*, Algarotti works with what are effectively thought experiments, thought experiments for the benefit of a marchioness who explicitly states her ignorance of mathematics:

> I must ingenuously confess, said the Marchioness, that though I have always regarded the *Mathematicians* with a singular Veneration, I do not yet understand what their Demonstrations are. However familiar they may at present be rendered, I do not comprehend them enough to find the solution of a Problem among the Patch-Boxes and Perfumes on my toilette.[64]

While the Countess of Twilight leads the children to the Marquis of Setstar's observatory 'that they might have the use of proper instruments,' Algarotti's narrator proposes a different mode of demonstration: he asks the marchioness to construct an experimental laboratory in her imagination.

The textual changes of the 1742 edition are similar in effect to those made to *Tom Telescope* in that they present a subtly different marchioness, a marchioness who is more than a simple recipient of the chevalier's knowledge, a significant change from the 1739 translation: 'She is permitted the occasional interjection, such as '"I see," said the marchioness, interrupting me,' while the narrator is teaching her. The 1742 edition is still a text that teaches Newtonian natural philosophy but the marchioness's response to the same prompt is '"Hold, hold, dear sir!" said the marchioness, disappointing me'.'[65]

Within the text, Newton is spoken of in reverential tones as the Chevalier works towards the book's climax, the presentation of the *Experimentum crucis*, 'the very Thing necessary to prove not only the different refrangibility, but every other principle in *Natural Philosophy*'.[66] The chevalier warms to his task:

> Till now you have only been conversant in the several Absurdities of those different Systems, which have successively owed their Rise to the Pride and vain Imagination of some Men, and

61 *Newton for the Ladies* (1739), I, B11v.
62 Fenelon, p. 53.
63 *Newton for the Ladies* (1739), I, B11r.
64 *Newton for the Ladies* (1742), II, C8r.
65 Miller, p. 15.
66 *Newton for the Ladies* (1742), II, C2r.

> their Reception to the Credulity of the People. Now *Newton* comes, who brings you Light and Truth, who speaks with Candour.[67]

This subtle apotheosis, accompanied by the commonplace analogy of light and knowledge, works in tandem with Miltonic citation to create the appropriate mind-picture for the marchioness. Having had the Newtonian system explained to her, the marchioness states that she understands 'thoroughly', before demanding:

> produce me some Proofs, I long for them […] at first I inclined to Des Cartes's Opinion, then I fell into that of Malbranche, and now (what with your observation-scheme) I am left without any system at all.[68]

No matter that she *understands*, she still wishes to be *shown*, just as Tom, having described an animal's reliance on air as 'self-evident', and 'needing no explanation' still places the rat in the air pump.[69] The chevalier replies: 'Madam, the Genius of Observation, depend on it, will soon make you ample Amends for your present chagrin.'[70] The way in which the chevalier makes the marchioness see demonstrates how much *Newton for the Ladies* differs from *Tom Telescope*'s utilitarian style, and a comparison between the 1739 and 1742 versions makes this all the more apparent. Carter's translation reads thus:

> Imagine yourself to be in a Place of *Milton*'s visible Darkness, or rather still darker a Place, if you will be absolutely deprived of all Light; and this shall be our Theatre of Reasoning and Observations.[71]

The 1742 text reads 'Be pleased to figure to yourself a Chamber entirely dark, a Chamber, in which, as *Milton* says, *reigns Darkness visible*. This shall be our Scene for the Search of Truth.'[72] This accentuation of Milton, coupled with the removal the word 'reason', intimates that it will be the imagination that is called on to engage with the experiment. The darkened chamber was a familiar Newtonian concept,[73] and Newton himself had written on the power of the imagination to create realistic images in the mind's eye:

> when the impresion of [Sun symbol] was not too strong upon my eye I could easily imagine severall shapes as if I saw them in the [Sun symbol]s place, whence perhaps may be gathered that the tendere{st} sight argues the clearest fantasie of things visible. & hence something of the nature of madnesse & dreame{s} may be gathered[74]

67 *Newton for the Ladies* (1742), II, B10r.
68 *Newton for the Ladies* (1742), II, B12r.
69 *Tom Telescope* (1798), F2r.
70 *Newton for the Ladies* (1742), II, B12r.
71 *Newton for the Ladies* (1739), II, C1v.
72 *Newton for the Ladies* (1742), II, B12r. Algarotti uses the same phrase in a letter to Lord Hervey in 1739, writing 'figure to yourself, my Lord, skeletons in rags' (Francesco Algarotti, *Letters from Count Algarotti to Lord Hervey*, 2 vols (Dublin: Saunders et al, 1770), I, p. 22).
73 See Newton's *Opticks*, Bk II, pt iv, p. 88.
74 *Questiones quaedam philosophiae,* add ms. 3996, CUL, p. 42. <http://www.newtonproject.sussex.ac.uk/view/texts/normalized/THEM00092> (accessed 28–10–09).

The imagination, it seems, can create the effect of seeing, and Newton connects the phenomenon with dreams, an experience the marchioness understands well:

> Philosophy and sound sleep do not agree very well together. My interrupted Dreams have transported me into the Region of optics, where I saw nothing but Prisms, Lenses, Rays differently refracted, coloured Images, in short, all those Experiments, and all the philosophical Apparatus which you described, arose successively in my Imagination like Visions and Fantomes. Whatever Charms these Things may have in themselves, I could never have imagined they would employ my Thoughts so strongly at a Time when it is not very customary to think of Philosophy.[75]

Algarotti draws on this imaginative power in both marchioness and reader, making the marchioness become the room within which Newton carried out his experiments with the prism: the 'genius of observation' is to take place entirely virtually, and verbally.

It is the phrase 'darkness visible' that sets off this imaginative chain of events, conjuring associations in the mind of his readership, who would have recognised phrase, poem and context. *Paradise Lost*, designed that Milton might discover the truth of God's universe, justify the ways of God to men, begins with this poetic rendering of chaos:

> A dungeon horrible, on all sides round
> as one great furnace flamed, yet from those flames
> no light, but rather darkness visible
> served only to discover sights of woe[76]

As Shelley realised in *A Defence of Poetry*, Milton's chaos connects thinking and being. When Satan comments that 'The mind is its own place, and in itself / Can make a Heav'n of Hell, a Hell of Heav'n',[77] in many ways foreshadows Newton's comments on the imagination, as well as reminding the reader of the biblical view of creation appropriated by Pope: 'And God said, Let there be Light: and there was light'.[78] Milton is clear that the importance of his argument necessitates his creating a new kind of poetry, 'things unattempted yet in prose or rhyme',[79] inviting the reader to create an internal, poetic world in which his argument can find itself.

It is this passage that fires the marchioness's imagination into transforming herself into a demonstrative stage, a stage of discovery that she will share with the reader. The darkness serves as a place of discovery as Newton, the actual light, and the light of knowledge are simultaneously allowed to enter, literally and figuratively enlightening her, as she imagines the experimental chamber, the light, the prism, the colours of the spectrum, and finally feels the sensation of witnessing Newton's great experiment.

75 Algarotti (1742), II, E2r–v.
76 John Milton, *Paradise Lost*, ed. by John Leonard (Harmondsworth: Penguin, 2000), I, ll. 61–4.
77 *Paradise Lost*, I, ll. 255–56.
78 Genesis, 1.3.
79 *Paradise Lost*, I, l. 16.

The marchioness functions both as her character and as surrogate reader, for as she figures this darkened chamber, so do we – the reader is effectively inserted into the text, and the experiment is thus witnessed, albeit virtually. The marchioness originally requested that she 'see some Experiment, which cannot possibly be explained by any other System than the *Newtonian*, and that I believe would satisfy me'.[80] She does not say 'I want you to tell me about', but 'I want to see', and by telling within the realms of the imagination, what is merely description is made real by the imagination. By *telling* in this specifically poetic fashion, the narrator re-constructs the experiment as surely as the experimental lecturers did, and thus ends up *showing* the experiment.

CONCLUSION

The changing nature of these two texts raises questions concerning their status and purpose. While *Tom Telescope* seems to have been edited in order to accentuate its resemblance to a 'real' experimental lecture series, *Newton for the Ladies* seems to have been edited in order to accentuate its poetic nature, a nature designed to conjure images in the imagination of its readers.

While the companion texts of Desaguliers et al were designed either to be read alongside demonstrations, or at least with the memory of a demonstration witnessed fresh in the mind, or to be read independent of the need for these machines, these two fictionalised texts work at conjuring the same powerful images through the use of the imagination in order that their audiences, who seem rarely to have been afforded the honour of witnessing these experimental demonstrations, can recreate, or even create, them within their heads. It may seem superfluous of Algarotti's narrator to authenticate his own narrative, saying that 'these effects must happen according to Sir *Isaac Newton*'s System; and these in reality do happen, as I myself have often had the Pleasure of seeing',[81] but if the marchioness is acting as a surrogate for the reader, the reader must surely understand that this is the author talking to them.

Both texts explore the complex relationship between writing, the imagination and reality, with Algarotti's narrator stating that poetry and Natural Philosophy both rely on observation, and observation leads us directly to Newton:

> we may affirm then, replied the Marchioness, that as every Thing which *Midas* touched was transformed to Gold, so every Thing that Sir *Isaac Newton* handled became Demonstration.[82]

In using poetry, and especially the sublime poetry of Milton, to allow the marchioness to explore the nature of light, Algarotti is also perhaps working alongside the increasingly popular poetic commonplace of the mind launching itself into space, the nocturnes and philosophic poems such as John Hughes' *The*

80 *Newton for the Ladies* (1739), II, C3r–v.
81 *Newton for the Ladies* (1739), II, B12v.
82 *Newton for the Ladies* (1739), II, C8v.

Ecstacy or Samuel Bowden's *A Poem Sacred to the Memory of Sir Isaac Newton,* which imagined themselves journeying through space, making a metaphor of the search for knowledge.[83] Algarotti is certainly foreshadowing Shelley and Coleridge's later aesthetic discourse in suggesting that poetry is akin to an act of creation, and it is the imagination which drives poetry. By harnessing the imagination, Algarotti negates the need for physical demonstration. *Tom Telescope* uses a different approach, that of increasingly technical description, but with the same end in sight, the re-creation of the experiment in the internal world of the reader through specific narrative strategies. The two texts differ in one simple but vitally important regard: *Tom Telescope* is designed to appeal to an audience which, as it grows, progressively acquires reason, and thus as it evolves as a text, it increasingly calls on the audience to engage their reason; *Newton for the Ladies*, on the other hand, appeals solely, and increasingly, to the audience's imagination. In both texts, however, the use of the internal audience, the audience within the narrative with which the reader identifies, means that the reader is in one sense left reading writing about learning by doing, and in another, very real sense, experiencing the experiments for themselves.

REFERENCES

Algarotti, Francesco, *Newton Explain'd for the Use of the Ladies*, trans. by Elizabeth Carter (London: E. Cave, 1739)

Algarotti, Francesco, *Newton Explain'd for the Use of the Ladies* (London: G. Hawkins, 1742)

Algarotti, Francesco, *Letters from Count Algarotti to Lord Hervey*, 2 vols (Dublin: Saunders et al, 1770)

Barney, Richard A., *Plots of Enlightenment: Education and the Novel in Eighteenth-Century England* (Stanford, California: Stanford University Press, 1999)

Calvecci, Elizabeth, 'A Witness Account of solar microscope projections: collective acts integrating across personal and historical memory', *BJHS*, 41 (2008), 369–84

Carter, Elizabeth, *A Series of Letters between Mrs Elizabeth Carter and Miss Catherine Talbot*, 1741–1787, 2 vols (J. Rivington: London, 1808)

Desaguliers, J. T., *Physico-Mechanical Lectures. Or, An Account of what is explain'd and demonstrated in the course of mechanical and experimental philosophy given by J. T. Desaguliers, etc.* (London: Printed for the Author, and Sold by him at his house in Channel-Row, Westminster, by Richard Bridger, 1717)

Desaguliers, J. T., *A Course of Experimental Philosophy*, 2 vols (London: Innys, Longman et al, 1745)

Douthwaite, Julia V., *The Wild Girl, Natural Man and the Monster* (Chicago: University of Chicago Press, 2002)

Feingold, Mordechai, *The Newtonian Moment: Isaac Newton and the Making of Modern Culture* (New York and Oxford: New York Public Library, Oxford University Press, 2005)

Fenelon, Francois, *Instructions for the Education of a Daughter* (London: Jonah Bowyer, 1707)

83 My thanks to Prof Annie Janowitz for this suggestion. See also William Powell Jones, 'Newton Further Demands the Muse', *Studies in English Literature*, 1500–1900, 3 (1963): 287–306.

Hauksbee, Francis, and William Whiston, *A Course of Mechanical, Optical, Hydrostatical and Pneumatical Experiments* (London: 1715)
Heering, Peter, 'The enlightened microscope: re-enactment and analysis of projections with eighteenth-century solar microscopes', *BJHS*, 41 (2008), 345–368
Locke, John, *Some Thoughts Concerning Education*, thirteenth edition (London: A. Millar et al, 1764)
Martin, Benjamin, *A Course of Lectures in Natural and Experimental Philosophy* (London: 1743)
Millburn, John R., 'An analysis of the subscribers to George Adams Junior's *Lectures on Natural and Experimental Philosophy* (1794)' (Aylesbury: for the author, 1985)
Miller, Laura, 'Publishers and Gendered Readership in the English Editions of *Il Newtonianismo per la Dame*' (forthcoming)
Milton, John, *Paradise Lost*, ed. by John Leonard (Harmondsworth: Penguin, 2000)
[Newbery, John] *The Newtonian System of Philosophy* (London: John Newbery, 1761)
[Newbery, John] *The Newtonian System of Philosophy* (London: Ogilvy et al, 1794)
[Newbery, John] *The Newtonian System of Philosophy* (London: Ogilvy et al, 1798)
Newton, Isaac, *Questiones quaedam philosophiae*, add ms. 3996, Cambridge University Library. <http://www.newtonproject.sussex.ac.uk/view/texts/normalized/THEM00092> (accessed 28–10–09).
Powell Jones, William, 'Newton Further Demands the Muse', *Studies in English Literature*, 1500–1900, 3 (1963), 287–306
Rees, Graham, ed., *The Oxford Francis Bacon*, 15 vols (Oxford: Clarendon Press, 2004), vol XI
Roscoe, S., *John Newbery and his Successors 1740–1814, A Bibliography* (Hertfordshire: Five Owls Press, 1973)
Secord, James A., 'Newton in the Nursery: Tom Telescope and the Philosophy of Tops and Balls, 1761–1838', *History of Science*, 23 (1985), 127–51
Shapin, S., and S. Schaffer, *Leviathan and the Air-Pump: Hobbes, Boyle, and the Experimental Life* (Princeton: Princeton University Press, 1995)
Shapin, Steven, *A Social History of Truth* (Chicago: University of Chicago Press, 1995)
Spedding, Ellis and Heath, eds, *The Works of Francis Bacon*, 7 vols (London: Longman et al, 1872)
Stanhope, E., ed., *Letters Written by the Late Right Honourable Philip Dormer Stanhope, Earl of Chesterfield*, 4 vols (Dublin: E. Lynch et al, 1775)
Stewart, Larry, *The Rise of Public Science: Rhetoric, Technology and Natural Philosophy in Newtonian Britain, 1660–1750* (Cambridge: Cambridge University Press, 1992)
Wollstonecraft, Mary, *A Vindication of the Rights of Woman* (London: J. Johnson, 1792)
Wu, Duncan, ed., *Romanticism: An Anthology*, 3rd edn (Malden: Blackwell, 2006).

THE ROLE OF CHEMISTRY TEXTBOOKS AND TEACHING INSTITUTIONS IN FRANCE AT THE BEGINNING OF THE NINETEENTH CENTURY IN THE CONTROVERSY ABOUT BERTHOLLET'S CHEMICAL AFFINITIES

Pere Grapí

ABSTRACT

The purpose of this paper is to show the extent to which French chemistry textbooks published during the first quarter of the nineteenth century reflected one of the conceptually most profound changes in chemistry. This was the new conception of chemical change introduced by the French chemist Claude-Louis Berthollet, which was founded on a new understanding of chemical affinity. The paper will primarily contrast Berthollet's ideas with the standard theory of chemical change founded on the theory of the elective affinities. The first part will show the relevance of that theory in French textbooks of chemistry in their educational context and, in the last part, the imprint of Berthollet's new affinities in chemistry textbooks will be analysed.

THE COMING OF A NEW UNDERSTANDING OF CHEMICAL CHANGE IN ITS EDUCATIONAL CONTEXT

Claude-Loius Berthollet (1748–1822) participated in the dissemination of Lavoisier's new chemistry. His works in the field of applied chemistry, for instance the development of a new bleaching process based on the use of chlorine, made a name for him among industrial chemists. His strict investigative methodology founded on a rigorous objectivity and the need of working out hypotheses to control experiences guided him to the construction of a new general theory of chemical affinities. However, Berthollet might also be known much more for a famous controversy that saw him in opposition to the chemist Joseph-Louis Proust regarding the proportions in the combination of chemical substances. Berthollet's belief was contrary to that of the fixed proportions of combination that Proust claimed to have proved. Proust's point of view proved to be more successful and was finally adopted by chemists.[1]

1 The best account of Berthollet's works and life is still the one by Goupil (1977).

At the beginning of the nineteenth century, Berthollet introduced an innovation that shook the world of chemistry and was considered subversive by practitioners of the discipline. Berthollet proposed an alternative view of chemical change to challenge the traditional conception founded on the system of elective affinities. This system had been firmly established and had been widely favoured by chemists as the standard one for interpreting chemical change since the early eighteenth century. "The opinions which Berthollet endeavoured to establish were not only inconsistent with those of Bergman, but utterly subversive of the whole science of chemistry, if he could have established them" (Thomson, 1831, Vol. I, p. 34).

According to Bergman's theory, the result of a chemical change was predetermined by the elective order established between two reacting substances in what were known as the tables of affinity; it reflected the belief of the savants of the eighteenth century in a concept of nature governed by strict and invariable laws. Throughout the eighteenth century these tables offered the possibility of 'seeing' the elective order of the affinities of certain substances and, consequently, were believed to provide visible evidence of that conception of chemical change. The tables were the real message for the assumed conception of chemical change.[2] Etienne-François Geoffroy's *Table de Rapports* (1718) was the prototype for a proliferation of tables of affinity during the eighteenth century. A table of affinity had to be interpreted by following its columns (Figure 1). At the head of each column is the symbol for the substance to which it refers. Below are the symbols for substances with which it reacts, arranged in their corresponding order of affinity so that the nearest substance exhibits greater affinity for the substance of reference and cannot be displaced by any of the substances lower down the column, but can remove any of them by combining with it.[3]

2 Alistair Duncan's (1962 and 1970) pioneering articles on chemical affinity in the eighteenth century are a mandatory reference. He offered a broader overview of the topic in his later book (1996). Chemical affinity as a unifying concept within chemical theory related to its nineteenth-century intellectual background in science, philosophy, and religion was treated by Trevor Levere (1971). The evolution of the concept of chemical affinity from the seventeenth century until its accommodation in the thermodynamics of irreversible processes was chronological analysed by Michelle Goupil (1991). The historian Mi Gyung Kim has reviewed the role played by the tradition of chemical affinities in the development of chemistry at the end of the eighteenth century, characterised by the works of Lavoisier and his school (Kim, 2003).

3 For instance, a substance A could decompose a compound BC (A + BC = AB + C) if, following the table of affinity, substance B showed a greater affinity for A than for C.

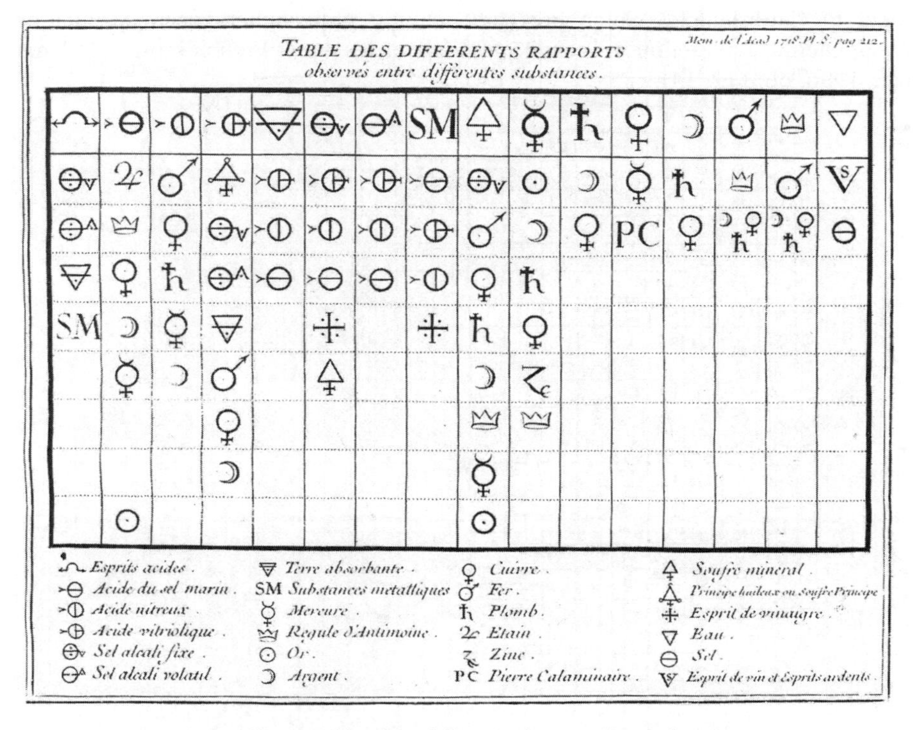

Fig. 1: Geoffroy's Table of the different relations observed in chemistry between different substances (1718).

After the publication of Geoffroy's table, 16 other tables of affinity and similar tables were drawn up before the publication of Torbern Bergman's tables of affinity in 1775 and 1783 (Figure 2). Bergman's tables were undoubtedly the most complete and successful tables among the practitioners of chemistry in the eighteenth century.[4] This long gap between the publication of Geoffroy and Bergman's tables has been associated with the resistance to Newtonian ideas and the attach-

4 The first version of Bergman's table was published by himself in 1775. Two other improved editions of that table were published in the Latin editions of Bergman's *Disquisitio de Attractionibus Electivis* (1775 and 1783). The first English translation of this text (*Dissertation on Elective Attractions*) appeared in 1785 and the French translation (*Traité des Affinités Chymiques ou Attractions Electives*) was published in 1788. The tables of affinity were not alone in supporting the diffusion and acceptance of the elective affinities. Ciphered figurative diagrams were also used to visualise chemical reactions (Crosland, 1959, pp. 65–90; Goupil, 1991, pp. 161–63, 188–89). Ursula Klein proposed an appraisal of the underlying conceptual structure of Geoffroy's table (1994 and 1995) and has studied jointly with Wolfgang Lefèvre the classificatory structures embedded in the affinity tables (Klein and Lefèvre, 2007, pp. 155–77). For an assessment of the tables of affinity as a visual summary of chemical reactions, see Kim (1992).

ment to Cartesianism, which prevented natural philosophers from speculating about chemical attraction in France and in countries under French influence (Duncan, 1996, pp. 112–14).

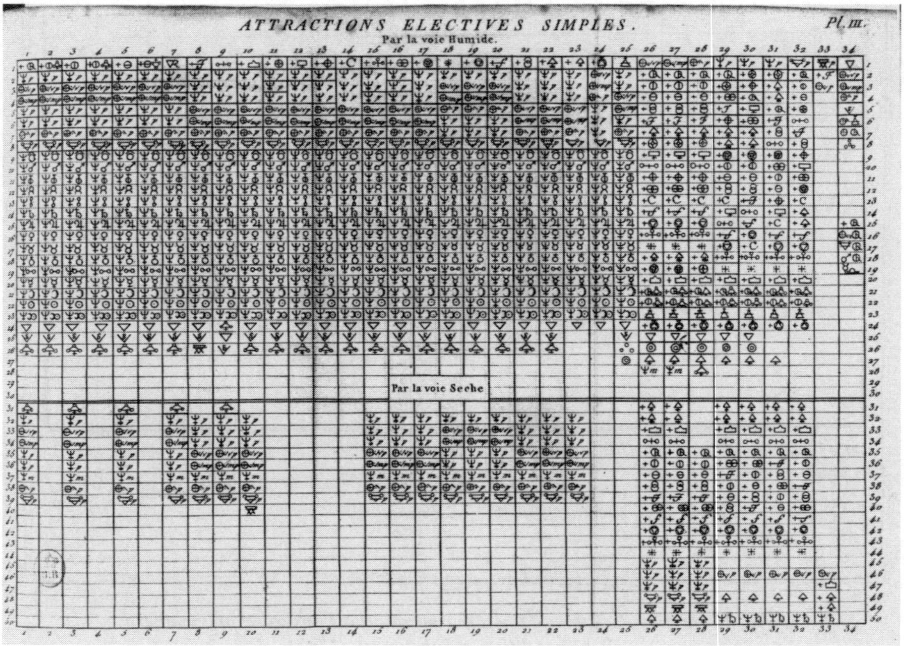

Fig. 2: The first 34 columns of Bergman's Table of Simple Elective Attractions (1788). The whole table had 59 columns and was divided into two parts—wet way and dry way—to account for the effect of heat on the affinities.

During the eighteenth century, chemical affinities became a coherent system that provided an explanation for chemical change, founded on the assumption that affinities were constant and elective. Chemical affinity was considered a Newtonian, short-range attraction that tended to link particles of different substances at such short ranges that the 'distance' factor could be assimilated by the factor of the 'shapes' of the particles.[5] It followed that the variety of the shapes of the particles (apart from heat) could cause the affinity to vary; therefore, the affinity between two substances had to be constant, because the shapes of the respective particles were constant. Accordingly, the system of elective affinities did not allow the quantity of a substance to disturb the order of the affinities established in the tables.

5 This approach was due to Buffon, who tried to generalise the laws of Newtonian attraction to chemical reactions (1765, pp. XII–XVI).

Berthollet's new system of affinities was the result of accurate observations and profound reflection on chemical change. Berthollet did not model his system on the formal physical-mathematical basis of short-range attractions, but rather around the phenomenon of neutralisation between acids and alkalis (Berthollet, 1803a, 1, pp. 68–69). He assigned to the new concept of 'chemical action' the leading role in deciding the course of a chemical change. This chemical action was understood as the tendency between two different substances to form a new combination, and was exerted according to the reciprocal affinities between substances, as well as according to their weight proportions. Disassociating the concept of 'affinity' from the leading role it had played in deciding the course of a chemical change was at the core of Berthollet's subversive ideas. For Berthollet, the affinities between substances did not have to be constant; for that reason, he had to provide another guiding assumption that could explain the variety of chemical changes and, above all, to explain the cases of chemical revertibility which challenged the system of elective affinities. The system of the elective affinities implied that the result of any chemical change was irrevocably determined and, consequently, that the end products of a chemical change could not combine again to reproduce the original substances. Chemical revertibility was thus theoretically forbidden. However, the practice of chemistry provided real instances of chemical revertibilty, which had become persistent anomalies at the end of eighteenth century (Grapí and Izquierdo, 1997, pp. 116–17).

Berthollet built his new interpretation of chemical change around the so-called laws of chemical action. The first law (the partition effect) established that when a substance acted on a compound, the subject of the combination was shared between the other two substances, depending on their affinities and their weight proportion.[6] The second law (the mass quantity effect) established that the quantity of a substance could substitute the effects of the affinity to produce the same degree of saturation. Berthollet regarded the phenomenon of chemical revertibility as a justification for his laws of chemical action and he considered the formation of compounds in variable proportions of combination as the very essence of those laws. This belief is an important point since it placed Berthollet in opposition to Proust, the advocate of fixed proportions of combination. Berthollet's prediction about the formation of compounds in variable proportions of combinations partially explains his inability to refute the theory of elective affinities once and for all. Indeed, the theory of elective affinities, which had dominated chemistry for nearly a hundred years, was abandoned without being satisfactorily replaced by another theory (Grapí, 2001, p. 117).

Along with the chemist Fourcroy, the mathematician and astronomer Pierre-Simon Laplace, and others, Berthollet was a member of what Lavoisier termed a "community of opinions", a kind of research school led by Lavoisier himself as the chef d'école. Lavoisier decided that the way forward lay in making chemistry

6 This law held that a chemical composition could occur without continuing to completion. Thus, when a substance A decomposed a compound BC, B was apportioned between A and C, and the species A, C, BC, and AB coexisted in the reacting medium.

a quantitative physical science, more akin to experimental physics than to natural history and pharmacy. Thus, Berthollet's emphasis on the relevance of quantities in chemical reactions was in the tradition of Lavoisier's new quantitative approach to chemistry. Lavoisier was sceptical regarding chemical affinity. In the 'Preliminary Discourse' of his *Traité élémenatire de chimie,* he dealt with the theoretical basis of his book by considering that affinities were still uncertain and could not provide a solid foundation for chemistry. In spite of this position, Lavoisier also contemplated affinities as the part of chemistry that might become an exact science (Lavoisier, 1789, pp. xiii–xiv). In this sense, Berthollet's research on affinities followed Lavoisier's tradition of approaching chemistry like experimental physics.

Berthollet published his system of affinities during the period between 1794 and 1803. This process was closely linked to two remarkable events: the chemistry courses at the short-lived École Normale of the year Ill (1794), and Napoleon Bonaparte's expedition to Egypt. Berthollet's exposition of affinities in the École Normale courses and the consequent publication of the *Séances*[7] marked the establishment of his own system. He finally presented his definitive ideas on chemical affinities in his 1803 book *Essai de statique chimique,* when the debate on his system of affinities had hardly even begun.[8]

Berthollet's conception of chemical change was framed within a wide and complex context in which domestic policy – economic, social, and educational – and foreign policy all played a part. A common denominator of this complex context was nitre, since the interest aroused by the production and refining of this salt in France at the end of the eighteenth century influenced, directly or indirectly, the factors that determined that context.[9] One of these factors was the establishment of a reliable chemical test to check the purity of nitre, the main component for the production of gunpowder at that time. In September 1793, Berthollet was appointed director of the nitre refinery of Saint Germain-des-Près in Paris, and it was there that an anomaly observed in the refining process of nitre became a key episode in the origin of Berthollet's affinities.[10] The problem of the nitre test was also present in Berthollet's chemistry courses at the École Normale when dealing with the anomalies of elective affinities. If, as Berthollet noted, the organisation of that course forced him to begin questioning the established princi-

7 The 12 lessons of Berthollet's chemistry course were published jointly with the lessons of the other courses at the École Normale in the *Séances*. The *Séances* were published again in 1808 to commemorate the re-foundation of the École Normale Supérieure. Berthollet's chemistry lessons (along with Hauy's physics lessons and Daubenton's natural history lessons) were published again in Guyon (2006).

8 It was between the publication of the *Séances* and the *Éssai* (Berthollet, 1803a) that the new affinities were constructed and assimilated. This process was reflected in the reading of the *Recherches sur les lois de l'affinité* at the Institute de France towards the end of 1799, in which Berthollet presented an advanced, mature explanation of his system of affinities.

9 For a reassessment of this context, see Grapí and Izquierdo (1997).

10 The refinery workers told Berthollet that, as the concentration of nitre in the washing waters increased, its capacity to dissolve more salt from the gross nitre diminished.

ples of chemistry, then the underlying theoretical considerations in the observations on the test for nitre might have been among the principles that required review (Berthollet, 1803b, pp. 288–89).

The École Normale was created by the Comité d'Instruction Publique in October 1794 to provide courses for people who would in turn become teachers in their own departmental écoles normales to train school teachers. The syllabus of the school was elementary for very few students, but highly advanced for the majority. As for Berthollet's chemistry course, it favoured deep reflection not only on chemical affinities but also on the methodology of the teaching of chemistry. Berthollet advised future teachers of the importance of making their students aware that the progress of science was unavoidably linked to the use of a rigorous methodology. He insisted on the need of teaching not only the theoretical content of chemistry, but also its public usefulness, and in this sense he encouraged teachers to pay visits to different workshops (distilleries, dye-works, etc.). Specially meaningful were his didactic guidelines for the teaching of the chemical affinities. In the first lesson – before expounding his seminal ideas on the new affinities – he explained to his students the best way of using visual resources for teaching the elective affinities: the table of affinities and the figurative diagrams. In an elementary course, he advocated the drawing up of a table of affinities by means of a wall chart developed throughout the course. His lessons revealed a true interest in the didactics of chemistry which has not been fully appreciated.[11] Berthollet's initial reflections on chemical affinities and their public presentation were, thus, also framed within the context of the construction of a national system of education in France.

THE LEADERSHIP OF ELECTIVE AFFINITIES IN THE TEXTBOOKS OF CHEMISTRY

Textbooks cannot be separated from the context in which they were used. Learning and teaching are processes that share a symbiotic relation. This reciprocal process has been regulated in schools, and it was precisely at the time of the French Revolution that education became a priority for the political authorities who needed to provide for the citizens' right to education. For an exploration of the role of chemistry textbooks in the controversy surrounding Berthollet's chemical affinities, an understanding of the aims of these textbooks in the educational context of post-revolutionary France is imperative: teaching institutions; chemistry teachers with their personal approach to the discipline and their own affiliation to a particular research school; the demands of the chemistry syllabus in

11 For a pedagogical assessment of Berthollet's chemistry course at the École Normale, see Bensaude-Vincent, Bret and Grapí (2006).

each institution or the intellectual ability of the students attending the courses.[12] In the discussion that follows, these elements will be taken into consideration in order to contextualise the teaching of chemical affinities in early nineteenth-century France, though they will not be scrutinised exhaustively.

In early nineteenth century France, apart from public institutions like the Muséum d'Histoire Naturelle and the Collège de France, important scientific research was also carried out by a private scientific society – La Société d'Arcueil, set up by Berthollet and Laplace at Arcueil in 1806. Maurice Crosland, the historian of science, pointed out that any assessment of French science at the beginning of the nineteenth century should distinguish between at least two main groups or schools: those who were committed to the Société d'Arcueil led by Berthollet and Laplace, and those who were attached to the Muséum d'Histoire Naturelle, led by the chemist Antoine-François Fourcroy. The Arcueil group promoted experimentation and discussion of problems with a physical-chemical approach, while Fourcroy's group dealt with chemistry as a discipline allied to pharmacy, natural history, and medicine (Crosland, 1967, pp. 221–23).

Fourcroy was a multifaceted character. His capacity as a researcher in chemistry and as a statesman matched his excellent communication skills as a teacher, and his commitment to politics and administration. From 1786 onwards, he was engaged in the establishment of a project to organise the teaching of chemistry. This project can be followed through his chemistry textbooks, above all the *Élémens d'Histoire Naturelle et de Chimie* (1789) and the *Philosophie chimique* (1792). This latter textbook proved to be very popular with beginners. It was recommended as a guide for first-year students at the École Polytechnique, and in 1809 it was also prescribed as a textbook for chemistry courses in lycées for secondary education. The axiomatic conception of chemistry in *Philosophie chimique* formed the basis for the organisation of the teaching of chemistry that Fourcroy presented later in his *Tableaux Synoptiques de Chimie* (1800). Fourcroy worked out his *Tableaux* guided by two criteria. First, chemical substances were arranged according to their complexity in eight classes, from simple substances to vegetable and animal substances. Second, Fourcroy assigned chemical affinity as the most suitable property to establish both distinctions and relations among those classes of substances. The perception that the knowledge of chemistry could be summarised in 12 tables provided a perspective that conveyed both Fourcroy's authority in the discipline, and the belief that chemistry was grounded on rationally organisable knowledge. These *Tableaux* were recommended for first-year students at the École Polytechnique and for teachers at central schools who were teaching chemistry for the first time. In 1806, 20 years after the beginning of the project, Fourcroy presented a definitive plan for the teaching of chemistry in the last edition of his *Philosophie chimique*. What was central to Fourcroy's programme for the teaching of chemistry was that the elective affinities were its

12 For a valuable study of the readers for whom French chemistry textbooks emerged as an independent genre during the first half of the nineteenth century, see Bensaude-Vincent, García-Belmar, and Bertomeu-Sánchez (2003).

backbone; they constituted the model for explanations of natural phenomena and appeared in the operations of the chemical arts.

Fourcroy was also involved in a project for the complete reform of the educational system, which ended in 1808 with the organisation of the different centres in the newly created Université Impériale. At that time Napoleon Bonaparte was the Emperor of France, and for him a politically established state needed a teaching body with fixed principles. This plan led to the creation of an imperial university as an institution devoted exclusively to public education for the whole empire. The experience that Fourcroy gained in this political episode helped him to work out his project for the organisation of the teaching of chemistry, while his conception of chemical phenomena also influenced his view of the educational system. "... since they [schools, special schools, and schools of application] all have close affinities, they must also have reciprocal influences on each other. Students of the first can become students of the second, emulation will double their efforts, and the good that must result from this competition will entirely affect public prosperity".[13]

Fourcroy's notion of chemical change ruled by affinities that converged to form new substances was latent in his idea of the change that various establishments of the Université Impériale, acting jointly, could generate in students, and thus provide useful citizens for the nation. In this way, Fourcroy's project for organising the teaching of chemistry was involved with the social and political processes that led to the establishment of a national educational system in France. It is in this context, therefore, that the leading role of Fourcroy's research school and of the elective affinities in the teaching of chemistry should be understood.

Fourcroy and his followers (Vauquelin, Guyton de Morveau, Brongniart, and Laugier) exerted tacit control over the teaching of chemistry in many of the higher education institutions during the first quarter of the nineteenth century. They assumed teaching posts at most of the important educational institutions like the Muséum d'Histoire Naturelle, the École de Pharmacie, the École de Médecine, and the École Polytechnique. However, as the century progressed, some of these chairs of chemistry came to be occupied by chemists associated with the Arcueil group In 1804 Louis Jacques Thenard was appointed Vauquelin's successor in the Collège de France when he took up a chair at the Muséum d'Histoire Naturelle. Joseph Louis Gay-Lussac replaced Fourcroy at the École Polytechnique in 1809 after his death, and Thenard replaced Guyton de Morveau in 1811 after his retirement.

Fourcroy's chemistry textbooks were widely used before Thenard's *Traité de chimie élémentaire, théorique et pratique* was published in 1813–14 (Thenard 1813). In his textbooks, Fourcroy defended his belief in elective affinities, pre-

13 "... puisqu'elles [lycées, écoles spéciales et écoles d'application] ont toutes des affinités intimes, elles doivent avoir aussi des influences réciproques les unes des autres. Les élèves des premières peuvent devenir les élèves de seconde, l'émulation doublera leurs efforts, et le bien qui doit résulter de ce concours rejaillira tout entier sur la prospérité publique" (Fourcroy in Hippeau, 1990). Translations into English hereafter are mine.

senting a perspective of chemistry in which elective affinities played a key role in interpreting chemical change. To this end, Fourcroy's *Élémens d'Histoire Naturelle et de Chimie* included a set of laws regulating the performance of chemical affinities,[14] supported by tables of affinity and figurative ciphered diagrams which displayed visible evidence of chemical change. These figurative diagrams (Figure 3) were intended to be a resource to facilitate the learning of the elective affinities to interpret chemical double decompositions; since they contained the names of substances according to the new nomenclature, they were also good visual propaganda both for the introduction of the new nomenclature and for the diffusion of the new antiphlogistic chemistry.

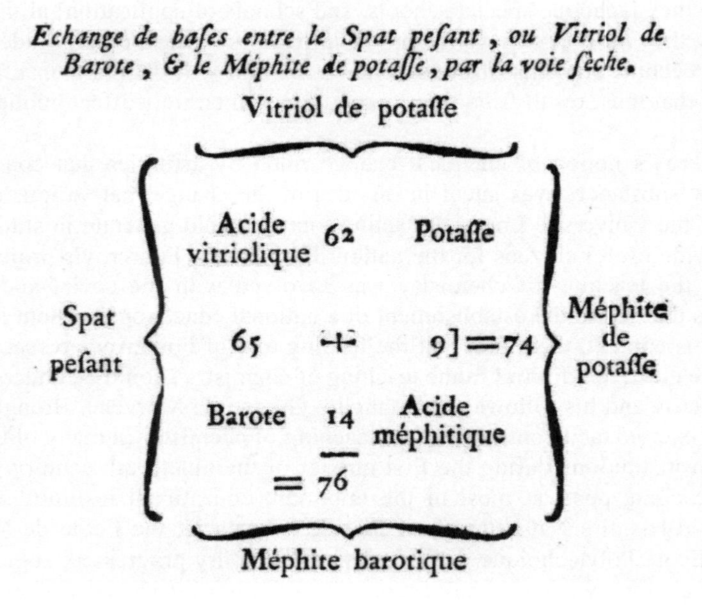

Fig. 3: Figurative ciphered diagrams of reciprocal decompositions between salts showing the corresponding degrees of attraction.

Fourcroy's textbooks became a mandatory reference for other contemporary textbooks that also presented a favourable exposition of elective affinities. This was the case, for instance, with Chaptal's textbook *Élémens de Chimie* (1790). Even though Chaptal was an industrial chemist little given to dealing with the theoretical features of chemistry, the last edition of this textbook which came out in 1803 recognised the challenge that Berthollet's innovation posed to the elective affini-

14 The number and the content of these laws varied significantly in Fourcroy's textbooks. For an analysis of the evolution of these laws in relation to Berthollet's laws of chemical action, see Grapí (2001, pp. 122, 125–26).

ties. "When chemistry was limited to the knowledge of some substances and was concerned only with certain facts, it was then possible to establish the tables of affinity … but these scales were built upon principles that have undergone modifications, the number of principles have increased, and we are forced to work on new bases".[15]

THE CONTRIBUTION OF THE FRENCH TEXTBOOK TRADITION TO THE PROPAGATION OF BERTHOLLET'S CHEMICAL AFFINITIES

The debate on Berthollet's chemical affinities from 1801 to 1817 did not result in a favourable acceptance of his system. The most important conclusion of this debate was Berthollet's own conviction that the exposition of his system of affinities had been misinterpreted.[16] However, during the first half of that period, Berthollet's affinities began to find an audience through chemistry textbooks. Therefore, the appearance of Berthollet's innovation in these textbooks deserves to be examined.[17]

Actually, it was in the first edition of Bouillon-Lagrange's *Manuel d'un Cours de Chimie* that Berthollet's law on the mass quantity effect was introduced for the first time (Bouillon-Lagrange, 1798, Vol. I, p. 14). The third edition of the textbook brought out in 1802 still favoured Berthollet's affinities, presenting them as a true alternative to the elective affinities. This book was used by the students of the École Polytechnique to recapitulate the experiments undertaken in Fourcroy's and Guyton's courses.[18] The book was soon recommended for students of

15 "Lorsque la chimie était borne à la connaissance de quelques substances, et qu'elle n'était occupée que de quelques faits, il était possible alors de dresser les tables d'affinité ... mais tous les principes sur lesquels on avait construit ces échelles, ont reçu des modifications, le nombre des principes s'est accru, et nous sommes obligés de travailler sur de nouvelles bases" (Chaptal, 1803, Vol. I, p. 30).

16 For a summary of the main obstacles that Berthollet's affinities had to face, see Grapí and Izquierdo (1997, p. 119).

17 This paper does not deal with the reception of Berthollet's affinities in British textbooks. However, it can be pointed out that British chemistry textbooks were initially more receptive to Berthollet's ideas than the French, but they were also quicker to oppose Berthollet's affinities when the emergence of chemical atomism discredited Berthollet's conjecture on the variable proportions of combination.

18 The case of the textbooks of chemistry used in the École Polytechnique is a complex one. By 1810 the Conseil de Perfectionnement of the school presented a list of the compulsory textbooks for first year students but no chemistry textbook was in this list (Bradley, 1976, p. 278). Up to the publication of Thenard's textbook in 1813, different textbooks of chemistry – as the ones mentioned in this paper – were recommended for the first year chemistry course (École Polytechnique, 1800–1801, p. 62; 1801–1802, p. 53).
Berthollet taught his new conception of chemical change in his own courses of chemistry at the École Polytechnique during the period 1801–1805. These courses were first intended to complement the students' practical instruction in chemistry, but they became courses dealing with advanced topics of chemistry that were beyond the majority of students who had only followed Fourcroy's elementary course of chemistry. Actually, Berthollet's courses were not

the École de Pharmacie, the central schools, and later the lycées (Bensaude-Vincent, García-Belmar, and Bertomeu-Sánchez, 2003, pp. 51–52).

The new organisation of secondary education in the lycées and collèges after the creation of the Université Impériale meant the loss of the prominent role that the sciences had played in the central schools' curriculum, which became progressively more literary. In addition, the fact that physics and chemistry were not subjects required for entering the École Polytechnique meant that their teaching was further neglected in the secondary schools. This new organisation of secondary education resulted in the publication of new science textbooks. René-Just Haüy's *Traité Élémentaire de Physique* and Pierre-Auguste Adet's *Leçons Élémentaires de Chimie* were prescribed as physics and chemistry textbooks for the lycées. From the second edition onwards, Haüy's textbook supported Berthollet's mass quantity effect but not his partition effect (Haüy, 1821, Vol. I, pp. 46–47). On the other hand, Adet's textbook introduced some issues of Berthollet's conception of chemical change; in particular, Berthollet's law of chemical action relative to the mass quantity effect was extended to all chemical changes (Adet, 1804, pp. 6–10). Despite the simplicity of the text and the absence of Berthollet' law on the partition effect, the book presented Berthollet's ideas without ambiguity.

Chaptal, the author of the previously mentioned *Élemens de chimie*, enjoyed a very close friendship with Berthollet, and during his time as Minister of the Interior he frequently attended meetings of the Société d'Arcueil. He returned to his home city of Montpellier after resigning as Minister, and there he began to write his *Chimie Appliquée aux Arts*. This text, in contrast with his previous *Élémens de chimie,* supported Berthollet's system more than any other text, as it presented his two laws of chemical action. What makes the theoretical background of *Chimie Appliquée aux Arts* innovative is precisely its support for Berthollet's affinities. However, this advocacy of Berthollet's ideas did not extend to the belief in the variable proportions of combination. For Chaptal, as for most chemists, admitting a degree of uncertainty in the proportions of combination of substances was equivalent to accepting that the results of chemical analyses might be accidental (Chaptal, 1807, Vol. III, p. 225).

The new higher education model implemented in France between 1806 and 1811 mainly left the role of examining lycées students intending to gain the Baccalauréat ès Sciences and Baccalauréat ès Lettres to the Faculties of Science and Letters. The training of professional experts was a duty of the grandes écoles (in a wide sense these were the École Polytechnique, the École Vétérinaire at Alfort, the École Normale Supérieure, and the faculties of medicine, law, and pharmacy). The level of science taught at the faculties of science was not high until the 1820 reassessment of the Baccalauréat raised the standards of science; future students of the faculties of medicine were required to hold this degree.

compulsory and a part of the attendees were chemistry teachers and in-service chemists. In 1805 Berthollet had to resign from these teaching duties when he realised that the content of his course did not already match the requirements of the school (Fourcy, 1828, p. 265).

From 1808, Louis-Jacques Thenard taught chemistry at the Faculty of Science at Paris. After 1804, Thenard had become a member of Berthollet's circle and was influenced by the physical approach to chemistry characteristic of the Société d'Arcueil, which is perceptible in his work on organic chemistry. Later, between 1813 and 1816, he published the first of the six editions of his best-selling chemistry textbook *Traité de Chimie Élémentaire, Théorique et Pratique.* Chemistry had undergone such important advances that the most respected textbooks were no longer able to serve as guides for those embarking on the study of chemistry. In his textbook, Thenard openly criticised the elective affinities and dismissed the tables of affinity as mere tables of decomposition. Furthermore, Thenard praised Berthollet's conception of chemical change for having overturned elective affinities, accepting the law of the chemical action on the mass quantity effect, but omitting any reference to the other law, the partition effect. It was because of his acceptance of the fixed proportions of combination that Thenard did not include Berthollet's law on the partition effect in his textbook. Assuming this law would have also implied acceptance of the validity of the variable proportions of combination, which were regarded as a serious obstacle to the results of chemical analyses.

The defence of Berthollet's affinities against elective affinities was particularly strong with regard to the reciprocal action between salts, and indeed this was the part of the book for which Berthollet's affinities were to be remembered in the future. Thenard's exposition of Berthollet's theory of uncrystallisable residua when dealing with the reciprocal action between soluble salts became a pronouncement in favour of Berthollet's affinities (Thenard, 1813–14, Vol. II, pp. 350–55). He presented Berthollet's experimental data on the reciprocal decomposition of salts using tables that made visible the mass quantity effect.[19] These tables (Figure 4) were not an exact counterpart of the affinity tables because they did not posses any predictive value; on the contrary, they transmitted the idea that the end products of a chemical change were not predetermined but dependent, both on the proportions of the reacting substances and on the experimental conditions.

Thenard's tables can be valued as visual aids showing the message of Berthollet's affinities. This was the surviving part of Berthollet's chemical laws of chemical action for which Berthollet's affinities were remembered. Therefore, Thenard and his textbook can be considered responsible for the fact that Berthollet's conclusion on the phenomena of reciprocal action between salts became finally acknowledged as the only Loi de Berthollet.

19 Actually Thenard used the same tables that Thomas Thomson had already worked out from Berthollet's experimental results (Thomson, 1802, Vol. III, pp. 224–27).

			ÉVAPORATION (a)		
Sels mêlés	**Proportions**	**Précipité**	**Sels provenant de la première**	**Sels provenant de la seconde**	**Eau mère**
Nitrate de chaux Sulfate de potasse	1 1	Sulfate de chaux	Nitrate de potasse Sulfate de chaux	Un peu de sulfate de potasse	Un petite quantité
Idem	1 2	Idem	Sulfate de potasse Sulfate de chaux	Nitrate de potasse Sulfate de potasse Sulfate de chaux	Un très-petite quantité
Idem	2 1	Idem	Sulfate de chaux Nitrate de potasse	Nitrate de potasse Très-peu de sulfate de chaux	Abondante (b)
Sulfate de soude Nitrate de chaux	1 1	Idem	Nitrate de soude	Nitrate de soude	Abondante (c)

(a) Après avoir soumis la dissolution à l'action du feu pendant un certain temps, on la laisse refroidir, afin d'en obtenir des cristaux ; puis on décante la liqueur surnageante, qu'on soumet de noveau à l'action du feu, etc. ; il en résulte donc des évaporations sucessives : ce sont ces évaporations qui sont désignées sous le nom d'*évaporations première, seconde*, etc.
(b) Composée de nitrate de chaux et de nitrate de potasse.
(c) Composée vraisemblablement de sulfate et de nitrtate de soude.

Fig. 4: A reproduction of one of Thenard's tables of the successive crystallisations between soluble salts.

BY WAY OF CONCLUSION: REASSESSING THE ROLE OF TEXTBOOKS IN A SCIENTIFIC CONTROVERSY

It is generally accepted that textbooks and scientific texts constitute the medium for instructing newcomers in a scientific discipline. The kind of scientific knowledge that textbooks seek to propagate was described by Thomas Kuhn, who said that textbooks incorporate and recount what the scientific community has taken as the body of accepted knowledge (Kuhn 1962, pp. 10, 47). Barry Barnes, in his discussion of Kuhn's contribution to the social sciences, argued that textbooks have not only underestimated certain perspectives that appeared unorthodox at one time, but have also avoided the acknowledged anomalies of normal science (Barnes 1982, p. 19). In contrast to these arguments, French chemistry textbooks at the beginning of the nineteenth century contributed significantly to the debate on Berthollet's chemical affinities, since they were able to reflect the most salient anomalies of the system of the elective affinities, as well as Berthollet's innovative ideas on chemical change.[20]

The forms in which Berthollet's affinities were presented in these textbooks were not completely loyal to Berthollet's ideas. Although the textbooks published after 1800 and in favour of the elective affinities alluded to Berthollet's conceptions, they effectively shaded them or adapted them to comply with the laws of

20 Barnes' claim stands basically for educational textbooks. Nevertheless, it does not mean that new findings and theories were not incorporated in French chemistry textbooks of the time. For instance, Fourcroy's textbooks were the first to publish Lavoisier's new chemistry but without debating its foundations in relation to its rival phlogiston chemistry. This debate was not incorporated in chemistry textbooks but in periodicals. In spite of this, certain chemistry textbooks contributed to the debate between elective affinities and Berthollet's affinities when the latter were still controversial among chemists, and this was unusual in the French chemistry textbook tradition.

elective affinities. Furthermore, none of the French textbooks committed to Berthollet's affinities offered a complete exposition of the elements of his system; they all neglected the law of chemical action that accounted for the partition effect, and none accepted the conjecture of the variable proportions of combination. The only scientific text published at that time that expounded Berthollet's two laws of chemical action was Chaptal's *Chimie Appliquée aux Arts,* a book that was not intended for beginners in chemistry. Above all, Thenard's *Traité* was the chemistry textbook for higher education that most clearly supported Berthollet's system of affinities, disseminating what would eventually be condensed into a single *Berthollet's law*.

Finally, the control that Fourcroy and his school exerted over many leading teaching institutions in France, his prolific textbook production, the influence of his manuals on other textbook writers, his activity in the field of educational politics, and his project for the teaching of chemistry in the context of the establishment of a national system of education are all factors that broaden our understanding of the prevalence of elective affinities in the French textbook tradition up to 1813. In the same context, chemistry textbooks represented an effective barrier for the implementation of Berthollet's affinities.

REFERENCES

Adet, P. A. 1804. *Leçons élémentaires de chimie, à l'usage des lycées*. Paris: Dentu.

Barnes, B. 1982. *T.S. Kuhn and Social Science*. New York: Columbia University Press.

Bensaude-Vincent, B., A. García-Belmar, A., and J. R. Bertomeu-Sánchez. 2003. *L'émergence d'une science des manuels. Les livres de chimie en France (1789–1852)*. Paris: Éditions des archives contemporaines.

Bensaude-Vincent, B., P. Bret, and P. Grapí. 2006. 'Leçons de chimie de Berthollet. Introduction', in E. E. Guyon, (dir.) *L'École Normale de l'an III. Leçons de physique, de chimie, d'histoire naturelle. Édition annotée des cours de Haüy, Berthollet et Daubenton*. Paris: Éditions Rue d'Ulm, pp. 239–52.

Bergman, T. O. 1788. *Affinités chimiques ou attractions électives*. Paris: Buisson.

Berthollet, C.L. 1803a. *Essai de statique chimique*, 2 vols. Paris: Firmin Didot.

———. 1803b. 'Essai de statique chimique', *Annales de chimie,* 46, pp. 288–93.

Bouillon-Lagrange, E. J. B. 1798–99. *Manuel d'un cours de chimie*, 2 vols. Paris: Bernard.

Bradley, M. 1976. 'An Early Science Library and the Provision of Textbooks: The École Polytechnique, 1974–1815', *Libri*, 26 (3), pp. 165–80.

Buffon, G. L. L. 1765. 'De la Nature, seconde vue', *Histoire Naturelle, générale et particulière, avec la description du cabinet du Roi*, 13 (I–XX).

Chaptal, J. A. C. 1790. *Élémens de chimie,* 3 vols. Montpellier: Picot.

———. 1803. *Élémens de chimie*, 3 vols. Paris: Détreville.

———. 1807. *Chimie appliquée aux arts,* 4 vols. Paris: Crapelet.

Crosland, M. 1959. 'The Use of Diagrams as Chemical "Equations" in the Lecture Notes of William Cullen and Joseph Black', *Annals of Science*, 15, pp. 75–90.

———. 1967. *The Society of Arcueil. A View of the French Science at the Time of Napoleon I,* London: Heinemann.

Duncan, A. M. 1962. 'Some Theoretical Aspects of Eighteenth Century Tables of Affinity', *Annals of Science*, 18, pp. 179–94, 217–32.

———. 1970. 'The Function of Affinity Tables and Lavoisier's List of Elements', *Ambix,* 17, pp. 28–42.
———. 1996. *Laws and Order in Eighteenth-century Chemistry*. Oxford: Clarendon Press.
École Polytechnique. 1800–1801; 1801–1802, *Rapport sur la situation de l'École Polytechnique présenté au Ministre de l'Interieur par le Conseil de Perfectionnement (an IX, an XI)*. Paris: Imprimerie de la République.
Fourcy, A. 1828, *Histoire de l'École Poluytechnique*. Paris: Chez l'auteur, à l'École Polytechnique.
Fourcroy, A.F. 1789, *Élémens d'histoire naturelle et de chimie*, 5 vols. Paris: Cuchet.
———. 1792. *Philosophie chimique ou vérités fondamentales de la chimie moderne*. Paris: Cl. Simon.
———. 1800. *Tableaux synoptiques de chimie*. Paris: Badouin.
———. 1990. 'Discours prononcé sur un projet de loi relatif à l'instruction publique', in C. Hippeau (ed), *L'instruction publique en France pendant la Révolution*. Paris: Klincksieck.
Geoffroy, E.F. 1718. 'Table de différens rapports observés en Chymie entre différentes substances', *Mémoires de l'Académie Royales des Sciences*, pp. 202–12.
Goupil-Sadoun, M. 1977. *Le chimiste Claude-Louis Berthollet (1748–1822). Sa vie, son œuvre*. Paris: J. Vrin.
Goupil, M. 1991. *Du Flou au Clair? Histoire de l'Affinité Chimique*. Paris: Éditions du CTHS.
Grapí, P. 2001. 'The Marginalization of Berthollet's Chemical Affinities in the French Textbook Tradition at the Beginning of the Nineteenth Century', *Annals of Science*, 58, pp. 111–135, 122–27.
Grapí, P. and M. Izquierdo. 1997. 'Berthollet's Conception of Chemical Change in Context', *Ambix,* 44, pp. 113–30.
Guyon, E. (dir): 2006. *L'École Normale de l'an III. Leçons de Physique, de Chimie, d'Histoire Naturelle. Édition annotée des cours de Haüy, Berthollet et Daubenton*. Paris: Éditions Rue d'Ulm.
Haüy, R. J. 1821. *Traité élémentaire de physique*, 3 vols. Paris: Huzard-Courcier.
Hippeau, C. (ed). 1990. *L'instruction publique en France pendant la Révolution,* Klincksieck, Paris.
Kim, M. Gyung. 1992. 'The Layers of Chemical Language, I: Constitution of Bodies v. Structure of Matter', *History of Science,* 30, pp. 60–96.
———. 2003. *Affinity, That Elusive Dream. A Genealogy of the Chemical Revolution*. Cambridge, Massachusetts and London: The MIT Press.
Klein, U. 1994. 'Origin of the Concept of Chemical Compound', *Science in Context*, 2, pp. 163–204.
———. 1995. 'E.F. Geoffroy's Table of Different *Rapports* Observed Between Different Chemical Substances – A Reinterpretation', *Ambix*, 42, pp. 79–100.
Klein, U. and W. Lefèvre. 2007. *Materials in Eighteenth-Century Science. A Historical Ontology*. Cambridge, Massachusetts and London: The MIT Press.
Kuhn, T. S. 1962. *The Structure of Scientific Revolutions*. Chicago: University of Chicago Press.
Lavoisier, A. L. 1789. *Traité élémentaire de chimie*. Paris: Cuchet.
Levere, T. 1971. *Affinity and Matter. Elements of Chemical Philosophy 1800 – 1865*. Oxford: Clarendon Press.
Séances des Écoles Normales Recueillies par les Sténographes et Revues par les Professeurs, 8 vols. Paris: Imprimerie du Cercle Social, 1795–97.
Thenard, L. J. 1813–1814. *Traité de chimie élémentaire, théorique et pratique*, 2 vols. Paris: Crochard.
Thomson, T. 1831. *A System of Chemistry of Inorganic Bodies*, 2 vols. London: Baldwin & Cradoch; Edinburgh: W. Blackwood.

INSTRUMENTS OF SCIENCE AND CITIZENSHIP: SCIENCE EDUCATION FOR DUTCH ORPHANS DURING THE LATE EIGHTEENTH CENTURY

Lissa L. Roberts

ABSTRACT

One of the two most extensive instrument collections in the Netherlands during the second half of the eighteenth century – rivaling the much better known collection at the University of Leiden – belonged to an orphanage in The Hague that was specially established to mold hand-picked orphans into productive citizens. (The other was housed at the Mennonite Seminary in Amsterdam, for use in the education of its students.) The educational program at this orphanage, one of three established by the *Fundatie van Renswoude*, grew out of a marriage between the socially-oriented generosity of the wealthy Baroness van Renswoude and the pedagogical vision of the institute's director and head teacher – a vision that fit with the larger movement of *oeconomic* patriotism. *Oeconomic* patriotism, similar to 'improvement' and *oeconomic* movements in other European countries and their colonies, sought to tie the investigation of nature to an improvement of society's material *and* moral well-being. Indeed, it was argued that these two facets of society should be viewed as inseparable from each other, distinguishing the movement from more modern conceptions of economics. While a number of the key figures in this Dutch movement also became prominent Patriots during the revolutionary period at the end of the century, fighting against the House of Orange, they did not have a monopoly on *oeconomic* ideas of societal improvement. This is demonstrated by the fact that an explicitly pro-Orangist society, *Mathesis Scientiarum Genitrix*, was organized in 1785 to teach science and mathematics to poor boys and orphans for very similar reasons: to turn them into productive and useful citizens. As was the case with the *Fundatie van Renswoude*, a collection of instruments was assembled to help make this possible. This story is of interest because it discusses a hitherto under-examined use to which science education was put during this period, by revealing the link between such programs and the highly charged question of citizenry.

INTRODUCTION

At the core of this essay are the stories of two institutions established in the Netherlands during the late eighteenth century to educate poor and orphaned boys. The first, made possible by the large bequest left by a wealthy noblewoman, was the *Fundatie van Renswoude*, which operated three orphanages in The Hague, Utrecht and Delft for specially chosen boys who demonstrated promise. Second was the privately organized Leiden society *Mathesis Scientiarum Genitrix* (MSG). The political orientations of these two institutions differed starkly during the highly charged decades which ended one century and began the next. While members of MSG demonstrated their allegiance to the House of Orange by asking the future King Willem I to be their official patron in 1785, a number of the *Fundatie*'s leading teachers and administrators agitated for the kinds of reform which were championed by the anti-Orange regime of the Batavian Republic. Nonetheless, they worked toward strikingly similar goals. Both placed practical mathematics and science at the centre of their programs' curriculum. And both did so with the express purpose of educating their charges to serve the public good. Placing the contextualized stories of the *Fundatie van Renswoude* and MSG next to each other thus offers a unique opportunity to explore how science education was appreciated across political lines in the Netherlands during a time which witnessed a full cycle from reform and revolution to war and restoration, not only in terms of stimulating knowledge production or material production for their own sakes, but also as a vehicle of socio-cultural renewal. To set the stage, this introduction will be followed by sections that discuss the context in which these stories should be placed.

In a recently published report, a European Commission committee argued for the urgency of reforming science education in the face of declining student interest. The report articulates three closely related reasons for the need to alter this trend. First, according to the report, the "[a]vailability of highly qualified science and technology professionals is a key factor for the establishment, import and success of high-tech industry in the European Union." More generally, it continues, citizens must possess increasing scientific literacy in order to understand and respond to the complex environmental, medical, economic and moral challenges that increasingly face society on both a local and global scale. Finally, the report presents science education as the key to Europe's cultural coherence. "… by giving … [citizens] the opportunity to develop critical thinking and scientific reasoning… science education helps fighting [sic] misjudgments and reinforcing [sic] our common culture based on rational thinking." (*Science Education Now*, 2007, pp. 6–7.)

Two closely related characteristics of this document make it of special interest in relation to this essay. The first is that it broadly ties science education to concern for issues ranging from economic growth and environmental sustainability to social stability, moral rectitude and cultural cohesion. Second is that its rhetoric characterizes science with the twin terms of rationality and instrumentality, and projects an image of European culture as innately rational and progress oriented.

What holds the report together, in other words, is a vision of science as the cornerstone of modern western culture which served and must continue to serve as the motor of both material *and* moral progress. It is not my purpose here to subject this portrayal to critique. Rather, I want to draw attention to the fact that this claim which links scientific education to socio-cultural concerns beyond economic and intellectual growth is strongly rooted in Enlightenment ideology.[1] From this perspective, then, focusing on scientific education only for its own sake or in relation to economic growth is just half the story.

This is not only a question of omission, however. By forgetting the moral concerns that lay at the heart of many science education programs and discussions during the eighteenth century and since, some historians have been led to misread the past. This is particularly so in the case of historians who seek to explain the Netherlands' belated process of industrialization, relative to Great Britain and Belgium, as rooted in Dutch society's declining interest and involvement in science and science education during the second half of the eighteenth century. A number of (especially economic) historians have accepted the view that modern technological and economic development are rooted in the spread and application of Newtonian mechanics, as most elegantly expressed in the work of Margaret Jacob. (Jacob, 1997, pp. 141–154)[2] This trend has led to both a narrow definition of 'science' and 'the culture of science', as well as to an equally narrow vision of the historical relation between material production and knowledge production. (Roberts et al, 2007) For while it is certainly the case that Dutch universities ceased to be the leading international centers for the study of experimental philosophy and medicine during the second half of the century, the investigation of nature continued to be valued in the Netherlands throughout the century, for reasons that went beyond the growth of scientific knowledge for its own sake, the goals of 'polite' amateur learning, and narrowly construed interests in material productivity and economic growth.

Indeed, a good deal of concern for the teaching of science, and education more generally in the Netherlands during the second half of the eighteenth century, was linked to what the Dutch called '*oeconomische patriotisme*', akin to what one might find in oeconomic and 'improvement' societies throughout Europe.[3] The word *oeconomie* did not actually find its way into Dutch dictionaries

1 That the coupling of scientific, social and moral progress was an outstanding theme during the Enlightenment, including by those who offered doubts, hardly needs a footnote. Diderot and d'Alembert's *Encyclopédie* stands out as perhaps its greatest monument.

2 For a critical variation, see Davids, 2005, pp. 330–336, in which he argues that the problem was not inattention to science, but insufficient interplay between 'science' and 'technology'.

3 There exists as yet no overarching study of this important Enlightenment movement, whose variegations were colored by local socio-political characteristics. Hence, for example, German variants developed within the context of government-sponsored cameralism, while Swiss and Dutch oeconomic societies were more engaged with the ideals and actualities of their local republican contexts and English 'improvers' operated in a context colored at least partially by the ways in which market-driven entrepreneurialism bridged the traditional gap between rural and urban forms of productivity. See, for example, Lowood, 1991; Wakefield,

until the eighteenth century. But its appearance should not be seen as signalling the birth of modern economic thinking. Rather, it drew on its etymological origins, linking market activities to a vision of society as a (domestic) household.[4] In such a situation, material productivity could not be considered as an autonomous goal; for the benefit of society, it had always to be accompanied by the goals of virtuous behavior and a sense of moral responsibility.[5]

This helps to explain why, across Europe and beyond, the eighteenth century is full of examples in which improvements in agriculture and manufacture alike were spurred by individuals and organizations who were not solely motivated by self-enrichment in the financial sense. More specific to this essay, it helps us understand why two of the largest scientific instrument and model collections in the Netherlands during the eighteenth century belonged to institutions which were not primarily interested in the advancement of science for its own sake and whose aims were other than (economically understood) profit maximization. The Mennonite Seminary in Amsterdam, which housed an enviable collection, included experimental philosophy in its curriculum so as to prepare Mennonite preachers for the task of linking an understanding of this world with concern for the world to come.[6] The other institution in question was an orphanage established in The Hague with an educational programme intended to nurture promising orphans whose talents would otherwise be lost to society. Along with fraternal institutions in Delft and Utrecht, this orphanage was funded by the *Fundatie van Renswoude*, which was capitalized by the bequest of a wealthy aristocratic widow, and staffed by a group of dedicated and socially engaged teachers.[7]

2009; Kapossy, 2007; Roberts, 2006; Mijnhardt, 1988. No study of this phenomenon would be complete without mentioning its global reach, both in terms of the ways in which global exchanges were taken up by European oeconomists and in terms of the establishment of oeconomic societies and activities outside Europe. On the first, see for example, Koerner, 1999.

4 Etymologically, the original Greek *oikos* and its Latin equivalent refer either to household management or ecclesiastical dispensation, explaining the initial Dutch translation as *huishoudkunde*. While the word *oeconomie* had acquired a broader meaning by the seventeenth century, referring to the orderly management of a larger community's resources (hence the Dutch *staathuishoudkunde*), it continued to carry with it a sense of the moral responsibility entailed in the organization and management of familial resources. See Shannot, 1736, p. 359; Arsy, 1682; Sewel, 1766, vol. I, p. 529 and vol. II, p. 350.
The introduction, definition and applied use of the term *oeconomie* in Dutch during the eighteenth century helps explain why the amorality of Adam Smith's economic views in *Wealth of Nations* originally found little fertile ground in the Netherlands, as opposed to his *Theory of Moral Sentiments*. See Kloek & Mijnhardt, 2001, p. 301.

5 Contrast this view with the otherwise informative work of the social and economic historian H. van den Eerenbeemt, who develops a claim that the Dutch Enlightenment was marked by a "marriage" between philanthropy and economics. Reference to marriage rather than (perceived) unity is to read a modern, liberal conception of 'economics' back into an inappropriate context. See e.g. Eerenbeemt, 1972 and 1977.

6 On the Mennonite instrument cabinet, see Zuidervaart, 2006.

7 For the most recent and complete study of the *fundatie*, see Gaemers, 2004.

What brings these two institutions together, beyond the fact that they invested in science education, is that they did so for reasons that help reveal how interest in and involvement with science were undergirded by a wide range of meanings attached to the concept of utility, recognized as a central tenet of the Dutch Enlightenment. (Roberts, 1999; Mijnhardt, 1987; Mijnhardt, 1998; Mijnhardt and Kloek, 2001) For some, studying nature was useful for what it taught about the creative greatness and goodness of god. For others, it was a necessary first step toward being able to control nature for material or social benefit. For some, an understanding of natural laws buttressed a drive to understand and reform the laws of society. Others sought to weave science education into a program that somehow tied meritocratic ideals to conservative political allegiance. In other words, thanks to the variegated understanding of and support for utility, the Dutch found it possible to disagree about politics while agreeing about the desirability and meritocratic goals of science education.[8]

This can perhaps best be seen by setting the *Fundatie van Renswoude*'s program alongside that of a contemporary organization established in Leiden, the amateur society *Mathesis Scientiarum Genitrix* (MSG), which also aimed to improve the lives of select orphans through the medium of science education. As mentioned at the beginning of this essay, these two organizations manifested rather different politics, but never wavered from their commitment to enrich society by educating its orphans. This study allows us, then, to go beyond the rhetoric of political opposition voiced during the turbulent years of revolution and war at the close of the eighteenth century, to examine a practical current of cultural continuity. Whether Orangist or Patriot, broad sectors of Dutch society could be counted on to support the goals of moral citizenship. For individuals such as the members of MSG and those involved with the *Fundatie van Renswoude* this entailed organizing and supporting educational opportunities for underprivileged children.

These two institutions, examined in the penultimate sections of this essay, reveal further how science and technology were harnessed to develop the talents of promising Dutch orphans for public benefit, thereby informing us about the place given science and technology in Dutch culture more generally. Based on a formula that tied meritocracy to social benefit through the medium of scientific and technological instruction, select orphans were taught mathematics, physics, architecture, instrument-making, the surgeon's art and various permutations of what we now broadly call engineering, both to realize their individual potential and to tap their talents for the good of the nation.

8 In this regard, the concept 'utility' can be considered as a boundary object. See Star and Griesemer, 1989: "Boundary objects are objects which are both plastic enough to adapt to local needs and the constraints of the several parties employing them, yet robust enough to maintain a common identity across sites...They have different meanings in different social worlds but their structure is common enough to more than one world to make them recognizable, a means of translation". p. 393. See also Daston, 1999.

While, as stated, oeconomic and improvement movements were widespread in Europe during the second half of the eighteenth century, the specifically Dutch history of having to contend with threats posed by the natural habitat left its mark on the movement's contours in the Netherlands. Dutch survival historically depended, that is to say, on shaping the physical environment and making the most of limited resources. (Lambert, 1971) If this meant centuries of innovative water management and land reclamation projects, it also required social and cultural cooperation, even when political agreement was not possible.[9] Some argued that it further required nourishing talent, wherever it might be found, for the sake of national moral and material progress. Hopeful that promising young minds could be productively cultivated through the application of science and technology, just as the Netherlands' originally savage landscape had been, these reformers tied science and technology to the meritocratic enhancement of Dutch culture and society.

MORAL CITIZENSHIP AND EDUCATION IN THE NETHERLANDS[10]

When Dutch reform advocates spoke of reinvigorating their 'fatherland' during the second half of the eighteenth century, they generally took this paternal metaphor to include a vision of the 'natural' social order and what was needed to maintain its health and stability in materially difficult times. The lower classes had a right to gainful employment, they argued, but also a duty to adhere to what many current-day politicians continue to refer to as 'Dutch norms and values'.[11] This meant first that the potential productivity of the poor – especially in urban areas that were suffering from growing overseas competition during the eighteenth century – should be tapped. If this was often expressed in positive terms as the key to alleviating poverty and re-establishing the Netherlands as an international centre of manufacture and trade, Hendrik Herman van den Heuvel – founder of the important reforming society *Oeconomische Tak* (the "Oeconomic Branch," founded in 1777 as an offshoot of the *Hollandsche Maatschappij der Wetenschappen* [Dutch Society of Science] to encourage material and moral recovery) and editor of the spectatorial journal *De Borger* – was more blunt. He strongly advised matching education to a child's social station, lest the poor come to think that they were too good to work with their hands. Too much education, he warned, was a threat to social stability. (*De Borger*, 2 (October 1778): 337–341.)[12]

9 "Few countries exist where the hand of man has exerted a greater formative influence in the shaping of the landscape". Lambert, 1971, preface.

10 For a general introduction to this topic, see Los, 2005, especially chapter 9.

11 Cornelis Ris, founder of the *Vaderlandsche Maatschappij van Reederij en Koophandel* (Fatherlandic Shipping and Commerce Company) in the town of Hoorn, for example, repeatedly described characteristics such as decency, humility, thrift, honesty and sincerity as traditional Dutch qualities. See, for example, Ris, 1777, pp. 16, 19.

12 Hake labels oeconomic patriots such as Van den Heuvel and Ris as 'conservative' for this stance, which leads him to argue more generally that oeconomic patriotism was a

In this, his views were consonant with those voiced by a number of 'enlightened' contemporaries throughout Europe. As discussed by Daniel Mornet in his classic study on the intellectual origins of the French Revolution, the French *philosophes*' ideal of popular education was predominantly to "make more useful men, not more thoughtful ones". Hence could a well-trained workforce augment the nation's prosperity without "inverting the order of things".[13]

Utility dictated for Van den Heuvel that the poor imbibe the values of hard work and civic virtue so as to render them materially productive and socio-politically passive. This can be seen, for example, in proposals that were floated within the *Oeconomische Tak* to train poor youth for service at sea. Traditionally a maritime nation, the Netherlands was losing ground to England militarily and commercially during the second half of the eighteenth century, partly due to lack of naval manpower. Some society members thought the answer was to organize an institute for maritime training, as English counterparts had done in 1756, but the society's central committee decided it was more efficient simply to offer monetary awards to boys who signed up for on-the-job training. (Titsingh, 1780) When such a school was nonetheless founded in Amsterdam by private subscription in 1785, the society did offer limited financial support. (Bierens de Haan, 1952, pp. 67–70) The emphasis remained fixed, however, on immediately perceived social benefit rather than popular enlightenment.

Other initiatives, set in motion by various municipalities and private parties during this period, tended to couple the establishment of workhouses for textile manufacture, as well as for products ranging from wallpaper to porcelain, with various incentives (either positive or punitive) to convert the 'idle' poor into productive workers. (Eerenbeemt, 1977) If domestic (a term that reminds us of the oeconomic equation of the nation with a familial household) output could thus be increased, reformers believed, Dutch society would once again flourish in terms of widespread material wealth and the kind of happiness found in moral rectitude; simply put, a return to the republic's 'golden age'. For the poor, however, the future looked somewhat less golden. Workdays in such institutions were long and often structured to include training programs for the young, who learned manual skills such as spinning and embroidery, along with basic reading, writing, arithmetic and the fundamentals of moral religion. The general goal was to train an army of citizens, through the propagation of industriousness, to accept their place in the social order while contributing to the moral and material good of the nation as a whole.[14]

conservative faction within the more general Patriot movement. Hake, 2004, p. 20. But, as stated below, Ris became so active in the Patriot movement that he had to flee his home town in 1787 when pro-Orangist forces gained political control of the area.

13 First quotation from Mornet, 1967, p. 421. Second quotation from François Philippe Gourdin, "De l'éducation physique et morale considerée relativement à la place que doivent occuper les Enfans dans l'ordre de la Société", (Lyon, 1779) cited in Chisick, 1981, p.173.

14 Reformers such as Cornelis Ris explicitly used the term 'citizen' to refer to the entire Dutch population. See Ris, 1777, passim.

This makes the educational and professional opportunities afforded orphans by the *Fundatie van Renswoude* and MSG, as will be discussed later in this essay, seem all the more striking. But, explaining the presence of these two institutions is not simply a matter of contrasting proclaimed opposites such as Enlightenment thinking and revolutionary ideology (both organizations were established too early, in any event, for this to be the case) or Orangist conservatism and Patriot progressivism. Ultimately, we can best understand them as manifesting a small but telling facet of the Netherland's complexly hewn culture of utility – one that tied science and technology to the narrowly applied ideals of meritocracy.

Even before turning to the *Fundatie* and MSG, however, we can see just how complex this culture was by recognizing the role played by the House of Orange in supporting efforts to revive Dutch productivity during the late eighteenth century, as well as after its establishment as the Netherlands' royal house after the fall of Napoleon. When the reforming society *Oeconomische Tak*, original home of Dutch oeconomic patriotism, was established in 1777 to stimulate a fusion of material and moral progress in the Netherlands, it did so with Stadholder Willem V as its official patron. Not only did he announce his intention henceforth to prefer domestic products over imports, a basic tenet of the society, he and his family put their money where their mouths were by supporting various ventures intended to produce and bring domestic goods to market. They became major stockholders, for example, in Cornelis Ris' *Vaderlandsche Maatschappij van Reederij en Koophandel* (Fatherlands Shipping and Commerce Company), which oversaw a network of production centers that specialized in products ranging from whaling ships to stockings and painted wallpaper as well as administering a school in which poor children were taught religious and moral responsibility, reading, writing, arithmetic and the sorts of basic manual skills with which they could be put to work.

Ris was pleased to have such highly-placed backing, but this didn't stop him from becoming active in the local Patriotic Society of Hoorn – so active, in fact, that he had to flee the city in 1787 following an Orangist victory there. (Kooijmans, 1985, p. 195.) The stadholder's family similarly bought shares in the porcelain factory and school set up in 1774 by the pastor Johannes de Mol in the small town of Loosdrecht. Further, Willem V made a point of paying official visits to sites that promised oeconomic progress, whether a semi-mechanized cloth factory in Amersfoort in 1777, a newly erected steam engine near Rotterdam in 1790 or a workhouse in Groningen in 1791. (Roberts, 2004; Eerenbeemt, 1977, pp. 40, 93, 98–114.)

When we combine this with the fact that Orangist financiers, such as the Hope family (owners of one of Europe's largest private banking organizations), organized and underwrote portions of the financing for such ventures, two things become apparent. First, no easy contrast or opposition between (oeconomic) Patriots and Orangists is possible in the realm of supporting oeconomic reform – an important vehicle for furthering the drive toward moral citizenship in the Netherlands. Second, the ability to share such socio-cultural goals across an apparent political divide and even cooperate in seeking their attainment did not end with the onset of revolution. Traces of this shared urge toward oeconomic reform could still be found in the Netherlands long after Batavians and Bonapartists alike were replaced by the establishment of a kingdom under Willem I (son of Stadholder Willem V). Consider, for example, the king's supportive stance regarding post-revolutionary Dutch industry and commerce, which can be seen as a partial outgrowth of the examples given by his father and 'oeconomic patriots' alike, as well as a partial continuation of policies previously enacted by King Louis Napoleon. Not only did King Willem I serve as official patron of MSG from its inception in 1785 (before he became king, that is), a position he resumed upon his return to the Netherlands in 1813 and continued to hold until his death in 1843. He was also a major stockholder in the *Nederlandsche Handel Maatschappij* (Dutch Trading Company – established to stimulate Dutch commerce, employment and overseas trade) and supported initiatives such as the series of weaving schools set up by Thomas Ainsworth in the early 1830s to provide employment and stimulate textile production in the Netherlands following separation with Belgium. (Coeverden, 1983) Clearly the historical circumstances – including economic (as opposed to oeconomic) policies and practices – were different by this time, but it makes more sense in this context to speak of the Dutch culture of utility having undergone a complex process of evolution than to refer to the revolutionary period as constituting a watershed.

AN ALTERNATIVE VIEW

As we have now seen, a number of prominent voices and initiatives in the Netherlands thus sought to mediate the challenges of progress and social control through programs that trained the poor to be materially productive, morally upright and socio-politically passive. If this was a fairly typical manifestation of Enlightenment thinking at the time, the physician and amateur historian Simon Stijl (1731–1804) offered a different perspective on the way in which educating the poor should serve the common good. In 1774 he published his widely read work, *De opkomst en bloei van de Republiek der Vereenigde Nederlanden* (*The Rise and Flourishing of the United Dutch Republic*), in which he offered a different view of the relationship between public education and social utility. Like many Dutch contemporaries, Stijl placed the broadly perceived problem of national decline in a moral context and argued that the source of his compatriots' woes was that they had been seduced by decadent French mores. Traditional values and industriousness had given way to materialism and idle sophistication. (Stijl, 1774, especially p. 687)[15] According to Stijl, the key to national recovery was education...but of a

15 For biographical background and a discussion of Stijl's "Enlightenment" style of history writing, see Smitkamp, 1950. While the explanation of Dutch moral decline through "Frenchification" was fairly common, it is interesting to note a variation on the theme offered by Stijl's conservative contemporary Elie Luzac. In addition to his more general social criticisms, Luzac claimed that Dutch natural philosophy was suffering from the rage for

certain kind. He advocated establishing a society to educate poor children, especially orphans.

To indicate what he had in mind, Stijl asked, "Why is mathematics, which is so useful for guiding human understanding, never counted among the characteristics of a good upbringing?"[16] Unlike his French counterparts, Stijl declared, he was not afraid of teaching children to think, no matter how unfortunate their circumstances. It was in society's best interest, he asserted, to help all its youth acquire the combined virtues of modesty, reflection and productivity. An upstanding cadre of lawyers, doctors, mathematicians, artists and craftsmen could thus be formed, who would bring credit to their fatherland as they helped to improve its material and moral conditions.[17] Rather than destabilize the social order, orphans gifted enough to achieve these goals would strengthen the society in which they lived.

EDUCATING ORPHANS

Before we finally turn to the special educational programs organized for orphans by the *Fundatie van Renswoude* and MSG, providing a sketch of the kind of life and learning experienced by most Dutch orphans is in order. The *Heilige Geest* orphanage in Leiden, from which students were recruited for MSG, provides a good example with which to begin. In 1777 nearby Leiden University's physics professor Jean Nicolas Sebastian Allamand had four model steam engines from the university's instrument collection put on display in the orphanage's courtyard for student lectures and public demonstrations. Despite the display's highly innovative character, no one at the orphanage recorded their reaction. Were orphans present at demonstrations? How did they react to university students in their midst? Were they also 'on display' when visitors came to see the orphanage and its advertised engines?[18] Meeting university students and tourists might have held up a sort of cultural mirror in which orphans could see their images reflected against the perception of those who observed them and their surroundings. But no

"French science" by which he meant the "superficiality" of the *Encyclopédistes* and the kind of *physique amusante* practiced by the Abbé Nollet. See, for example, various remarks in his *Nederlandsche Letter-Courant,* vol. III (1760), *passim.*

16 "Waarom wordt de Wiskunst, die so nuttig is om aan het verstand eene geregelde leiding te geeven, nooit onder de punten van een fraaie opvoeding geteld?" Stijl, 1774, p. 689. For trends in society membership and the distinction between dilettante and reforming societies, see Mijnhardt, 1988. For Dutch spectatorial literature, see Johannes, 1995.

17 Contrast Stijl's plan with more usual programs for the "education" of orphans, which often amounted to no more than exploiting their labour. Groenveld, Dekker & Willemse, 1997; Hoker, 1982.

18 We should remember here both that an article about the placement of one of the machines in the courtyard appeared on the front page of the *Leidse Courant*, 22 June 1777 and that visits to institutions such as orphanages and asylums were common forms of entertainment. See Roberts, 2006.

recorded testimony exists to tell us how orphans perceived this complicated set of cultural relations.

We do, however, know how orphans were educated at the *Heilige Geest* orphanage and in the Netherlands generally.[19] First, like all Dutch orphans, they were required to wear uniforms which were coloured to match the coat of arms of the city that housed their institution. *Heilige Geest* orphans thus dressed in red with blue, black and white accessories, and a patch on the left sleeve to represent Leiden's municipal shield. (Mieris, 1784, vol. III, p. 105) Whether inside or outside the orphanage, then, the status of Dutch orphans was clearly marked. Daily life and future prospects were similarly regimented, making their place and expectations equally clear.

Indeed, regimentation was a key pedagogical tool. Consider this primary rule of conduct: "that during the day and evening school time, no one may leave his place without the teacher's consent, nor is anyone allowed to ask questions or recite their lessons out loud, but rather to be so quiet that nothing can be heard". This disciplined atmosphere coupled with an overriding emphasis on religion (all the books in the orphanage library, for example, dealt with religious subjects) were meant to prepare orphans for a life of docile acceptance.[20]

The first thing they had to accept was a fixed schedule. Younger children had lessons every weekday from nine until twelve and two until five; lessons on Saturdays went from nine until eleven. The time was spent reading and writing, mostly in relation to catechism, and basic arithmetic. Three afternoons a week were deemed enough for older children who worked doing laundry, sewing and the like. Classes for older boys apprenticed in crafts as textile manufacturing, plumbing, smithing, painting and glass making, were held in their workshops. And for a final group of older girls whose work kept them busy all day, school was held at night from seven to nine. Lessons in penmanship, number recognition, and Christian morality remained an organized and organizing part of every orphan's life as long as they lived under the orphanage's care.[21]

19 For a good introductory overview, see Groenveld, Dekker & Willemse, 1997, especially pp. 194-215.

20 "Dat geduurende de dag en Avondschooltijd, niemand van zijn plaats mag opstaan zonder consent van de meester, als ook dat niemand vermag iets 't zij vragens of lessen verstaanbaar te leeren, maar so stil dat er niets van gehoord kan worden". *Archief Heilige Geest Weeshuis*, Leiden municipal archives, #3743. The catalogue of library books is found in #3744. Not everyone internalized this lesson, of course. The records of the orphanage contain numerous accounts of runaways and other disciplinary problems.

21 Ibid., #3743 and #3865. The only exception to these requirements was for orphans who were placed in service, either military, naval or with the West or East Indies Companies. See #221 and #3887.

MATHESIS SCIENTIARUM GENITRIX

If anything, such a disciplined environment seems more conducive to deadening imagination and curiosity than to fostering them. But in 1786, potential salvation came for a small number of orphans at the *Heilige Geest* Orphanage. A local society dedicated to teaching working class children mathematics offered to educate the orphanage's most promising (male) students on a yearly basis at its own expense.[22] Though its founders neither belonged to the *Oeconomische Tak* nor cited Simon Stijl, their society's history straddled what the one practiced and the other preached. The *Genootschap der beschouwende en werkdadige wiskunde, onder de zinspreuk: "Mathesis Scientiarum Genitrix"* (Society for theoretical and practical mathematics, under the motto "mathematics gives birth to science/knowledge") was established in Leiden in 1785, thanks especially to the efforts of Pieter van Campen, surveyor, mathematics teacher, wine gauger, captain in the *Lydsche Oranje Schutterij* (Leiden Orangist Militia), and author of books ranging in subject from algebra and mechanics to the construction of self-powered water mills and (bad) poetry. Many of the society's members shared his Orangist sympathies, leading them to seek (successfully) the patronage of the prince who ultimately became King Willem I. Under his official protection, they set out to serve their fatherland by instructing "*de kinderen des volks*" in mathematics and its practical applications. In 1786, they decided to include a small number of orphans in the society's student population.[23]

Originally, the society offered a mixed program. Students from Leiden University and the city's middle class were trained to help teach fortification design, the mathematical analysis of projectiles, and navigational techniques to working class boys in the hopes of making them fit for military service or life at sea. If this seems in line with the *Oeconomische Tak*'s orientation, the curriculum offered to orphans promised more. Instead of being channelled toward a career in the army or at sea, orphans were introduced more broadly to the principles of mathematics, architecture, experimental physics, and drawing. Too little remains in the archives to indicate whether the primary goal of this program was to train skilled craftsmen or expand the orphans' horizons. A class schedule from 1792 which does still exist includes courses in arithmetic, seamanship, mechanical engineering (*werktuigkunde*), drawing, geography, architecture, experimental physics and algebra. We know further that society members sought to maintain the program's educational level by organizing a subsidiary society in which they

22 For more general discussion of mathematical education at this time, see Beckers, 2003.

23 For biographical information on Pieter van Campen, see *Nieuw Nederlandsch Biografisch Woordenboek*, 1924, vol. 6, columns 260-261. Other founders included Arnoldus van Gennep, law student at University of Leiden, J. Reinier, theology student, Pieter Rijk, bricklayer and teacher, A. de Bruine de Neve, medical student, and Betholomeus van den Broek, sculptor and teacher. Van den Broek taught drawing and draughting for the society for some fifty years before retiring. The quoted phrase is from Bemmelen, 1910, p. 41. For the agreement between the society and the orphanage, see *Archief Heilige Geest Weeshuis*, #3757.

practiced operating scientific instruments and built machine models that could then be shared with the children. Finally, courses were held for a time in the same room as that which housed the artist Frans van Mieris' *Maatschappij van Schilder- en Teekenkunde* (Society for the Arts of Painting and Drawing). One can only imagine the impact of such encounters and surroundings.[24]

In 1802 the orphans' curriculum was adopted for everyone and university and middle class students were no longer trained to teach their lower-class piers. (Dijk, 1885, p. 17) MSG's Orangist members might have been less inclined to train sailors and soldiers for the Batavian Republic, but local politics undoubtedly also had something to do with this newly chosen direction. As early as 1788 Allamand's successor at Leiden University, C.H. Damen, complained about the society's program and pretensions. Threatening action to protect his own academic turf and clientele, Damen also criticized society members for thinking they could effectively teach military and navigational arts without having experience themselves and without including a practicum in their lessons. No war was ever won or sea ever crossed by reading a book, he averred. (p. 14)[25]

Damen was less damning of the courses offered by the society to a select group of orphans. It never hurt anyone to learn how things work and it might actually happen that an orphan would thereby be led to rise above his circumstances. But even here Damen remained less than optimistic. He argued that few orphans had the attitude and aptitude needed to succeed as an architect or millwright. As if to echo his words, the only surviving letters from the society's teachers to the orphanage from this period are filled with disparaging remarks about orphans' lack of ability and commitment.[26]

This did not keep the society from carrying on with its self-appointed mission, however, nor from attaining both a good reputation and social standing for its work. Public examinations, for example, advertised students' ability to calculate complicated sums in their heads. Students learned from both published textbooks and handbooks written specially by society members. They studied experimental physics and mechanical engineering (*werktuigkunde*) through experimental demonstrations with the society's collection of instruments and mechanical models, as well as by building their own model machines. As the nineteenth century progressed, MSG carried on, firmly anchored by its dual allegiance to monarchy and meritocracy.[27]

24 For the class schedule see Peeperkorn, 1985, p. 73. The sub-society mentioned was originally named the *Wis- en werktuigkundig genootschap onder de spreuk: kunst word door oefening aangekweekt.* It continued to exist until 1840 as the *Physisch Genootschap*. For mention of the MSG's relation with Frans van Mieris, see Ibid., p. 27.

25 For more details, see *Archief Koninklijk Genootschap 'Mathesis Scientiarum Genitrix'*, Leiden municipal archives.

26 *Archief Heilige Geest Weeshuis*, #3758.

27 See Peeperkorn, 1985 for nineteenth and twentieth century.

THE FUNDATIE VAN RENSWOUDE

The evidence that Damen drew on for his somber picture of orphans' educational achievements came from his (however biased) familiarity with the *Fundatie van Renswoude*. In 1754, Maria Duyst van Voorhout, vrijvrouwe (baroness) van Renswoude died, leaving two legacies.[28] The first was a sum of 500,000 guilders to provide advanced education for select orphans from the Municipal Orphanage of The Hague, the Reformed Orphanage of Delft, and the *Stadsambachtskinderhuis* of Utrecht. The second is the historical presence of a Dutch noblewoman who collaborated with natural philosophers and practiced a kind of philanthropy usually associated with late eighteenth-century reforming societies.[29]

Together these legacies point to how complicated the socio-political map of eighteenth-century Dutch culture was. Simon Stijl's patriotism led him in the late eighteenth century to combine his call for popular education with a Montesquieuian program of mixed government in which Stadholder, regents and "the people" formed a productively balanced whole. (Smitskamp, 1950, p. 216) Members of *Mathesis Scientiarum Genitrix*, on the other hand, linked Orangist sympathies with Enlightenment ideals even during revolutionary times. Decades earlier, the Vrijvrouwe van Renswoude shared the disdain of a number of her fellow regent-aristocrats for the House of Orange while she showed her intellectual and economic independence by endowing a unique set of educational institutions for orphans rather than leaving the whole of her fortune to relatives. (Langenbach, 1995)

Van Voorhout was explicit about the sort of education she wanted specially chosen orphans to receive. They were to be trained in "mathematics and physics, drawing or painting, sculpting or carving, building strong dikes to protect our land from heavy inundations of water, and similar liberal arts."[30] In practice this meant that the educational program was divided in two. Internally, *Fundatie* pupils followed a curriculum which included mathematics instruction (*mathesis*), drawing and catechism lessons. Pupils further studied foreign languages, penmanship, current events and geography. Outside the foundation, pupils received individualized training, in keeping with the career for which they were being prepared. In sum, their education was intended to prepare them for useful lives as watchmakers, instrument makers, surveyors, engineers, architects, surgeons, artists, teachers and

28 For biographical information, see Booy & Engel, 1985, pp. 19-22.

29 The Vrijvrouwe van Renswoude apparently met Antonie Van Leeuwenhoek while growing up in Delft and retained a friendship with him in later years, though the only letters still extent to evidence a relationship are those written by Van Leeuwenhoek to her husband. These letters contain details of microscopic observations made in her home. See *Alle de Brieven van Antonie van Leeuwenhoek* volume 10 (1979), p. 161. See also Wall, 1760, p. 18.

30 "...Mathesis, teekenen of schilderkonst, Beeldhouwen of Beeldsnijden, oeffeningen in sware Dijkagien tot behoudinge van ons Landt tegens sware overstromingen van 't water, of dergelijke Libere Konsten". This portion of her testament is quoted in Langenbach, 1991, p. 10. I leave the questions of artistic and surgical training largely aside (which do not alter my argument) because of space limitations.

military officers. A series of portraits made of the Foundation's first students in Utrecht, captured the promise this education bestowed on the boys by portraying them with the instruments of their chosen trade in hand.[31] So was Jan Weteling, who specialized in navigation and surveying and who, after spending time in service of the Dutch East Indies Company, became a school master in later life, pictured holding an astrolabe. (pp. 37–39) Jan Wormerus Raven, whose study in surveying and engineering ultimately led to his appointment as inspector general of Water Management (*inspecteur-generaal van Waterstaat*) and mayor of the Zeeland town of Sas van Gent, posed with a Holland circle in his hands. (pp. 56–58) And Hendrik Groenendaal, whose training as an instrument maker included a temporary apprenticeship in Paris and who later worked making instruments for Martinus van Marum and Teylers Museum, chose to be portrayed with a microscope he had made. (pp. 84–86)

The foundation was also interested in raising its students' social and cultural horizons. Foundation orphans wore a specially designed uniform to distinguish them from those who remained in the institutions from which they came. Beside academic courses, they learned dancing, the niceties of polite conversation and table manners. And students were encouraged to learn about the creative arts and to read as extensively as possible. The idea was that they should they not only serve society, but that they should politely fit into it as well.

These were boys who would never have attended university on their own, yet they were exposed to many of the same authors and instruments as university students. An inventory of the Delft library from 1790, for example, lists the titles of some 200 study books. Students were further supported to purchase their own books. Books in Dutch, Latin, French, and German on subjects including natural history, mathematics, astronomy, mechanics, military arts, fortification building, geography, architecture, shipbuilding, machine building, navigation, and surveying appeared in the library and student collections, indicating that students had a broad range of interests and working knowledge of more than one language.[32]

The physics cabinets of the three foundations were well enough equipped to rival, if not surpass, that of Leiden University. In 1787, for example, the Delft foundation bought a large portion of Leiden Professor Allamand's private instrument collection when he died. An inventory taken immediately after the purchase lists some 265 instruments spread over eight categories: mechanical and machine models; hydrostatic and hydraulic instruments; instruments for experimenting with air, fire and magnetism, light, and electricity; instruments for astronomy and navigation as well as sundials; and, finally, measuring devices and artillery instruments.[33]

31 Ibid. This book is based on an exhibition held in January 1991 at the Centraal Museum, Utrecht.

32 See Booy & Engel, 1985, appendix 1 (pp. 275 - 283) for a complete list of titles.

33 Ibid., appendix 2 (pp. 285 - 290) lists the entire inventory of the Delft collection. For The Hague, see Florijn, 1802, p. 47.

In The Hague, the situation was even more lavish. An inventory of its collection from 1802 fills seventy five pages and includes 667 separate entries, some of which were items purchased from well known London instrument makers and others crafted by foundation students. (Florijn, 1802) One of the entries, for example, described what must have been an impressive copper and ivory orrery made by the London instrument maker E. Wright, but powered by a mechanism made by one of the orphans, Johannes Anthonie van Baarda. (p. 6.) Others included a wide range of measuring devices, globes made by the surveyor and geographer to Queen Anne, John Senex, an octant from John Hadley's hand, several balances and devices for demonstrating physical principles which were discussed by 's Gravesande, Nollet, Desaguliers and Musschenbroek (relevant page numbers are given in the inventory). So too were detailed working models of various machines, mills and constructions such as bridges and building elements part of the collection as were a number of apparatus for use in the study of water management. Both English and domestically made air pumps were present, along with various items for studying air(s), sound and heat. Of special interest in this last category were two items discussed by 's Gravesande – an aeolipile and a steam-driven wagon – both of which demonstrated the basic principles and power of controlled steam. Optical instruments also abounded as did devices for demonstrating static electricity and gravity, making this a highly exemplary collection both in terms of its size and breadth of coverage. Foundation students were indeed a lucky lot.

The institution's governors initially budgeted 3000 guilders annually (and more, as needed) for the purchase of instruments. The amount tapered off as the instrument collection grew, but results remained impressive. To compare, Leiden University spent some 10,000 guilders on instruments for the university physics cabinet for the period 1742–1811, leaving it with 410 items. During the same period, about 25,000 guilders were spent on the Teylers Museum collection in Haarlem, where Martinus van Marum sought to turn the museum into an international center for natural philosophical research. Given these figures, the Swedish astronomer and diarist Bengt Ferrner had cause to predict that, within a few years of his visit in 1759, the foundation in The Hague would house the best instrument collection in the Netherlands.[34]

34 For The Hague, see Kernkamp, 1910, pp. 492-494. It is worth quoting a sizeable excerpt: "Twaalf van de vlugste jongelingen uit het Burgerweeshuis worden hier opgenomen om te worden onderwezen in Frans, tekenen, schermen en dansen, maar vooral in mathematica en fysica, in welke laatste vakken zij onderwezen worden door een zekere la Faille, die, te oordelen naar het gesprek, dat ik met hem voerde, een geschikt en bekwaam man is voor dat doel. Alle theorie wordt zoveel mogelijk bekrachtigd door experimenten, waarvoor allerlei soorten van instrumenten gekocht worden, zonder er op te letten, of zij duur zijn of niet. La Faille heeft slechts te bestellen bij de beste instrumentmakers, die hij kent, en op voorwaarde, dat het instrument zo goed wordt, dat het volkomen beantwoordt aan zijn doel en de goedkeuring verwerft van de professoren Musschenbroek en Lulofs in Leiden, mogen zij zoveel daarvoor bedingen, dat zij met voordeel hun tijd en hun opmerkzaamheid daaraan kunnen wijden en geen vrees behoeven te koesteren, dat men hun afdingt. Jaarlijks is een som

The precise details of how Van Voorhout's instructions were to be carried out were left to local committees. Hence, the institutions in Delft, Utrecht and The Hague differed from each other in a variety of ways that at least partially reflected the differences among the three orphanages from which Van Renswoude charges came. But they shared a basis in mathematics and experimental physics, a special emphasis on fields that related to water and land management along with the making of precision instruments, and an apprenticeship system that placed students in the best possible place to learn their chosen trade. These facts are revealing in a number of ways. They tell us what the governing regents considered necessary to prepare talented boys for a life combining personal development and public service. We see also their commitment to bolstering the Dutch tradition of seeking security and prosperity through engineering – whether of the land- or waterscape.

Thanks to the enlightened largesse of their benefactress, then, *Fundatie van Renswoude* students were exposed to the latest innovations in the arts and sciences. If they pursued this striking opportunity, they stood to realize the utilitarian and meritocratic ideal that authors such as Stijl wrote about and so many others dreamed of. And, contrary to what Damen reported in his disparaging letters to MSG, a number of orphans did precisely this, obtaining university degrees in a few cases and rising through the ranks of organizations such as the East Indies Company, private engineering firms and public water boards. Teachers might have complained about some students' lack of diligence and some students might not have completed the program, but many of those who did went on to enjoy productive careers. Of the forty two pupils admitted to the Utrecht program between 1761 and 1795, for example, twenty four are known to have finished their education and to have gone on to establish careers. In some cases, the contextual challenges of economic downturn and revolution proved too great, but there were enough success stories to make this a remarkable episode.[35]

As we will see, the pool from which the Utrecht program drew its charges was the weakest of the three. Not surprisingly, then, completion rates and career paths of students in Delft and The Hague evidence a higher degree of success. Of the fifty orphans who entered the Delft foundation by the end of the eighteenth century, for example, only eight did not complete their education successfully – one because of premature death. And while a small number (4) of Delft's graduates went on to work in various capacities on VOC ships and foreign colonies, most

van 3000 gulden bestemd voor het aankopen van instrumenten, zodat hier binnen weinig jaren de grootste verzameling bijeen zal zijn, die men ergens findt; bovendien is er meer geld beschikbaar voor instrumenten, wanneer in een of ander jaar meer dan 3000 gulden nodig is. La Faille deed in onze tegenwoordigheid enige proeven met de instrumenten, die hij had, welker aantal reeds zo groot is, dat het hier te wijdlopig zou worden om ze op te sommen, want er zijn er reeds voor een bedrag van meer dan 10,000 gulden. Alle proeven gingen precies, zoals zij volgens de theorie moesten gaan".
For instrumental purchases at Leiden and Haarlem, see Clercq, 1987, p. 171.

35 For an overview of student successes, see Langenbach, 1991, for Utrecht; Booy & Engel, 1985, for Delft, and Hardenberg, 1964, for The Hague.

managed to practice their occupations in the Netherlands. These included carpenters and furniture makers, jewellers, surgeons, millwrights, teachers and surveyors. (Booy and Engel, 1985, pp. 291–295)

ORPHAN EDUCATION AND THE URGE TO REFORM

Successful *Fundatie* students owed much to their teachers, especially the *mathesis* (pure and applied mathematics and physics) instructors who served as unofficial directors of the three schools and who inspired their students both pedagogically and by example. From the start, these teachers worked to give substance to the meritocratic vision later articulated by Stijl that coupled personal development with national rebirth. As evidenced by the inaugural speech given in Delft by the man who would go on to be head teacher there for thirty years, this vision provided the very context in which the foundation's mission would be pursued. Johannes van der Wall used the occasion to make two points. He claimed first that physical labor alone could do little to advance society. Only an education stressing the arts and sciences could spur useful inventions and social happiness. Second, he praised the foundation's decision to establish separate schools for the most promising orphans rather than trying to teach the same program to all. Not every child is blessed with equal potential, he explained. Those with greater abilities deserve to be nurtured for the good of society. (Wall, 1760, pp. 21–22, 35–36)[36]

Before coming to Delft, Van der Wall taught structural engineering (*bouwkunde*) to military officers as he wrote a doctoral dissertation on navigation (*ars navigandi*) at Leiden University. Though tempted by offers to become professor at the universities of Leiden and Utrecht, Van der Wall stayed in Delft until his death in 1787. In addition to his work for the foundation, he examined captains and pilots for the East Indies Company, gave public lectures on mathematics and astronomy for the city of Delft and served as the city's inspector of public works. These tasks never eroded his dedication to his students, however. When good textbooks were unavailable, for example, Van der Wall wrote his own, on subjects ranging from practical geometry, mill construction and navigation to physics, geography and human anatomy.[37] Further, he arranged the best apprenticeships possible for his students and followed their progress closely. In one case, he mentored the young man who would become his own successor, making sure that the foundation supported Abraham van Bemmelen until he completed his studies with Allamand at Leiden University. (Booy & Engel, 1985, pp. 58–64) In another

36 It is striking to notice that Van der Wall repeatedly made use of metaphors drawn from water-management to speak of the foundation' s mission.

37 Van der Wall apparently custom-made handbooks for his students that fit with their lessons in ways that available published texts could not do. They were never themselves published, nor have examples survived. What we know of them comes from reports Van der Wall made to the orphanage's board of regents. See especially "Rapport Van der Wall" (7/11/1783), *Archief van de Fundatie van Renswoude* #60.

case, Van der Wall shepherded Frederick Willem Conrad through an apprenticeship as surveyor, leading to his employment by the agency responsible for overseeing the Netherlands' rivers (the predecessor of today's Department of Water Management (*Rijkswaterstaat*)). Conrad ultimately became the agency's inspector general, following in the footsteps of Leiden University's physics professor Johan Lulofs. (Winter, 1988, p. 103)

Jacob Baart de la Faille showed similar dedication and ability in The Hague. While foundation director, he also taught mathematics to The Hague bureaucrats, showing his students by example how to combine theory with practice. Baart de la Faille *fils*, who taught by his father's side before becoming professor of mathematics and physics at the University of Groningen, paid homage to his father by posthumously publishing his book *Verhandeling over de rekenkunsten* (Lectures on Mathematics) in 1778 and by teaching his own courses in Groningen in a way that stressed practical application as much as theory. (pp. 101–105; Hardenberg, 1964, p. 162)

Utrecht was a different situation; so is the story of its head teacher. The *Stadsambachtskinderhuis* (Municipal Workers' Orphanage), from which Utrecht foundation students came, was established to house children who had nowhere else to go. Local Catholic orphanages wouldn't take them because they were born out of wedlock or were not fully Catholic. The more prestigious *Burgerweeshuis* (Burger Orphanage) would not accept them because their parents weren't municipal citizens. These were children from whom little was expected and to whom little was traditionally given. Unsurprisingly, none of them seemed appropriate candidates for the foundation when the first round of examinations was held. To solve this problem, a preparatory school was established to ready potential students for the foundation. Its director had to work closely with the foundation's head teacher to insure that things proceeded smoothly and effectively. In the case of the first director and head teacher, Dirk de West and Laurens Praalder, the results were striking both within and beyond the foundation.[38]

In 1773 Praalder used his spare time to help establish a society dedicated to applying the arts and sciences to social needs, the *Genootschap 'Besteedt den tijd met Konst en Vlijt'* (Society 'give time to art and industry').[39] De West became one of the society's first members. Not everyone was pleased with the society's unorthodox and un-hierarchical organization, however. Members seeking higher status for themselves and the society applied for and received a provincial patent that officially recognized them as the *Provinciaals Utrechtsch Genootschap* (Provincial Society of Utrecht). Not only was the society reorganized hierarchically, it redirected its purpose along more traditional lines. Praalder's vision of social im-

38 Langenbach, 1991, p. 102, makes the important point that, while the students in Utrecht overall seem to have achieved less than their fellows in Delft and The Hague, the success of the foundation in Utrecht has to be measured against the initial circumstances in which these orphans found themselves.

39 For a quotation from the society's constitution, see Singels, 1923, p. 30. For details regarding goals and membership structure see Mijnhardt, 1988, p. 107.

provement through public education was replaced by the desire for dilettantish gratification.[40] He and De West left the group.

They and their views soon found a new home however. This time it was De West who helped found the Utrecht chapter of the *Maatschappij tot Nut van 't Algemeen* (Society for Public Welfare), a soon-to-be nationwide society dedicated to the same ideals that motivated him and Praalder at the foundation and, previously, at the *Genootschap 'Besteedt den tijd met Konst en Vlijt'.* Initially a small society, *Het Nut* (as it was popularly called) developed into a network of local chapters where participants shared local interests and an increasingly national orientation that called for nationwide educational reform. The society's success can be gauged in two ways. First, it had fifty-two chapters by the turn of the century. Second, it provided the intellectual force behind a series of laws that organized elementary education on a national basis in the Netherlands beginning in 1801. (Lenders, 1988, especially pp. 24–48)

Members of *Het Nut* were motivated by their commitment to creating a nation of moral citizens. (Mijnhardt, 1988, pp. 259–294) We can recognize this retrospectively as a political goal in that it rested on a nationalist vision of the Netherlands – a vision in which all citizens received equal treatment under the law. In order to achieve this end, though, the society (like many other late eighteenth-century Dutch societies) declared itself officially apolitical. Discussions of party politics and partisan religion were barred from its meetings, which opened its membership to otherwise outspoken Patriots, loyal Orangists and men who were pragmatic enough to ride the rough waves of contemporary politics in order to get things done. (p. 266)

And get things done they did. A number of popular books on topics ranging from arithmetic and child-rearing to physics and natural history were published thanks to society efforts. And by cultivating good relationships with government officials, *Nut* members gained appointments as educational inspectors, overseeing a fledgling national school system based on their recommendations. The system's structure was laid out in a government commissioned report, written and published by *Het Nut* in 1796. The report called for a two-tiered program. A general curriculum was proposed for the nation as a whole, while a more advanced course of study was to be offered to those with "special" capacities. (*Algemeene Denkbeelden*, 1796) The society embraced equal treatment under the law, therefore, but did not advocate social homogenization. Its proposal was explicitly meritocratic. Only those showing promise would be nurtured beyond the minimum. All students would be taught basic literacy, numeracy and morality; only a few would study science and languages. (p. 94) As Van der Wall had stated years before at the *Fundatie van Renswoude's* inauguration, society was moved by scientific enlightenment and not the brute force of physical labor. But only a few were ca-

40 We can see this by contrasting the work that Praalder and De West did at the foundation with the content of pedagogical guides written by another of the society's prominent members. W.E. de Perponcher (member of the Utrecht nobility) directed both his books (1774 & 1782) exclusively to members of the upper classes.

pable of providing that intellectual power. The rest would serve the national good by internalizing the precepts of Christian morality and following their superiors' lead.

If the Netherlands remained a hierarchical society throughout the long eighteenth century by intellectual accord as well as practical circumstances, its educational programs and ideals helped give it form and direction. This should not be read, however, as an endorsement of the view that it was a society in which a growing gap separated the world of learning and culture from the practical concerns of work and technical development.[41] Dutch universities continued to serve a variety of purposes under the unifying banner of utility while private initiative supported various schemes for redressing what was broadly perceived as the twin plagues of material and moral decline. Among these private initiatives were the amateur society *Mathesis Scientiarum Genitrix* and the *Fundatie van Renswoude*, both of whose programs reflected a facet of the Dutch Enlightenment's utilitarian ideal of disciplining nature and society for the public good. By paying special attention to morals, experimental physics, mechanics, applied mathematics and engineering, they prepared their orphans to serve a society of which they were made a part.

To round out this picture, two important qualifications must be added. First, Dutch faith in meritocracy seems less the outcome of a rise in individualism than the result of a commitment to the Netherlands as a national community. In this context, science and technology provided a medium for putting Dutch resources to profitable use, whether through engineering the landscape or educating orphans who showed promise. This form of 'nationalism' was not a product of revolution, but a longer-standing Dutch tradition that had itself grown out of centuries of practical experience. It was, therefore, a sort of hybrid that mediated between an ideal and idealized national community on one hand and the realities of localism on the other. This helps to explain the co-existing pursuit of and commitment to reform, on one side, and resistance to the centralizing urges of programs put forth by both the Batavian and Napoleonic regimes.

Second, Dutch culture in the second half of the eighteenth century nurtured the myth of apolitical sociability, whereby various groups – aristocrats and middle class, Orangists and Patriots – could participate in the achievement of the same or similar goals despite their overt political differences. If revolution and war had a major impact on such activities, it was primarily to create financial distress and instability. Throughout it all, however, the concept of utility – liable to a range of overlapping meanings and applications – leant cultural continuity where social and political interests could not.[42]

41 Compare Davids, 1990 with Frijhoff, 1981, pp. 287-288.
42 See note 8 above.

POSTSCRIPT

Both MSG and the Fundatie van Renswoude continue to exist to this day, though in rather different forms than that met in this essay. In general this has to do with, on one hand, historical developments in the field of Dutch education and, on the other hand, with the history of orphanages from which they drew their student-charges. As the nineteenth century progressed, secondary schools and institutions of higher education – including those that offered scientific technical courses in their curricula – increasingly appeared and became accessible to a broader range of young people. In conjunction with such broad developments in educational opportunities, programs run within orphanages ultimately became redundant. Faced with this situation, the *Fundatie van Renswoude*'s three governing boards decided to take joint action in 1913. After establishing that they would not legally be held in violation of the foundation's original purpose, as laid out in the Vrijevrouw van Renswoude's will, they redirected the foundation's money to providing scholarships for promising students (not only orphans) who wanted to attend university.

Those responsible for MSG responded to the changing environment in a rather different way, both in terms of the kind of education offered and in terms of funding. As the nineteenth century progressed, surrounding technological and industrial developments increased the need for daytime training of boys to be augmented by courses in mathematics, physics and various sorts of engineering. To meet this demand, MSG organized an evening educational program, increasingly encouraged by the support of Leiden's cultural and intellectual establishment. By the end of the century, a program in electrical engineering was established, under the personal guidance of Leiden professor H. Kamerlingh Onnes. This longstanding link with the university both fed MSG's reputation and provided future opportunities for MSG students. By 1921, more than 450 students attended its evening school.

Following the end of World War Two a day school was also started to answer Leiden's ever-growing demand for technical studies. By 1966 almost 1100 students were in attendance. Quite understandably, such dynamic developments could not have been financed by private subscription alone. Neither was it seen as desirable to maintain the schools as privately administered institutions, given the various government regulations to which they had to adhere. Alongside the historical transition from a small program, overseen by a private society (sixty-seven students received instruction in 1793) to a multi-tiered educational institution, it was almost inevitable that private funding would be replaced by government financing. Though it no longer sports its full Latin name, the fully public MSG Technical High School continues to offer technical education to teenagers in Leiden.

REFERENCES

1796, *Algemeene Denkbeelden over het Nationaal Onderwijs, ingeleverd in den jaare 1796, van ween de vergadering van hoofdbestuurders der Nederlandse Maatschappij: Tot Nut van 't Algemeen*, Amsterdam.
1924, *Nieuw Nederlandsch Biografisch Woordenboek*, A.W. Sijthoff's Uitgevers-Maatschappij, Leiden.
2007, *Science Education Now: A Renewed Pedagogy for the Future of Europe*, Office for Official Publications of the European Communities, Luxembourg.
Archief Heilige Geest Weeshuis, Leiden municipal archive.
Archief Koninklijk Genootschap 'Mathesis Scientiarum Genitrix', Leiden municipal archive.
Archief van de Fundatie van Renswoude Archive, Delft municipal archive.
Arsy, J. L. d': 1682, *Le Grand Dictionaire François-Flamende*, J.J. Schipper, Amsterdam.
Beckers, D.: 2003, *Het Despotisme der Mathesis: Opkomst van de Propaedeutische Functie van de Wiskunde in Nederland 1750–1850*, Uitgeverij Verloren, Hilversum.
Bemmelen, J.M. van: 1910, "Rede gehouden up de 106de Algemeene Vergadering, 23 April 1891", *Jubileum: Uitgave en Programma der Feesten ter Gelegenheid van het 125 Jarig Bestaan van het Genootschap Mathesis Scientiarum Genitrix te Leiden*, MSG, Leiden.
Bierens de Haan, J.: 1952, *Van Oeconomische Tak tot Nederlandsche Maatschappij voor Nijverheid en Handel*, Tjeenk Willink and Son, Haarlem.
Booy, E.P. de & Engel, J.: 1985, *Van Erfenis tot Studiebeurs. De Fundatie van de Vrijvrouwe van Renswoude te Delft. Opleiding van Wezen tot de Vrije Kunsten in de 18de en 19de Eeuw. De Fundatiehuizen. Bursalen in Deze Eeuw*, Fundatie van Renswoude, Delft.
Chisick, H.: 1981, *The Limits of Reform in the Enlightenment: Attitudes Toward the Education of the Lower Classes in Eighteenth-Century France,* Princeton, Princeton University Press.
Clercq, P. de: 1987, "In de Schaduw van 's Gravesande. Het Leids Physisch Kabinet in de Tweede Helft van de Achttiende Eeuw", *Tijdschrift voor de Geschiedenis der Geneeskunde, Natuurwetenschappen, Wiskunde en Techniek* 10: 149–173.
Coeverden, F.W.W.H. van: 1983, *Goor en Thomas Ainsworth. Uit de Geschiedenis van Twentsche Texteilnijverheid*, Commissie Ainsworth-herdenking, Goor.
Daston, L.: 1999, "Afterword: The Ethos of Enlightenment". In W. Clark, J. Golinski & S. Schaffer (eds.) *The Sciences in Enlightened Europe*, University of Chicago Press, Chicago, pp. 495–504.
Davids, C.A.: 1990, "Universiteiten, Illustre Scholen en de Verspreiden van Technische Kennis in Nederland, Eind 16e – Begin 19e Eeuw", *Batavia Academica* VIII: 3–34.
Davids, C.A.: 2005, "Shifts of Technological Leadership in Early Modern Europe", in C.A. Davids & Lucassen, J. (eds.), *A Miracle Mirrored: The Dutch Republic in European Perspective*, Cambridge University Press, Cambridge.
De Borger.
Desaguliers, J.T.: 1751, *De Natuurkunde uit Ondervindingen opgemaakt*, Amsterdam.
Dijk, J.A. van: 1885, *Toespraak: Het Eeuwfeest van het Genootschap Mathesis Scientiarum Genitrix te Leiden*, P.W.M. Trap, Leiden.
Eerenbeemt, H.F.J.M. van den: 1972, "Het Huwelijk Tussen Filantropie en Economie: een Patriotse en Bataafse Illusie", *Economisch- en Social-Historisch Jaarboek* 35: 28–64.
Eerenbeemt, H.F.J.M. van den: 1977, *Armoede en Arbeidsdwang: Werkenrichtingen voor 'Onnutte' Nederlanders in de Republiek 1760–1795*, Martinus Nijhoff, The Hague.
Florijn, J.: 1802, *Naamlijst en Korte Beschrijving van alle de Wis- en Natuurkundige Werktuigen, behoorende aan de Stichting van Wylen de Hoog Edele Welgeboorene Vrouwe, Vryvrouwe van Renswoude, in het Haagsche Burger Weeshuis*, The Hague.
Frijhoff, W.: 1981, *La societé Néerlandaise et ses gradués, 1575–1814*, Rodopi, Amsterdam.
Gaemers, C.: 2004, *Nalatenschap als Toekomst. De Fundaties van de Vrijvrouwe van Renswoude 1754–1810*, Walberg Press, Zutphen.

Gravesande, W. 's: 1746, *Elemens de physique, demontré et confirmé par des Experiences*, Leiden.

Groenveld, S., Dekker, J.J.H. & Willemse, Th.R.M. (eds.): 1997, *Wezen en Boefjes. Zes Eeuwen Zorg in Wees- en Kinderhuizen*, Verloren, Hilversum.

Hake, B.: 2004, "Between Patriotism and Nationalism: Johan Hendrik Swildens and the 'Pedagogy of the Patriotic Virtues' in the United Dutch Provinces During the 1780s and 1790s", *History of Education* 33: 11–38.

Hardenberg, H.: 1964, *Het Burgerweeshuis voor Nederlands Hervormden te 's Gravenhage 1564–1964*, Zuid-Hollandsche Boek- en Handelsdrukkerij, The Hague.

Hoker, M. d': 1982, "Arm Onderwijs voor Arme Kinderen. Bijdrage tot de Studie van het Lager Beroepsonderwijs in de Zuidelijke Nederlanden Tijdens de Achttiende Eeuw", Werkgroep Achttiende Eeuw, *Onderwijs en Opvoeding in de Achttiende Eeuw: Verslag van het Symposium, Doesburg 1982*, Holland University Press, Amstedam, pp. 103–111.

Jacob, M.: 1997, *Scientific Culture and the Making of the Industrial West*, Oxford University Press, Oxford.

Johannes, G.J.: 1995, *De Barometer van de Smaak: Tijdschriften in Nederland 1770–1830*, SDU Publishers, The Hague.

Kapossy, B.: 2007, "Republican Political economy", *History of European Ideas* 33: 377–389.

Kernkamp, G.W.: 1910, *Bengt Ferrner's Dagboek van Zijne Reis door Nederland in 1759 (Bijdragen en Mededeelingen van Het Historisch Genootschap* XXXI).

Kloek, J. & Mijnhardt, W.: 2001, 1800: *Blauwdrukken voor een Samenleving*, SDU, The Hague.

Koerner, L.: 1999, *Linnaeus: Nature and Nation*, Harvard University Press, Cambridge.

Kooijmans, L.: 1985, *Onder Regenten: De Elite in een Hollandse Stad, Hoorn 1700–1780*, De Bataafsche Leeuw, Amsterdam.

Lambert, A.: 1971, *The Making of the Dutch Landscape*, Seminar Press, London.

Langenbach, M.: 1991, *Onbekend Talent. Leerlingen van de Utrechtse Fundatie van Renswoude, 1761–1795*, Walburg Press, Zutphen.

Langenbach, M.: 1995, "Een Eigenzinnige Dame: Maria Duijst van Voorhout, Vrijevrouwe van Renswoude", *Jaarboek Oud-Utrecht*: 85–112.

Lenders, J.: 1988, *De Burger en de Volksschool: Culturele en Mentale Achtergronden van een Onderwijshervorming, Nederland 1780–1850*, SUN, Nijmegen.

Los, W.: 2005, *Opvoeding tot Mens en Burger: Pedagogiek als Cultuurkritiek in Nederland in the Achttiende Eeuw*, Verloren, Hilversum.

Lowood, H.: 1991, *Patriotism, Profit, and the Promotion of Science in the German Enlightenment: The Economic and Scientific Societies, 1760–1815*, Garland, London & New York.

Luzac, E. (ed.), *Nederlandsche Letter-Courant*.

Mieris, F. van: 1784, *Beschryving der stad Leyden*, Leiden., Amsterdam.

Mijnhardt, W.: 1988, *Tot Heil van 't Menschdom. Culturele Genootschappen in Nederland, 1750–1815*, Rodopi, Amsterdam.

Mijnhardt, W.: 1998, "The Dutch Republic as a Town", *Eighteenth Century Studies* 31: 345–349.

Mijnhardt, W. and J. Kloek: 2001, *1800: Blauwdrukken voor een Samenleving*, Sdu Publishers, The Hague.

Mornet, D.: 1967, *Les Origines intellectuelles de la Révolution française*, A. Colin, Paris.

Musschenbroek, P van: 1762, *Introductio ad Philosophiam Naturalem*, Leiden.

Nollet, J.A.: 1759, *Natuurkindige Lessen, door Proefneemingen bevestigd,* Amsterdam.

Palm, L. C. (ed.): 1979, *Alle de Brieven van Antonie van Leeuwenhoek*, vol. 10, Swets & Zeitlinger, Lisse.

Peeperkorn, L.: 1985, *Het Genootschap Mathesis Scientiarum Genitrix*, Private printing, Leiden.

Perponcher, W.E. de: 1774, *Instruction d'un père à son fils*, Utrecht.

Perponcher, W.E. de: 1782, *Onderwijs voor kinderen*, Utrecht.

Ris, C.: 1777, *Ontwerp ter proeve ter oprigtinge eener Vaderlandsche maatschappy van Reedery en Koophandel, ter liefde van 't Gemeenebest*, De Erve Houttuin, Amsterdam.

Roberts, L.: 1999, "Going Dutch: Situating Science in the Dutch Enlightenment". In .W. Clark, J. Golinski & S. Schaffer (eds.) *The Sciences in Enlightened Europe*, University of Chicago Press, Chicago, pp. 350–388.

Roberts, L.: 2004, "The Arcadian Apparatus: The Introduction of Steam Engines into the Dutch Landscape", *Technology and Culture* 45: 251–276.

Roberts, L.: 2006, "P.J. Kasteleyn and the Oeconomics of Chemistry", *Ambix* 53: 255–272.

Roberts, L.: 2007, "Devices Without Borders: What an Eighteenth-Century Display of Steam Engines can Teach us About 'Public' and 'Popular' Science", *Science and Education* 16: 561–572.

Roberts, L., S. Schaffer and P. Dear (eds.), *The Mindful Hand: Inquiry and Invention from the Late Renaissance to Early Industrialisation*, Edita KNAW, Amsterdam, 2007.

Sewel, W.: 1766, *Volkomen Woordenboek der Nederduitsche en Engelsche Taalen*, Kornelis de Veer, Amsterdam.

Shannot, S.: 1736, *Nieuw Woordboek der Nederlantsche en Latynsche Tale*, Joannes van Braam, Dordrecht.

Singels, N.J.: 1923, *Uit de Geschiedenis van het P. U. G. (Provinciaals Utrechtsch Genootschap) 1773–1923*, A. Oosthoek, Utrecht.

Smitkamp, H.: 1950, "Simon Stijl als Verlicht Geschiedschrijver", *Bijdragen voor de Geschiedenis der Nederlanden* VI: 199–217.

Star, S.L. & Griesemer, J.R.: 1989, "Institutional Ecology, 'Translations' and Boundary Objects: Amateurs and Professionals in Berkeley's Museum of Vertebrate Zoology, 1907–1939", *Social Studies of Science* 19: 387–420.

Stijl, S.: 1774, *De Opkomst en Bloei van de Republiek der Vereenigde Nederlanden*, Amsterdam.

Titsingh, G.: 1780, *Bedenkingen over de schaarsheid van Zeevaarend Volk in het gemeen, en het verval onzer nationale zeevaart in het bizonder*, Amsterdam.

Wakefield, A.: 2009, *The Disordered Police State: German Cameralism as Science and Practice*, University of Chicago Press, Chicago.

Wall, J. van der: 1760, *Redenvoering ter plegtige inwijinge van het Kunst- en Kweekschool, volgens de laatste wille van wylen de Hoog Welgeboorne Vrouwe, Maria Duyst van Voorhout, vryvrouwe van Renswoude*, Delft.

Winter, P.J. van: 1988, *Hoger Beroepsonderwijs Avant la Lettre. Bemoeiingen met de Forming van Landmeters en Ingenieurs by de Nederlandse Universiteiten van de Zeventiende en Achttiende Eeuw*, Noord-Hollandsche Uitgevers Maatschappij, Amsterdam.

Zuidervaart, H.: 2006, "'Meest Alle van Best Mahoniehout Vervaardigd'. Het Natuurfilosofisch Instrumentenkabinet van de Doopsgezinde Kweekschool te Amsterdam, 1761–1828", *Gewina* 29: 81–112.

THE SCIENTIFIC CULTURE IN EIGHTEENTH TO NINETEENTH CENTURY GREEK SPEAKING COMMUNITIES: EXPERIMENTS AND TEXTBOOKS

Constantine Skordoulis, Gianna Katsiampoura and Efthymios Nicolaidis

ABSTRACT

The scope of this paper is to describe how the Greek scholars of the late eighteenth and the early nineteenth centuries related to the political and intellectual developments in Europe during the Enlightenment through scientific instruments and experimental physics. We review the organization of Greek education during the Ottoman domination, focusing on the secondary schools and the scholars who introduced the experimental methodology in the late eighteenth century. In our analysis, the role of the scholar–teacher and his textbook are central to the adaptation and appropriation of the new scientific culture based on experiments and the experimental method. Furthermore, our analysis of the late eighteenth and early nineteenth century Greek physics textbooks focuses on what we term 'iconic experiments' (experiments performed by famous European scientists), since they constitute one of the main channels through which European experimental tradition was transferred to the Greek speaking regions. Throughout this paper, the teaching of the new science of 'Newtonianism' is associated with the ideas of the Enlightenment and the French Revolution, and is considered to have paved the way for the liberation of the Greek population from the Ottoman Empire.

INTRODUCTION

Scientific instruments in modern Greece appeared in the late seventeenth century when Chrysanthos Notaras (1663–1731) built some basic astronomical instruments, including a quadrant astrolabe (for his life and work, see Stathi, 1999). This type of astrolabe was a cheaper version of the classic astrolabe and was well known to the Ottomans. It originated from the Arabs and the first person who introduced it to the West was Profatius Tibbon (1236–1305), a Jew from Provence; thus the name given to the instrument was 'astrolabe of Profatius'. Notaras' instruments are the first for which there was profound evidence and they were preserved even around the 1930s, but since then they have been missing. Around 1700, Notaras purchased some survey instruments during his educational trips to

Italy and France, as one might infer from his book *Introductio ad Geographiam et Sphaeram* (1716). Later, he also purchased telescopes for the entertainment of the Greek princes of the Romanian provinces of the Ottoman Empire.

These instruments along with a few others that were available in the Greek area in the eighteenth century were exceptions in an educational landscape where experiment and observation were absent. The Byzantine tradition did not favour the acquisition of scientific instruments. Byzantine mathematical sciences supported the learning of complex calculations, but disregarded experiments and observations in favour of formal reasoning. The University of Padua, the main seat of study for Greeks in the seventeenth and eighteenth centuries, resisted the advent of the new physics and hence the experiment before the establishment of the Teatro di filosofia sperimentale by Giovanni Poleni in 1739 (Vlahakis, 1998).

During the late eighteenth century, and especially the years after the French Revolution and the changes that took place in the Greek speaking communities of the Ottoman and Austro-Hungarian empires, experimental physics made its way into Greek secondary schools (Henderson, 1971). There were experimental laboratories in the schools of Ioannina, Milies, Chios, Smyrna, Kydonies, Constantinople, Bucharest, and Jassy. The articles of the teachers in the journal *Hermes the Scholar* and travellers' reports give evidence for the scientific instruments and laboratories. Very few of these instruments have survived, such as those of two partisans of modern Greek Enlightenment, Anthimos Gazis and Theophilos Kairis. Gazis sent instruments to teach physics and geography to the school of Milies from Vienna, and Kairis purchased a telescope during his stay in Paris in 1810 (Vlahakis, 1999). These instruments were used mainly for qualitative demonstration experiments and were intertwined with a certain ideology: the introduction of the new science as a product of the Enlightenment. In this framework, experiments were performed both for the students and the public. Their purpose, apart from teaching, was to impress the audience so it could recognise the importance of science, which was considered in enlightened Europe as the champion of all human activities.

The scope of this paper is to describe the following relation: how the Greek scholars of the late eighteenth and the early nineteenth centuries related to the political and intellectual developments in Europe of the Enlightenment through scientific instruments and experimental physics. Our research is situated within a wider research area examining the reception of scientific ideas from countries of the European scientific centre to countries of the European scientific periphery (see on that theme, Blay and Nicolaidis, 2001; Skordoulis, 2009).

To examine issues of reception requires discussing the ways in which ideas and methods that originate in a specific cultural and historical context are introduced into a different context, with different intellectual traditions and educational institutions (see Dialetis, Gavroglu and Patiniotis, 1999; Mendoza, Nicolaidis, and Vandersmissen, 1999). In this respect, examining the role of the scholar-teacher is crucial. The scholars of the periphery were not simply disseminating ideas and methods acquired from the centres of Europe, but they acted upon them and adapted them to the local setting. The role of the scholar-teacher and his textbook were central to the adaptation and appropriation of the new scientific culture. The 'experiment' and the new experimental method was the most important constituent of this culture.

What we argue in this paper is that the introduction of the new mode of scientific thinking based on experiment was a necessary precondition for the connection of the national aspirations of the Greek middle classes with the Enlightenment project, and that the teaching of the new science of 'Newtonianism' became a political action associated with the ideas of the Enlightenment and the French Revolution, and was thought of as paving the way for the coming national liberation. Our paper gives a general outline of the introduction of the experimental method in Greece, and of the physics textbooks published during the late eighteenth and early nineteenth centuries by Greek scholars for citizens living in what is now mainland Greece and among the Greek speaking communities in what was then the Ottoman and Austro-Hungarian empires.

The paper is organised as follows: in the first section, we describe the socio-historical context for appreciation better understanding of the opposition to the new experimental method and its political and educational function. In the second section, we give an account of the educational institutions in mainland Greece and among the diaspora (the Greek communities that were founded in Western Europe by the Byzantines after the Ottoman conquest, for example, in Venice, and later by Greek merchants who fled the Ottoman empire, for example, in Vienna, Trieste, Odessa, etc.) where the new experimental method was introduced in physics instruction, and also of the teachers who introduced the new curricula. We discuss the didactical use of the experiments and its importance in educating the population in a new way of thinking, paving the way for the course of national liberation that was to follow. In the third section, we classify the published physics textbooks and analyse the experiments described in five of the most widely used textbooks published in the period between 1766 and 1812 in Vienna, Leipzig, and Venice. Our analysis is based on a thematic classification of the experiments described therein, with emphasis on what we term 'iconic experiments' (that is, experiments performed by scientists that were already famous in the period of interest).

SOCIO-HISTORICAL BACKGROUND

After the conquest of Constantinople by the Ottomans in 1453, Sultan Mohamed II appointed a new patriarch to the Orthodox Church, granting him jurisdiction over many aspects of religious and civil life of the Christian populations of the empire. The Patriarchate was integrated as a state institution in the Ottoman Empire. That was a well-calculated manoeuvre in order to reinforce the schism between the Catholic and Orthodox Churches initiated in 1054. One of the most important privileges granted to the Patriarchate was the responsibility for the education of the Christian population.[1] Till the seventeenth century, the curriculum aimed at strengthening the Orthodox Christian faith and preparing the students, whose number was limited, to become members of the clergy and to hold administrative positions in the institutions of the Patriarchate. During the eighteenth century, however, the content of what was taught was not solely determined by the Church; rather, it was the outcome of the social relations of power, the often diverging priorities and conflicting interests of the religious hierarchy and of the various social groups with significant economic activities. It is therefore necessary to give a brief exposition of the agendas of the religious and social groups involved.

In order to understand the Enlightenment's impact on Ottoman Balkan society, we should remember that in the pre-1820s Ottoman Balkans, most of the urban strata, mercantile groups, and religious and secular elites were either ethnic Greeks or those enculturated into the Greek ethnie. Both the peasantry and the literate and urban Greek Orthodox groups were "Greek" in the sense of being Orthodox. After 1750, the influence of the western Enlightenment led to secularisation, liberalism, and an undermining of the religious world view of the Eastern Church. With the French Revolution, this trend intensified. Greek Orthodox intellectuals reconceptualised the Orthodox Rum millet. They argued for a new, secular "Hellenic" national identity (see Roudometof, 1998).

At the beginning of the sixteenth century, the education of the Greek speaking population was mainly religious and centred around the education of the officials of the clergy. Later in the seventeenth century the Patriarch Kyrillos Loukaris (1570–1638) appointed the Neo-Aristotelian philosopher Theophilos Korydaleas (1574–1646) as the director of the Patriarchal Academy. Korydaleas introduced the study of Aristotle's physics in the Academy and wrote extensive comments on *Physics* and on *Generation and Corruption*, initiating an educational policy aimed at shaping a cultural identity of the Orthodox population distinct both from Muslims and Catholics, and in direct continuity with the Ancients. Within this framework, secondary schools were founded, scholarships were granted to students to study at the universities of Western Europe mainly in Italy, and the publication of books was financed (Nicolaïdis, 2001).

In the late seventeenth century, a new social stratum appeared in the scene. This was a bourgeois group of Greek origin called 'the Phanariots'[2] who were involved in commercial and financial activities, and who wanted to increase their capital through an affiliation with the state institutions of the Ottoman Empire. Thus the Phanariots served as high-ranking officials in the Orthodox Patriarchate

1 The Ottoman Empire based its domination on the organisation of 'millets', that is to say, a system for controlling non-Muslim populations by delegating major responsibilities to their religious leaders who were appointed by the Sultan. The first millet to be created was that of the Orthodox Christian Church, followed by the Armenian millet and the Jewish millet. The millets had their own laws, collected their own taxes in lieu of their loyalty to the empire, and managed their own educational system.

2 The Phanariots took this name from the neighbourhood of Phanari in Constantinople where the Orthodox Patriarchate is still located and where most of them lived.

and in the Sublime Porte (the Court of the Sultan). They were appointed as diplomats by the Sultan and, most important, as Governors of the Danubian counties of Wallachia and Moldavia, developing the already influential cultural Greek presence in these provinces (Demaras, 1977). The Phanariots were situated within the framework of 'enlightened despotism' and sought the modernisation of the Ottoman state. In their eyes, this modernisation was associated with the improvement of their position within the Ottoman state structures; therefore, they opposed any idea of a Greek nation state. At the same time, as they matured as a social substratum, they disputed the authority of the Patriarchate as the sole mediator between the Sultan and the Greek population. They especially disputed the sole authority of the Patriarchate over educational matters and so, in their areas of governance, they founded the 'Hegemonic Academies' (Colleges) of Bucharest and Jassy and favoured the introduction of Newtonian physics in their curricula.

However, the situation changed drastically during the years both preceding and following the French Revolution. The majority of the Phanariots were of the view that the Revolution jeopardised their political agenda of 'enlightened despotism' and their prospects of gaining progressively greater influence within the Ottoman Empire. The French Revolution was portrayed as the result of the philosophical ideas of the Enlightenment and, consequently, the Phanariots' conviction and affection for the Enlightenment and subsequently for Newtonianism started to decline.[3]

The third social factor that influenced educational matters in that period was the new class of Greek merchants and craftsmen flourishing in what is now mainland Greece and also in diaspora, i.e. the Greek communities that were founded in Western Europe by the Byzantines after the Ottoman conquest e.g. in Venice and later by Greek merchants that fled of the Ottoman Empire e.g. in Vienna, Trieste, Odessa etc. (Kardasis, 2001; Kitromilides, 1999). This social class, which had no access to the institutions of the Ottoman state, favoured the foundation of a new nation state, and became the driving force behind the revolution for Greek national independence in 1821.

In the two centuries preceding the Greek national revolution of 1821, this new social class of merchants and craftsmen supported financially the foundation of new schools, and the introduction of the new philosophical and scientific trends of European thought (Bokaris and Koutalis, 2008). This class sought to be educated in the 'new scientific spirit'; at the same time, the new scientific spirit introduced to the Greek speaking communities of merchants and craftsmen was connected with radical political ideas influenced by the European Enlightenment and the French Revolution (Kitromilides, 1990 and 1999). Furthermore, the intellectuals originating from or who sided with this class were inspired by the Enlightenment,

3 For an analysis of the controversial and ambivalent positions of the Phanariots, see Papachristou (1992).

and viewed science education as a vehicle that would lead to the liberation of a people living under Ottoman imperial rule.[4]

Our analysis of policies concerning education in the Greek speaking communities of the Ottoman empire is based on the tripole schema of social power relations outlined earlier; in the level of ideas this can be described as a conflict between two intellectual currents: (*a*) the followers of neo-Aristotelianism initiated by Korydaleas in the seventeenth century and (*b*) the followers of 'Newtonianism' as the new science came to be known in the period of interest of this paper.[5]

In the last decades, historians in Greece have introduced the already established term 'Neo-Hellenic Enlightenment' (Demaras, 1977) in order to identify the introduction and adaptation of the ideas of the Scientific Revolution and the Enlightenment in the Greek intellectual space. It is beyond doubt that the new science with its foundation on experimental method lead to a decisive break with scholasticism and metaphysical superstition associated with clergy and feudal rule (Iliou, 1975).

SCHOOLS AND TEACHERS IN THE LATE EIGHTEENTH AND EARLY NINETEENTH CENTURIES

Historians of science and education have shown that in a number of schools for secondary and higher education for the Greek communities of the Ottoman Empire and the diaspora, in the late eighteenth and early nineteenth centuries, there were organised science teaching laboratories and demonstration experiments formed an integral part of physics teaching (see Karas, 1997 and 2003). In this section, we give an account of the schools of higher education and the teachers who introduced the new experimental method in the teaching of physics in the curricula of these schools. It should be noted that at this point in our period of study, the number of schools for higher education was increasing. Till about the middle of the eighteenth century the majority of the student population belonged to the weaker economic classes destined for careers in teaching, the clergy or as lower administration officials, that is, professions with meagre income. In contrast, the children of the upper social classes had private tutorials and then they were sent to study in European universities. What changed in the late eighteenth century, in the period when the influence of the Enlightenment ideas was at its

4 A sound example of an intellectual of this kind is Rigas Velestinlis who wrote *Physics Selections*, published in Vienna in 1790. He was influenced by the ideas of the Enlightenment and especially by the French movement of the Encyclopédie (Paris 1751-65).

5 At this point a remark on methodology should be made: our methodology is based on the analysis of the relations of social classes and institutions of the Greek population in the Ottoman Empire. But one has to note that social classes are not homogeneous formations and that psychological, religious, and other ideological factors influence the behaviour of individual members, besides socio-economic interests.

peak, was the social origin of the students in the schools – it was expanding and students from all social strata attended the schools.

One has to take into account that till about the middle of the eighteenth century most of the teachers/scholars, even the dissidents, were members of the clergy. Therefore, all differentiations and ideological disputes were taking place within the institution of the Orthodox Church. Scholars, like Methodios Anthrakitis (1660–1749) who was a clergyman, were excommunicated by the Orthodox Church due to their adherence to the new philosophical ideas (in Anthrakitis' case, the French philosophers Nicolas Malebranche and René Descartes).[6]

By the middle of the eighteenth century, the figure of the teacher-priest was replaced by the figure of a professional scholar whose mission was to provide secular education. The teaching of the works of the priests of the Orthodox Church, of Aristotle, and of practical arithmetic was replaced by a curriculum influenced by the new ideas of the Enlightenment, serving the social, political, and ideological priorities of the communities which supported the schools financially. In fact, the curriculum of each school was the outcome of the social power relations between the local communities and the Patriarchate. The new generation of professional scholars had also realised that they could not teach the new scientific ideas of the Enlightenment in the same way as the old curriculum. So a new teaching methodology was necessary, based on experiment and observation.

The first school that introduced physics courses based on the new experimental method was the Athonias School in the monastic community of the Holy Mountain Athos (Aggelou, 1988). This school was founded in 1750; after Eugenios Voulgaris[7] was appointed as the Director in 1753, the school started to flourish, attracting not only students who intended to pursue clerical studies, but also others seeking a higher education with the total number of students amounting to over 200 in some cases. The reason for this success was the fame accompanying Eugenios Voulgaris as a teacher and the desire of the literate population to

6 Methodios Anthrakitis, after being excommunicated, was forced to burn his own books in the courtyard of the Patriarchate in Phanari. When reinstated, he taught in the neo-Aristotelian tradition.

7 Eugenios Voulgaris (1716-1806) received his education in his native place, the island of Corfu, under Vikentios Damodos, an important scholar of the period, and then in the island of Zakynthos under Antonio Katifaro, a professor of the Flagginian School of Venice. He continued his studies in the School of Ioannina (a wealthy commercial town of north western Greece) under Athanassios Psalidas. After he became a priest, in 1737 or 1738, he went to Italy in order to study theology, philosophy, European languages, and natural sciences. Before taking up the directorship of the Athonias School, Voulgaris was the director of the Maroutsis School of Ioannina (founded in 1742 by Greek merchants of the community of Venice) where he designed a curriculum introducing natural sciences for the first time. For a short period, he was also responsible for the School of Kozani (founded in 1745 by the community of Greek merchants living in the region of what is now Hungary). After leaving Athonias School, he taught at the Patriarchal Academy in Constantinople. He spent the last years of his life, after 1770, in the Royal Court of Catherine the Great in St Petersburg as Archbishop of Slavonia and Chersonesos. In 1788, he was elected a Foreign Member of the Royal Society of London during the Presidency of Sir Joseph Banks.

be acquainted with the new science of Newton (Patiniotis, 2007; Petrou, 2001).[8] Voulgaris was a very experienced teacher and introduced a curriculum that laid special emphasis on mathematics, physics, and cosmography based on the experimental methodology. His teaching notes compiled in the form of a manuscript formed the basis of his textbook that was published later in 1805 in Vienna. In 1759, his strong adherence to the Newtonian ideas did not find favour with the religious hierarchy of Athos, and he was forced to give up the directorship of the school after severe disagreements over the content of his teaching. Gradually, the school lost its fame and was closed in the first decade of the nineteenth century.

In 1780, in the Academy of Bucharest, Konstantinos Vardalachos[9] was a senior teacher of philosophy, mathematics, and experimental physics. He reorganised the school, promoting natural sciences as the main teaching subject and introducing instruments of physics and chemistry in the instruction. Research has shown that every Saturday the students had to devote one hour to the laboratory (Camariano-Cioran, 1974).

In the Hegemonic Academy of Jassy in what is now Romania, teaching experiments were performed by Iosipos Moisiodax and Nikiphoros Theotokis.[10] As a headmaster, Theotokis reorganised the academy around the middle of 1760. In his second period in office (1774), he came in conflict with the neo-Aristotelian cycles of the city due to his insistence in teaching the 'new philosophy' and he was forced to leave the school (Camariano-Cioran, 1974).

In the Kaplaneios School of Ioannina, in north western Greece, the 'experimental physics' of Vardalachos was taught by Christophoros Philitas and Athanasios Psalidas (1797–1820). The latter prepared his own 'general physics' teaching notes based on Horvath's *Physica Generalis*.

In the Patriarchal Academy of Constantinople, the headmasters Dorotheos Proios (1804–1807), Stefanos Dougas (1809–1810), and Konstantinos Koumas (1814–15) taught experimental physics and performed experiments during lectures. But they faced severe opposition for their teaching practice and they were driven to resign from their teaching posts.

The Academy of Kydonies (now Aivalik in Asia Minor, Turkey) was founded in 1790. From 1800 up to 1821, two well-known scientific figures acted as directors: Benjamin Lesvios (1800–1812) and Theophilos Kairis (1812–1821). Both of them had written textbooks on physics, which they taught during their time of service. Teaching experiments were performed since 1796.

8 Petrou's article is written in Greek but an English translation by the author is available at the Royal Society of London.

9 Konstantinos Vardalachos (1775-1830) studied in Padua and Pisa. He taught the physical sciences in Bucharest, Chios, and Odessa.

10 Nikiphoros Theotokis (1731-1800) studied in Padua and Bologna where he worked with Giovanni Poleni and also made astronomical observations with Zanotti. He taught the physical sciences in Constantinople, Jassy, and Vienna. After 1775, he joined Voulgaris in Russia, where he became the Archbishop.

A course on experimental physics with experiments was taught for a short period in the school of the island of Chios, which was founded in 1815. In 1816, Konstantinos Vardalachos was appointed there as a teacher.

The 'Gymnasium of Sciences' (Iliou, 1975) was founded in the city of Smyrna (what is now called Izmir in the mainland of Asia Minor in what now is Turkey) in 1808 with Konstantinos Koumas[11] as its director. The school was connected with the developing merchant class of the city. In 1810, it changed its name to 'Philological Gymnasium' in contrast to the 'Evangelical School' that remained in the old tradition, and organised experiments in physics and science laboratories. Experiments were performed in public for audiences other than the students of the school. In 1819, the school was attacked and partly destroyed by opponents of the experimental method.

In 1813, the school of Milies in Pilion (the mountain home of the centaurs according to mythology) was founded with Gregorios Konstantas as the headmaster and Anthimos Gazis as a senior teacher. Research in the archives of the school has shown that experimental physics and chemistry were taught there, and that a rich collection of scientific instruments has been brought to this school to be used for teaching purposes (Andritsaki-Fotiadi, 1993).

A year before, in 1813, the Greek community of Zemlin (now Zenum), a city then belonging to the Austro-Hungarian Empire (near Beograd of what is now Serbia) founded a school where Demetrius Darvaris[12] introduced the teaching of experimental physics.

In the school of Ambelakia in Thessaly (central Greece), Stephanos Dougas donated a number of scientific instruments with the aim of establishing an Institution for Higher Education. This project was not successful due to political developments and shortage of funds.[13]

DIDACTICAL APPROACH

We do not posses concrete information about the exact curricula of these schools concerning the physical sciences. It seems that every teacher appointed followed his own syllabus. What we know is that physical sciences were taught for six to

11 Konstantinos Koumas (1777-1836) studied in Vienna. He taught in Smyrna and Constantinople. After the Greek national revolution of 1821, he fled to Vienna where he was arrested. After his release, he spent most of his time with scientific writing. He acquired a Doctorate from the University of Leipzig and became a member of the Academies of Berlin and Munich.

12 Demetrius Darvaris (1757-1833) studied in Vienna, Halle, and Leipzig. He taught in Zenum and Vienna.

13 For a detailed account and further reading on schools and curricula, see the work of Karas (2003) and Nicolaïdis (2001). See also various publications of the Institute of Neohellenic Research, NHRF, Athens (presented in the *Newsletter for the History of Science in Southeastern Europe*, http://www.hpdst.gr/publications/newsletter).

eight hours a week in the Academies of Bucharest[14] and Jassy, five to six hours weekly in Kydonies and Ioannina, three to four hours weekly in Constantinople, etc. Experimental apparatuses and instruments were bought by rich merchants in Western Europe and donated to the schools of the Greek speaking communities. Due to the high cost, there was the possibility of only a single experimental apparatus for each kind of experiment to be installed in a school.

The performance of the experiments and the role of the experiment in the teaching process was, in most cases, the sole responsibility of the teacher and, to a lesser degree, of the curriculum that was approved by the Board of each school. Most of the experiments were demonstration experiments, performed by the teacher in front of the students with the sole purpose of the verification of a theory that had been taught.

Students were observers only and not experimenters. What was missing in these experiments was the process of measurement. This meant that students were not trained in the mathematical processing of their findings (the quantification of the observation), missing the most important feature of the new methodology introduced by the Scientific Revolution: the connection in the last analysis of the experiment with mathematics.

In the textbooks used by the students, there were no pictures and drawings of the apparatuses or of the experimental set ups described therein. There were only some drawings (etchings, gravures) in the last pages of the textbooks for some instruments or experimental set ups. Despite these drawings, the majority of experiments described were more or less ‘thought experiments’. The students had to use their imagination in order to construct a mental representation of the experimental set up. This was where the role of the teacher became indispensable. The teacher had a central role in producing additional material in order to facilitate students’ understanding of the experiment.

THE TEXTBOOKS

During the period from the late eighteenth to the early nineteenth century, a large number of science textbooks were published. These textbooks were widely circulated among students and the general public and they played a very important role, not only in teaching but also in familiarising students with the experimental method in schools where there were no organised laboratories and the students were taught about the experiments solely through the textbooks (Karas, 2009; Xenakis, 1994).

In our study, we examine only the physics textbooks published in this period. Besides the physics textbooks, there were textbooks of astronomy, chemistry, geography, botany/natural history, and a very large number of mathematics. M.

14 For example, the curriculum of the Hegemonic Academy of Bucharest was designed in 1707 by Chrysanthos Notaras. The students were introduced to philosophy following independent courses in logic, natural philosophy, and metaphysics.

Patiniotis (2006) proceeded to classify 135 scientific and philosophical textbooks published during the eighteenth century. He classified the textbooks according to their content which is identified with the various disciplines in the structure of the eighteenth century curricula of the schools.

We proceed in a different manner. We have selected only the printed physics textbooks of the late eighteenth and early nineteenth century and we have distinguished the following three types: (*a*) translations of physics textbooks of European authors (Petrou, 2006); (*b*) physics textbooks based to a large extend on similar textbooks by European authors; and (*c*) physics textbooks written by Greek authors sometimes comprising their teaching notes.

The first category of physics textbooks includes:

a) Benjamin Martin (*Grammatica delle Scienze Filosofiche*, Italian edition published in Venice 1795), translated by Anthimos Gazis and published in Vienna (1799),
b) Johann Heinrich Helmuth (*Volksnaturlehre zur Dampfung des Aberglaubens*, 6th edition published in Venice 1810), translated with the editorial assistance of Spyridon Vlantis and published in Venice (1810),

The second category includes the following authors:

a) Nikiphoros Theotokis, *Elements of Physics*, in two volumes published in Leipzig, 1766 and 1767. This textbook is based to a very large extent on Abbe Nollet's Italian edition *Lezioni di Fisica Sperimentalle* (Venice, 1747) and Pieter van Musschenbroek's *Elementae Physicae* (Naples, 1751).
b) Eugenios Voulgaris, *Philosophers' Favourites* (Vienna, 1805), which is also based on van Musschenbroek's *Elementae Physicae* (Naples, 1751).
c) Rigas Velestinlis, *Physics Selections* (Vienna, 1790) based on the *Encyclopédie* published in Paris, 1751–65 and Geneva, 1778–1779.
d) Konstantinos Koumas, *Elementary Treatise on Mathematics and Physics* (Vienna, 1807), based on Jean Claude Fontaine *Cours encyclopédique et élémentaire de mathématique et de physique* (Vienna, 1800).

The third category comprises physics textbooks written exclusively by Greek authors that sometimes are edited versions of their teaching notes from the lectures in the schools where they taught. In this group we have selected the physics textbooks of three prominent teachers who used the new experimental method in their course of teaching:

a) Konstantinos Koumas, *Synopsis of Physics* published in Vienna (1812),
b) Konstantinos Vardalachos, *Experimental Physics*, published in Vienna (1812), and
c) Demetrius Darvaris, *Physics* (three volumes) published also in Vienna (1812).

TEACHING EXPERIMENTS IN THE TEXTBOOKS THEMATIC CLASSIFICATION

We have chosen two textbooks from the second category (Theotokis, 1766; Voulgaris, 1805) and the three mentioned in the third (Koumas, 1812; Darvaris, 1812; Vardalachos, 1812). The main reasons for selecting these textbooks were that they had a wide circulation in printed form, they had been used in teaching, and they were written by Greek authors (not translated) since our study focuses on how the Greek scholars distributed in the periphery the ideas and methods acquired from the centres of Europe.[15] We proceed to a thematic classification of the experiments described therein: mechanics, mechanics of fluids, acoustics and optics, heat, electricity and magnetism.

In the mechanics section, the most common experiments described have to do with the study of the pendulum, the free fall in the presence of air or in vacuum, the motion in an inclined plane, the study of friction, the study of the momentum, and the various forms of impact (elastic, plastic) between spheres. There are also experiments to demonstrate the view that the elements are the undecomposable constituents of material bodies, ultimately composed of particles of various sorts and sizes which cannot be resolved in any known way.

In the sections of fluid mechanics (properties of liquids and gases), there are experiments concerning atmospheric pressure, cohesion forces, the shape of the surface of liquids in a vessel, the laws of flux, elasticity, and incompressibility of fluids.

In the acoustics and optics section, there are experiments about the nature and properties of sound, analysis and synthesis of white light, and also experiments in geometrical optics with lenses and mirrors.

In the section on heat, most of the experiments concern expansion and contraction, determination of the boiling point of various substances, melting and freezing, and conduction of heat.

Finally, in the electricity and magnetism section, there are experiments on magnetisation by induction, demonstration of the distribution of magnetic lines around the poles of a magnet, electrification by friction, charging and discharging of a capacitor ('Leyden jar'), electrolysis, detection of atmospheric electricity, animal electricity, construction of a voltaic pile, etc.

ICONIC EXPERIMENTS

A very interesting feature of the textbooks is the extensive reference to experiments that were considered landmarks in the development of classical physics, and which had been performed by famous physicists of those days. In the textbooks, special emphasis is given to the name of the scientist who performed the

15 The majority of Greek scientific textbooks were printed between 1770 and 1820. See Pappas and Karas, 1987.

experiment, for example, "the experiment of Boyle", etc. In the following section we give an indicative account of the iconic experiments described in the textbooks considered.

In the mechanics section, there are descriptions of the free fall experiments by Pieter van Musschenbroek, Marin Mersenne, Giovanni Poleni, and Giambattista Riccioli. There are also references to the free fall experiments by Galileo and Newton.

We can also find experiments on friction by John Theophilus Desaguliers and Pieter van Musschenbroek, the experiments on the constitution of matter by Anton van Levenhook and Johann Christophorus Sturm advancing the corpuscular theories, and also the experiments by Willem Jacob Gravesand, G. Poleni, and G. Riccioli in order to prove that the energy of a moving body ("vis-viva" is the term used in the original text) is not proportional to its velocity but to the square of its velocity.

In the mechanics of fluids sections, extensive descriptions are given about the experiments of Robert Boyle, Evangelista Torricelli, Daniel Bernoulli, and others regarding the experimental demonstration of the laws of flux, the dependence of hydrostatic pressure on the height of the liquid inside the vessel, etc. There are also descriptions of the experiments of Blaise Pascal (on the dependence of pressure on altitude), the experiments of Giovanni Poleni showing that the speed of a liquid flowing from the bottom of a vessel depends on the friction between the liquid and the vessel walls, the experiments of Jean Senebier on the elasticity of water, the historical experiment of Otto von Guericke with the 'Magdeburg hemispheres', Edme Mariotte's experiment to prove that the quantity of liquid that flows from the vessel depends on the height of liquid in the vessel, Jean Picard's experiment on the dependence of flow from the geometry of the vessel, etc.

Extensive references are also made to the experiments performed in the Academy of Florence. From the ancient figures, there are references to the works of Archimedes,[16] especially his Screw and his method to determine the percentage of gold in the crown of the king of Syracuse, and of the findings of Sextus Julius Frontinus (40–103 AD).

In the optics section, there is extensive reference to Newton's experiment for the analysis and synthesis of white light, Newton's rings, on Christian Wolff's observations of the human blood with a microscope, on the observation of the absorption of solar rays by cloths of different colours by Benjamin Franklin, and also William Herschel's experiment to prove that light and caloric are of different nature.

In the section on heat, there is an account of Fahrenheit's experiments for the calibration of various types of thermometers, the experiments in the Academy of

16 We underline here the reference to Archimedes in order to show that both neo-Aristotelian scholars and adherents of Newtonianism built their work in continuity with ancient Greek thought. In the works of Newtonian scholars there are references to Archimedes, with the scope of highlighting the importance of experiment and practise which was a characteristic of the method of Archimedes.

Florence (the dependence of the freezing point on pressure), Musschenbroek's experiment on the linear expansion of a metal rod when heated, and Alessandro Volta's experiment on water cooling and water freezing during evaporation.

In the electricity and magnetism section, there are references to Musschenbroek's technique for creating new magnets, Abbe Nollet's experiment on electricity and the human body and on capacitor charging by an electrical machine, the electric discharge by Ludolf of Berlin (1744), experiments on electrification by William Gilbert and Grey, the experiment of the magic 'horseshoe' by Benjamin Franklin and also his experiment for the detection of the electricity of the atmosphere.

The electrolysis of potassium, sodium, barium, calcium, and magnesium as discovered by Sir Humphry Davy (1807) is also presented. There is also extensive reference to the works of Luigi Galvani (animal electricity) and Volta (circuits with battery).

The reference to the experiments of these famous scientists served a dual purpose. On the one hand, the name of the scientist acted as a means to persuade the students about the validity of the theory presented. On the other hand, the European experimental tradition was transferred to the Greek speaking regions and the literate population was getting acquainted with the European experimental tradition.

EPILOGUE

Despite the difficulties and the drawbacks mentioned, one has to appreciate the contribution of the new experimental methodology in altering the Aristotelian style of thinking among the educated population of the Greek speaking communities. Such a process of change was not at all easy given the strong affiliation to the 'glorious past' of classical antiquity. The introduction of the new mode of thinking was a necessary precondition for the connection of the national aspirations of the class of merchants and craftsmen with the Enlightenment project.

In many cases the experimental methodology caused reactions not only from the official Church, but also from the group of teachers who sided with the Aristotelian tradition initiated by Korydaleas in the seventeenth century and/or other conservative members of the wider community. One has always to recall that the class of merchants and craftsmen that was politically oriented in the creation of a new nation state, opposed the Church hierarchy and the bureaucracy of the Phanariots who wanted to contain national aspirations within the policy of the modernisation of the Ottoman state. Thus, the introduction of the new experimental method was caught in political turmoil, and the teaching of the new science of 'Newtonianism' became a political action associated with the ideas of Enlightenment and the French Revolution and was thought of as paving the way for the coming national liberation.

REFERENCES

Aggelou, A. 1988. *Όψεις του νεοελληνικού διαφωτισμού (Aspects of Modern Greek Enlightenment)*. Athens: Hermes.

Andritsaki–Fotiadi, D. 1993. *Κατάλογος παλαιότυπων της ιστορικής βιβλιοθήκης των Μηλιών (List of Old Editions in the Historical Library of Milies, Pilion)*. Athens: Hellenic Ministry of Education and Religious Affairs and Historic Library of Milies.

Blay, M. and E. Nicolaïdis 2001. *L'Europe des sciences : constitution d'un espace scientifique*. Ed. Seuil, Paris: Seuil.

Bokaris, E. P. and V. Koutalis. 2008. 'The "System of Chymists" and the "Newtonian Dream" in the Greek Speaking Communities in the 17th – 18th Centuries', *Science & Education*, 17 pp. 641–61.

Camariano-Cioran, A. 1974. *Les Académies princières de Bucarest et de Jassy et leurs professeurs*. Thessaloniki: Institute of Balkan Studies.

Demaras, K. Th. 1977. *Νεοελληνικός Διαφωτισμός (Neohellenic Enlightenment)*, Second Edition. Athens: Hermes.

Dialetis, D., K. Gavroglu and M. Patiniotis. 1999. 'Sciences in the Greek Speaking Regions during the Seventeenth and Eighteenth Centuries: The Process of Appropriation and the Dynamics of Reception and Resistance', in K. Gavroglu K. (ed.), *The Sciences in the European Periphery During the Enlightenment*. Dordrecht: Kluwer Academic Publishers [Archimedes2], pp. 41–71.

Henderson, G. P. 1971. *The Revival of Greek Thought 1620–1830*. London/Edinburgh: Scottish Academic Press.

Iliou, Ph. 1975. 'Luttes socials et movement des Lumières à Smyrne en 1819', in *Structure sociale et développement culturel des villes sud-est européennes et adriatiques aux XVIIe–XVIIIe siècles*, Bucarest: Association Internationale d'Études du Sud-Est Européen.

Karas, Y. (ed.). 2003. *Ιστορία και φιλοσοφία των επιστημών στον ελληνικό χώρο, 17ος–19ος αι., (History and Philosophy of Sciences in the Hellenic space, 17th–19th Centuries)*. Athens: N.H.R.F. –Metaichmio.

———. (ed.). 1997. *Οι επιστήμες στον ελληνικό χώρο (The Sciences in the Greek Space)*. Athens: N.H.R.F. –Trochalia.

———.2009. *Οι περιπέτειες της νεοελληνικής επιστήμης (The Adventures of Neohellenic Science)*, Second Edition. Athens: Hellenic Physical Society.

Kardasis. V. A. 2001. *Diaspora Merchants in the Black Sea: The Greeks in Southern Russia, 1775–1861*. Maryland and Oxford: Lexington Books.

Kitromilides, P. M. 1990. 'The Idea of Science in the Modern Greek Enlightenment', in P. Nicolakopoulos (ed.), *Greek Studies in the Philosophy and History of Science*. Dordrecht: Kluwer Academic Publishers.

———. 1999. *Tradition, Enlightenment and Revolution*, PhD Dissertation, Harvard University, 1978. Also published in Greek: *Νεοελληνικός Διαφωτισμός: οι πολιτικές και κοινωνικές ιδέες*. Athens: MIET.

Mendoza, C-L., E. Nicolaidis and J. Vandersmissen (eds). 1999. *The Spread of the Scientific Revolution in the European Periphery, Latin America and East Asia*, Proceedings of the XXth International Congress of History of Science, Vol. V. Turnhout: Brepols.

Nicolaïdis, E. 2001. 'L'intégration des Balkans dans L'espace Scientifique Européen', in M. Blay and E. Nicolaïdis, *L'Europe des Sciences: Constitution d'un Espace Scientifique*. Paris: Seuil, pp. 353–400.

Notaras, Chrysanthos. 1716. *Εισαγωγή εις τα Γεωγραφικά και Σφαιρικά* (*Introductio ad Geographiam et Sphaeram*). Paris.

Papachristou, P. 1992. *The Three Faces of the Phanariots: An Inquiry into the Role and Motivations of the Greek Nobility under Ottoman Rule 1638–1821*, MA Thesis, Simon Fraser University, Canada.

Pappas, V. and Y. Karas. 1987. 'The Printed Book of Physics: The Dissemination of Scientific Thought in Greece, 1750–1821, before the Greek Revolution', *Annals of Science*, 44 (3), May, pp. 237–44.

Patiniotis, M.2007. 'Periphery Reassessed: Eugenios Voulgaris Converses with Isaac Newton', *The British Journal for the History of Science*, 40, pp. 471–90.

———. 2006. 'Textbooks at the Crossroads: Scientific and Philosophical Textbooks in 18th Century Greek Education', *Science & Education*, 15, pp. 801–22.

Petrou, G. 2006. 'Translation Studies and the History of Science: The Greek Textbooks of the 18th Century', *Science & Education*, 15, pp. 823–40.

———. 2001. 'Eugenios Voulgaris (1716–1806) and the Royal Society of London', *Neusis*, 10, pp. 181–98 (in Greek).

Roudometof, V. 1998. 'From Rum Millet to Greek Nation: Enlightenment, Secularization, and National Identity in Ottoman Balkan Society 1453–1821', *Journal of Modern Greek Studies*, 16, pp. 11–48.

Skordoulis, C. D. 2009. 'Hellenic Studies in History, Philosophy of Science and Science Teaching: New Perspectives', *Science & Education*, 18, pp. 1193–97.

Stathi, Penelope. 1999. *Χρύσανθος Νοταράς, Πατριάρχης Ιεροσολύμων, πρόδρομος του Νεοελληνικού Διαφωτισμού* (*Chrysanthos Notaras, Patriarch of Jerusalem, Precursor of the Greek Enlightenment*). Athens: Association of the Athens Alumni of the Great School of the Nation.

Vlahakis, G. N. 1999. 'The Greek Enlightenment in Science: *Hermes the Scholar* and its Contribution to Science in Early Nineteenth-century Greece', *History of Science*, 37, pp. 319–45.

———. 1998. 'The Introduction of Classical Physics in Greece: The Role of the Italian Universities and Publications', *History of Universities*, XIV, pp.157–79.

Voulgaris, Eugenios. 1805. *Philosophers' Favourites* (*Τα αρέσκοντα τοις φιλοσόφοις*). Vienna: George Vendotis Hellenic Editions

Xenakis, Ch. Th. 1994. *The Experiment as Methodology of Research and Knowledge in the Works of Prerevolutionary Period Greek Scholars*, Unpublished PhD Dissertation, University of Ioannina (in Greek).

THE MAGIC LANTERN FOR SCIENTIFIC ENLIGHTENMENT AND ENTERTAINMENT

Willem Hackmann

The Magic Lantern, which three centuries since served no better purpose in the hands of pretended wizards and necromancers, than to induce a belief in their possession of supernatural powers, now occupies a deservedly high position as a Scientific Instrument, capable of affording rational amusement to the young, instruction to "children of a larger growth", and in the hands of the philosopher, a means of demonstrating some of the most beautiful phenomena in science.[1]

INTRODUCTION

Audiences today are accustomed to the sophisticated computerized visual displays produced by software such as PowerPoint, which have become indispensable aids in the teaching of science. However, these techniques have evolved over centuries, and have their roots in the Renaissance science of optics.

The magic lantern, the precursor of PowerPoint presentations, was developed as an optical device by natural philosophers, and was popularized by itinerant showmen from the 1720s onwards. This device quickly moved from the relative obscurity of the laboratory and the cabinet of curiosities to the public arena, and by the late 18th century was recognized as a useful aid in science education. It was this dual capability of the magic lantern both to entertain and to act as a visual aid in describing the latest scientific discoveries that made its use so fruitful in disseminating and popularizing science.

However, the study of optics (especially optical devices) was often associated with magical tricks and illusions, and for that reason there was sometimes resistance to incorporate this material into what was considered to be 'legitimate' science. According to an often quoted story, the Aristotelian physicist at the University of Pisa, Guilio Libri, refused to look through Galileo's telescope in 1610.[2] When Libri died shortly thereafter, Galileo quipped that he would have another opportunity to see the satellites of Jupiter on his way up to heaven.

1 'A Mere Phanton', The Magic Lantern. *How to Buy and How to Use It, Also How to Raise a Ghost* (London, 1876 edition), p. 7.

2 John Joseph Fahie, *Galileo. His Life and Work. With Portraits and Illustrations* (London: John Murray; Elibron Classics edition, 1903), p. 101.

Fig. 1: The lecturer with his English triple (triunial) lantern. The heyday of the triple lantern was in the 1880s, by the mid1890s it was going out of style. Highly difficult to use as each lantern unit was controlled by manual valves operating the oxy-hydrogen 'limelight' jets. The fatal blow was the coming of motion pictures in the 1900s. In this lantern the limelight has been replaced by halogen lamps. The lantern effects were produced by a variety of slides: from simple dissolves with multiple slides to single and double 'slipping slides', lever slides and several versions of mechanical rackwork slides, such as the 'Wheel of Life' shown here. The lecturer recreated by means of PowerPoint and digital projector the scientific effects that could be achieved, thereby merging the old and the new ways of 'Learning by Doing'.

The early history of the lantern is closely related to that other 'magical' optical device: the portable camera obscura. Indeed, the magic lantern has been called 'the camera obscura in reverse'. In the camera obscura, a brightly lid scene is reproduced by means of a simple lens either within the dark body of the apparatus or cast onto a translucent glass screen, while in the case of the magic lantern, a scene painted on a glass slide is illuminated from within the apparatus and projected onto a distant white screen. This can only be achieved in the latter case with a fairly complicated optical arrangement, which took time to evolve and reached its zenith in the 19th century (Fig. 1).[3]

3 The most useful modern source for many of the aspects discussed in this paper is the groundbreaking volume by David Robinson, Stephen Herbert and Richard Crangle (eds), *Encyclopaedia of the Magic Lantern* (London: The Magic Lantern Society, 2001). Other useful works are Herman Hecht (ed. Ann Hecht), *Pre-cinema History: an encyclopaedia and*

BRIEF HISTORY OF THE INVENTION OF THE MAGIC LANTERN

The original camera obscura as a darkened room or enclosed box with a hole in one side is a very ancient device that came into prominence in the 13th and 14th centuries for observing eclipses.[4] From the 15th century onwards the camera obscura[5] was used as an aid for artists especially those involved with book illustrations, and in the study of perspective and geometrical optics.[6] As a means of 'mechanically' copying images it shared this function with the pantograph invented about this time.

The camera obscura only became a useful scientific tool after the addition of a lens to concentrate the image, but its early development has strong links with the invention of the telescope, followed by that of the microscope – all image-forming devices that had a great impact on visualization in natural philosophy.[7]

annotated bibliography of the moving image before 1896 (London: Bowker Saur/BFI, 1993); Deac Rossell, *Laterna Magica – Magic Lantern Vol. 1* (Füslin Verlag, 2008); Laurent Mannoni, Werner Neke, Marina Warner (eds), *Eyes, Lies and Illusions: Drawn from the Werner Nekes Collection* (London: Hayward Gallery in association with Lund Humphries, 2004). This astonishing collection has also been compiled on six DVDs: I. 'Was geschah wirklich zwischen den Bildern?'; II 'Durchsehekunst'; III. 'Belebte Bilder'; IV. 'Vieltausendschau'; V. ild-Raum'; VI. 'Wundertrommel' (commentary in German , French and English), of which III and VI are the most relevant to this paper. See http://wernernekes.de/00_shop1/index.php?p=productsList&iPage=2

4 In the 13th century Roger Bacon described the use of a camera obscura for the safe observation of solar eclipses, after which this technique is described by many scholars. Francesco Maurolico, or Maurolycus, describes making solar observations in a darkened room in his (*Cosmographia,* 1535); the same procedure is described by Erasmus Reinhold in his edition of G. Purbach's *Theoricae Novae Planetarum* (1542); while his pupil Rainer Gemma Frisius, used it for the observation of the solar eclipse of January 1544 at Louvain, making measurements and drawings, described in his *De Radio Astronomico et Geometrico* (1545). He proposed that his technique can be used for observation of the moon and stars. The same arrangement was used by Copernicus, Tycho Brahe, M. Moestlin and his pupil Kepler (who applied it in 1607 to the observation of a transit of Mercury).

5 John Hammond, *The Camera Obscura: A Chronicle* (Bristol: Institute of Physics Publishing, 1981); 'Camera Obscura', *Encycl. Britannica*, 11th ed., 1910), pp. 104–107.

6 David Hockney, S*ecret Knowledge: Rediscovering the Lost Techniques of the Old Masters* (London: Thames and Hudson, 2001), and the physicist argue that the realism in Renaissance art was due to optical aids (the camera obscura). For the ensuing debate see Sven Dupré, 'Introduction. The Hockney–Falco Thesis: Constraints and Opportunities', *Early Science and Medicine*, **10**–2 (2005), pp. 126–135; Antonio Criminisi and David G. Stork, 'Did the great masters use optical projections while painting? Perspective comparison of paintings and photographs of Renaissance chandeliers,' in J. Kittler, M. Petrou and M. S. Nixon (eds), *Proceedings of the 17th International Conference on Pattern Recognition*, Volume IV (2004), pp. 645–648.

7 Willem Hackmann, 'Natural Philosophy Textbook Illustrations 1600–1800', in Renato G. Mazzolini (ed.), *Non-Verbal Communication in Science Prior to 1900* (Florence: Leo S. Olschiki, 1993), pp. 169–196; Wolfgang Lefèvre, Jürgen Renn, Urs Schoepflin (eds), *The Power of Images in Early Modern Science* (Birkäuser Verlag, 2003).

As is the case of all these optical instruments, their precise origins are a mystery. In a sense this hardly matters for our purpose. Certainly their origins have as much to do with natural magic as with natural philosophy, and with the urge to invent extraordinary mechanical devices with which to impress the Renaissance courts. Thus, Giovanni da Fontana (*c.*1395–1455) describes a portable lantern intended to strike fear into the enemy by projecting a picture of the devil painted on its glass window, among the devices for conducting medieval warfare in his *Bellicorum instrumentorum liber* (1420).

Fontana's manuscript drawing is optically correct as he shows the projected image aligned the same way as it is painted, but as his lantern (or lamp) has no projection lens, the image it produced would not have been very sharp, but it would have served its purpose of terrifying the enemy. There is no particular evidence that Fontana invented the projecting lantern which he calls here an 'Apparentia Nocturnal', but he does describe the basic concept of the magic lantern, that is the projection of an image.[8]

More than a century later, Giovanni Battista della Porta describes in the first edition of his *Magiae naturalis* (1558) a simple method, which was probably already widely known, of projecting a design sketched on the surface of a mirror reflected by the sun's rays on to a white wall or screen. This was essentially a camera obscura with a convex mirror. His book was published in 1658 in English as *Natural Magick*. Thirty-one years later, in the 2nd edition of his *Magiae naturalis* (1589) he discloses that he had replaced the concave mirror by a convex lens some years ago but had kept it a secret. In this interval the use of the lens had already been described by a Venetian nobleman, Daniello Barbaro, in his book *La Pratica della perspettiva* (1568).[9]

In the 1590s Giovanni Battista della Porta describes a means of aiding sight through the combinations of a convex and a concave lens separated by some distance (and with the concave stronger than the convex to yield sharper images), but their *potential* capacity to view objects beyond the range of natural sight was not recognized until a decade later by spectacle makers in the Netherlands.[5] The commercial exploitation of the telescope started in 1608, and was immediately taken up by natural philosophers. Within a year or so both Thomas Harriot in England and Galileo in Italy had directed their instruments to the moon, and by the end of 1611 Christoph Scheiner had turned his telescope into a camera obscura for sun-spot observations.[10] The principal early scientific use of the camera obscura was in astronomy and the close affinity of the camera obscura to the telescope is demonstrated by two early camera obscuras in the collection of the Mu-

8 Willem Tebra, 'The Magic Lantern of Giovanni da Fontana', *The New Magic Lantern Journal* (*NMLJ*), **2**–2 (1982), pp 10–11. Fontana's work is often simply referred to as *Liber instrumentorum*.

9 Daniello Barbaro, *La Pratica della Perspettiva* (Venice: Camillo and Rutilio Borgominieri, 2nd ed., 1569), p. 192.

10 In 1611 the projection method (using the new telescope as a camera obscura) was used in observing sun-spots by Johann Fabricius, Galileo and Christopher Scheiner, and became a common projection technique.

seum of the History of Science in Oxford: a portable instrument similar to a telescope and a reflex version with lenses in a telescopic tube, both dating from the early 1700s.[11]

A consequence of the telescope was the invention of the compound microscope by Galileo as reported by the Scottish mathematician John Wedderburn in 1610. Galileo's microscope was an unwieldy affair, and its development and commercial exploitation fall outside the scope of this brief paper. By the early 1620s portable microscopes were commercially available and Christiaan Huygens' father purchased such an instrument in London in March 1622.

The experience of developing the lens-system for the telescope and the microscope fed into that other 'magical' technology that of projecting images. In the first edition of *Ars magna lucis et umbrae* ('The Great Art of Light and Shadow'), published in 1646, the Jesuit priest Athanasius Kircher rearranged della Porta's lens camera obscura so that it could project images using sunlight or candlelight. It required only a little rearrangement of these basic elements to create the magic lantern, which was described and depicted by Kircher in the second edition of his *Ars magna*, published in 1671.

There has been a great deal of controversy in the literature about Kircher's two illustrations showing his magic lantern, as they show the unusual arrangement of having no projection lens to focus the images. In this respect they are similar optically to Giovanni da Fontana's 'terror lantern' of two centuries previously, even to the extent that the images of the slides are upright as the images on the screen, which is optically correct for this kind of arrangement.[12]

Kircher does not claim to be the inventor of the magic lantern – that claim was made for him by successive commentators. By this time the lantern had already been described by the Dutch natural philosopher Christiaan Huygens and by the Danish mathematician Thomas Rasmussen Walgenstein (who is mentioned by Kircher). Either could be the inventor of the magic lantern. Huygens probably constructed his first lantern in 1659 (or perhaps a little earlier, thus, 12 years before Kircher's engravings), and Walgenstein at around the same time. Huygens and Walgenstein may have met as early as1649, but they certainly met in Leiden in 1657/8.[13]

11 Museum of the History of Science, Oxford: hand-held Camera Obscura, *c.* 1700, inv. no.83805 and small Reflex Camera Obscura, *c.* 1710, inv. no. 75945.

12 Willem Wagenaar has suggested ('The Origins of the Lantern', *NMLJ*, **1**–3, pp. 10–12) that the illustrations show point-source projectors. Herman Hecht ('The History of Projection 1', *NMLJ*, **6**–1, pp. 1–4) suggested that when Kircher published the second edition of his book (1671), he had somehow to claim that he thought of the lantern first, and that his lantern was not only much better but much bigger as well. But Kircher did not publish details until 12 years after Huygens had a lantern and six years after Samuel Pepys had bought one. See *Encyclopaedia of the Magic Lantern* (note 3), pp. 152–3; Jocelyn Godwin, *Athanasius Kircher's Theatre of the World* (London: Thames and Hudson, 2009), pp. 211–213.

13 Hecht (note 12), p. 3; the entry on Huygens in *Encyclopaedia of the Magic Lantern* (note 3), p. 142. For Huygen's sketch of the optical system of his lantern, see his *Oeuvres complètes* (Den Haag, 1888–1950), vol. XIII, p. 786. The controversy who was the *actual* inventor of

Huygens was ambivalent about the lantern as for him it pandered to the public's interest in magic and extraordinary spectacles, while he was keen to promote the method of ratiocination in natural philosophy or scientific thought. Walgenstein, on the other hand, certainly exploited his 'lantern of fear' to the full and got the device widely known, so that by the late 1670s and early 1680s the first practical representations of the lantern were widely disseminated in the books of Johann Christoph Sturm and Johann Zahn.[14]

DEVELOPMENT OF THE SCIENTIFIC LANTERN

The magic lantern is not intrinsically 'scientific', what makes it so is the intention of the operator. The same is true, of course, of any of these 'passive' optical devices. In the course of the 18th century the magic lantern was increasingly noticed in general books on science and optics. Of particular importance was the textbook *Physices elementa mathematica* (Leiden, 1720–1) by the Dutch professor of physics Willem Jacob's Gravesande, which not only produced the first coherent course of university physics (natural philosophy) based on the methodology of Isaac Newton, but also described a magic lantern powerful enough to be used in the lecture theatre. This lantern was made in collaboration with the famous Leiden instrument maker Jan van Musschenbroek. The strength of the illumination was enhanced by a concave mirror reflector. Otherwise sunlight could be used by mounting the lantern's objective system in front of a hole in a window-shutter. This lantern's sophisticated optics would not be surpassed until well into the 19th century.[15]

Two other 18th-century optical projection instruments closely associated to the magic lantern were the solar and lucernal microscopes. Both clearly have their roots in the camera obscura and magic lantern. In 1636 Daniel Schwenter, professor of mathematics and oriental languages at Altdorf invented the 'scioptic ball', made of lignum vitae with two long-focus lenses, which when mounted in an aperture converted a room into a large camera obscura. He devised it for drawing panoramic views but it was soon used for observing astronomical events. We have seen that 's Gravesande followed a similar line of development in the 1720s when he simply used sunlight as an illuminant for his lecture theatre's lantern on bright sunny days. This was soon to be taken one step further. By replacing this lens

the lantern has by no means been settled, see Deac Rossell, 'The True Inventor of the Magic Lantern', *NMLJ*, **9**–1 (2001), pp. 7–8.

14 Johannes Zahn describes twelve different lantern models in his *Oculus artificialis teledioptricus* (Würzburg, 1685). For the iconography of the lantern, see David Robinson, *The Lantern Image. Iconography of the Magic Lantern 1420–1880* (The Magic Lantern Society, 1993); and the two supplements of 1997 and 2009.

15 Peter de Clercq, *The Leiden Cabinet of Physics. A Descriptive Catalogue* (Leiden: Museum Boerhaave Communication 271, 1997), pp.108–111. On the history of the Musschenbroek workshop, see Peter de Clercq, *At the Sign of the Oriental Lamp. The Musschenbroek Workshop in Leiden* (Rotterdam: Erasmus Publishing, 1997).

system with a simple compound microscope, the 'solar microscope' was invented, probably by Daniel Gabriel Fahrenheit in the early 1730s, although usually attributed to A.J.N. Lieberkühn of Berlin in 1738.[16]

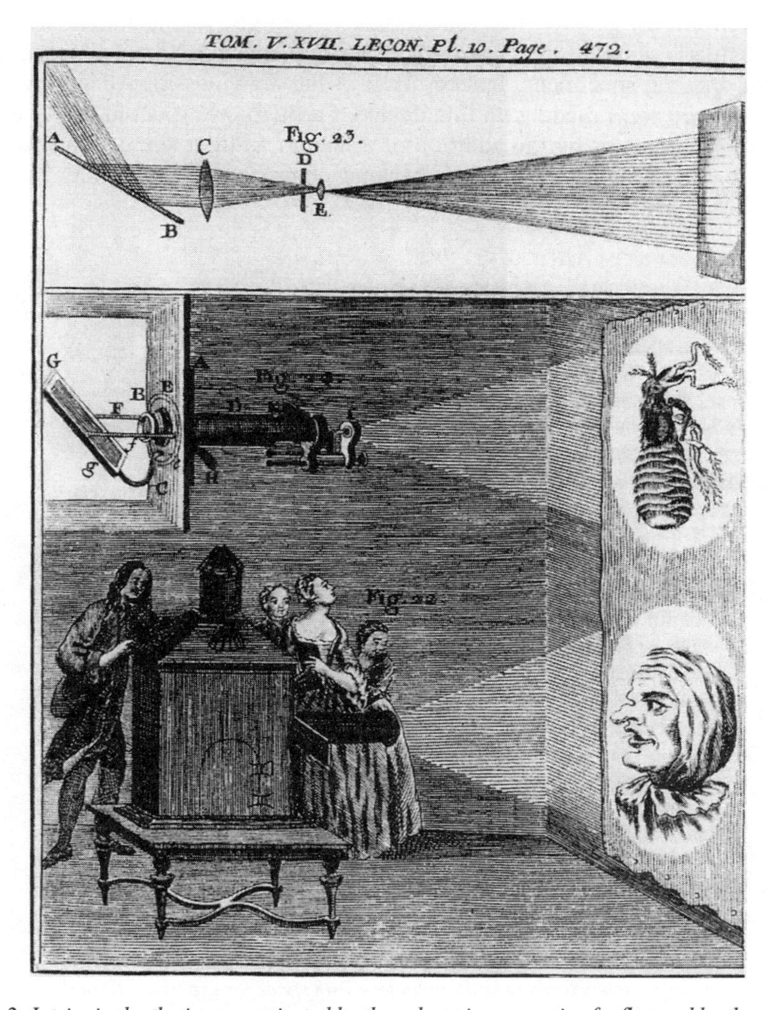

Fig. 2: Intriguingly, the image projected by the solar microscope is of a flea and by the magic lantern is of a grotesque face. The image shown by 's Gravesande's lantern of 1720, too, is another 'monster' image, this time of the devil (although other slides of this lantern that have survived are of animals and landscapes). Perhaps this reflects subconsciously the division which then prevailed that microscopes and telescopes were 'scientific' and the magic lantern was for 'spectacles'. Engraving from Abbé Nollet's Leçons de physique expérimentale (1755).

16 G. L'E. Turner, *Collecting Microscopes* (London: Studio Vista and Christie's, 1981), p. 95.

In Jean Antoine's Nollet's *Leçons de physique expérimentale* (1755) the drawing power of both the magic lantern and the solar microscope to awe and to instruct have been cleverly associated by the large images projected on the wall of the grotesque face and the hairy insect (Fig. 2). The lucernal microscope was invented by the well-known London instrument maker, George Adams Junior in 1787. It was particularly suitable for viewing both opaque and transparent objects, and as an aid in drawing specimens. Indeed, most of the drawings for Adams' *Essay on the Microscope* were made with this device. Furthermore, it could easily be converted into a projector by the addition of a lantern as light source, and such lanterns with microscopic attachments became commonplace from the 1820s onwards.[17]

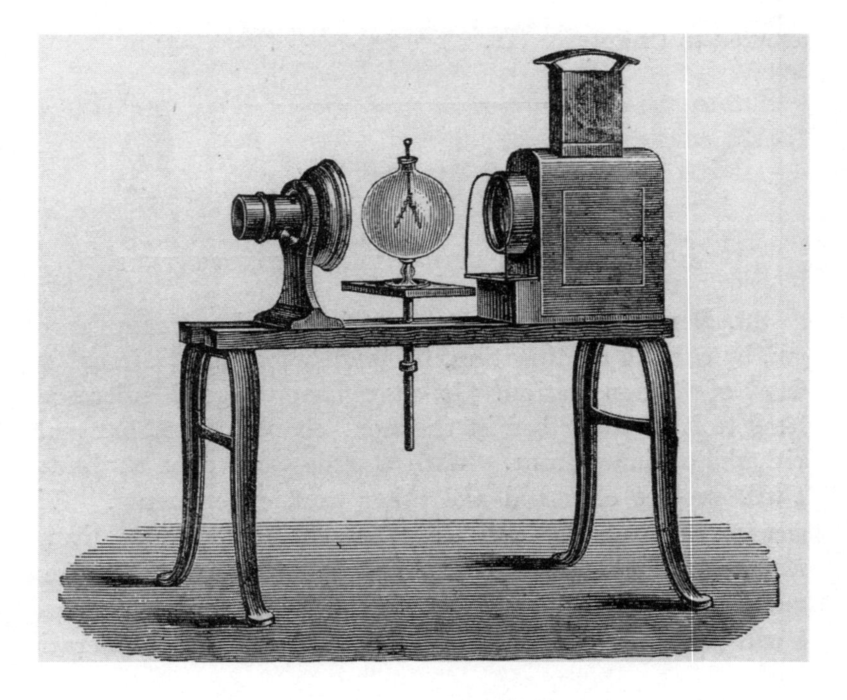

Fig. 3: A simple science demonstration lantern with space for experiments, in this case showing a gold-leaf electroscope, 1890s.The model steam engine piston shown in Fig. 6 would replace the gold-leaf electroscope in this demonstration.

17 Adam's lucernal was an improved variation produced in Germany and described by M.F. Ledermüller in 1762. It was illuminated by an oil lamp and was a combination of the magic lantern and the solar microscope, see Turner, *ibid.*, pp. 96–99.

The heyday of the 'scientific lantern' was the mid to late 19th century. Several factors were in place that aided this development. There were considerable improvements in the optics of the lantern, in the mode of illumination and in the manufacture of slides. The lack of a suitable light source had hindered the use of the lantern for large public audiences until several new powerful illuminants became available: an improved Argand lamp, limelight, the paraffin burner, and by the turn of the 19th century, electric arc light.[18]

The technical and scientific advances made at this time that began to impinge on society, stimulated public interest, and science was introduced into secondary education, and for mature students. Specialized scientific lanterns were produced for this burgeoning market. These were, on the whole, lanterns with an open area in front of the condenser (Fig. 3), which enabled experiments to be demonstrated and projected from there, and with specialized objectives for specific scientific demonstrations, such as the spectroscope and polariscope. Some also had a separate optical system in which the light could be projected upwards before being projected forwards (the precursor of the modern overhead projector).[19]

CREATING 'SCIENTIFIC EFFECTS' WITH THE MAGIC LANTERN

The telescope and microscope allowed the observer to see objects beyond common sight and therefore beyond common sense experience. Whole new worlds opened up which had to be interpreted and which had a huge impact on the imagination. Manipulating these instruments was generally a private experience but when the technology became available to project these images this experience could be shared with large audiences. The magic lantern enhanced this process, and helped to make the magic of science universal.

Christiaan Huygens constructed the lantern for family entertainment, and made sketches of 'moving pictures' (such as a skeleton that removes its head when moved in rapid sequence that would remain a stock figure in magic lantern shows).[20] He became reluctant to be associated with this device. As a serious student of natural philosophy, natural magic had no place in his scheme of things. Indeed, he undoubtedly regarded the 'lantern of fear' an obstacle to scientific pro-

18 Terence Rees, *Theatre Lighting in the Age of Gas (London: The Society for Theatre Research*, 1978).

19 For a brief history of the scientific lantern, see the entry 'Science Lantern', *Encyclopaedia of the Magic Lantern* (note 3), pp. 269–270. See also Lewis Wright, *Optical Projection. A Treatise on the Use of the Lantern in Exhibition and Scientific Demonstration* (London, 1891). I have also used the 2nd edition (1906) and the posthumous edition (1920).

20 This was drawn by him in 1659 and was probably inspired by the dancing skeleton of the *Danse Macabre*, which was popular at that time. The original is in Leiden University Library: ms HUG 10 fol. 76v, reproduced in Laurent Mannoni and Donata Pesenti Campagnoni, *Lanterne magique et film peint. 400 Ans de Cinéma* (Paris: Éditions de la Martinière and La Cinèmathèque Française, 2009), pp. 29–35.

gress and a trivializer of the scientific experience. Educators such as 's Gravesande, Petrus van Musschenbroek and the Abbé Nollet appreciated that the power of the extraordinary images of the lantern to awe and to amuse could be used to educational advantage, and were among the first to establish this genre. In his physics textbook *Beginselen der natuurkunde, beschreven ter dienste van der landgenooten* (1736), Petrus van Musschenbroek described lantern slides with moving images, such as a mill with turning sails, a woman bowing, a rope dancer, a drinker raising a goblet, and a hat and wig being lifted from the head of a bald man.[21] From these humble beginnings came the extraordinary diversity of animated scientific slides that became such a popular aid to teaching in the 19th century.

Initially the predominant aim of the magic lantern was entertainment rather than serious scientific study, as is made clear by such titles as William Hooper's *Rational Recreations: in Which the Principles of Numbers and Natural Philosophy are Clearly and Copiously Elucidated, by a Series of Easy, Entertaining Interesting Experiments* (1774). In this sense the magic lantern could be treated in the same way as the projecting solar microscope of which the serious microscopist Dr Goring complained in 1827:

> The image of the common solar microscope may be considered a mere shadow, fit only to amuse women and children.[22]

Throughout the 18th century the magic lantern was primarily used in the context of natural philosophy as a light source to project the massively enlarged images of natural objects produced by the projection microscope, and this continued well into the 19th century. According to one contemporary ironic comment:

> To see a flea as large as a camel must gratify every flea-bitten observer, by inducing a satisfaction and sense of security at not having been devoured by the attack of such animals.[23]

The second function of the lantern was to project images on prepared glass slides to accompany the science lecture or discourse. In this sense the lantern had the

21 Musschenbroek described 5 ways of causing movement of the image, four involved some form of lever arrangement and one employed the endless-cord and pulley mechanism to rotate the glass disc on which the image was painted, such as the sails of a windmill, see John Barnes, 'The Projected Image. A Short History of Magic Lantern Slides', *NMLJ*, **3**–3 (1985), p. 3 (pp. 2–7). But Musschenbroek was not the first, see note 41.

22 Quoted by Turner (note 16), p. 98.

23 *Ibid.*, pp. 98–99. According to Turner, by the end of 19th century the element of the fairground show had disappeared, and the Negretti & Zambra catalogue, for instance, was pointing out the educational value of the magic lantern with microscope attachment. Earlier optical instrument makers traded on the public's thirst for 'scientific' spectacles, as Philip Carpenter did with his lucernal projections in his 'Great Microcosm' exhibition in the 1820s, which inspired the famous cartoon 'Monster Soup commonly called Thames Water' (1828) by William Heath; see Stuart Talbot, ''The Perfect Projectionist': Philip Carpenter, 24 Regent Street, London', in *Bull. Scientific Instrument Society*, No. 88 (March 2006), pp. 17–20, reprinted in *NMLJ*, **10**–3 2007), pp. 49–51. For a contemporary description, see *Arcana of Science and Art or An Annual Register of Useful Inventions and Improvements*, Vol. 1 (London, 1828), pp. 14–15.

same function as illustrations in textbooks, that was, to expand the information visually (Fig. 4, colour plates).

The early images were quite primitive but as the lantern evolved technically, in particular the optics and the illuminants, this influenced the way the images were painted and manufactured. Hand-painted images were adequate for the low-level illumination of the lantern up to the mid-19th century. From then on limelight, multiple projection and dissolving views required extremely detailed transparent paintings in oils, water colours and dyes. Some extraordinary results were achieved. By the latter part of the 19th century, to satisfy demand, large numbers of photographic and lithographic slides were produced, but these fall outside the scope of this paper.[24]

The third function of the lantern was to create scientific effects and project these onto the screen. These were in effect scaled-down versions of lecture-demonstration apparatus so that they could now be performed within the space created between the condenser and the lenses of the scientific lantern. A performer of the stature of Michael Faraday continued in the footsteps of his predecessors in using the magic lantern in his scientific discourses at the Royal Institution. His lantern featured in the *Illustrated London News* of 1845 (Fig. 5) was conventional as the specialized scientific lantern has not yet been developed. But the lecture-demonstration techniques which he popularized could be adapted for the scientific lantern when this became available. Soon thereafter the specialized scientific lantern was developed as the teaching of science became more widespread.

24 With the development of limelight, multiple projection and dissolving views by these more sophisticated lanterns the slides had to be correspondingly more detailed to be in register. The standard size was between 3¼ and 4 inches on very thin crown glass. The painting mediums were oil paints, water colours or dyes. The binding agency was clear shellac varnish or albumen with water paints. Only a restricted colour palette could be used. Outstanding 19th century slide painters were: W.R. Hill (Royal Polytechnic Institution), H.M.J. Underhill, and Desch (France) and Paul Hofmann (Germany). Photographic slides were introduced in the 1850s and 60s. Perran Garnier, *A Manual of Painting on Glass for the Magic Lantern* (London: J. Barnard & Son, n.d., *c.* 1899). See also Dennis Crompton, David Henry, Stephen Herbert (eds), *Magic Images: The art of the hand-painted and photographic lantern slides* (London: Magic Lantern Society, 1990).

Fig. 5: Faraday lecturing on magnetism and light at the Royal Institution, Illustrated London News (1845). He was assisted by Charles Darker, an optician who in 1866 invented the lantern kaleidoscope (see note 45). The painter J.M.W. Turner was strongly affected by one of Faraday's lectures in which he displayed the magnetic field, as projected in this engraving.

'SCIENTIFIC' LANTERN SLIDES FOR LECTURE-DEMONSTRATIONS AND RECREATIONAL SCIENCE

The increasing dependence of the European economy on industrialization had its impact on science education, in particular in Germany, followed by the USA, France and England. The Great Exhibition of 1851 in London, and subsequent international exhibitions demonstrated the importance of a technical literate workforce. By the 1870s science facilities were expanded in universities and secondary education, and special science-demonstration apparatus was developed for teaching the new science courses. In England, for instance, a Royal Commission on Scientific Instruction and the Advancement of Science was set up in 1870 under the chairmanship of the Duke of Devonshire. To their report of 1872 was appended a catalogue of apparatus needed for teaching elementary science, with prices and illustrations.[25] Four years later the Special Loan Collection of Scientific Apparatus was established at South Kensington, the precursor of the present-day Science Museum in London.[26]

A wide range of lantern teaching demonstration apparatus was developed for schools and colleges: chemical tank slides for demonstrating precipitation and crystallization; slides with glass tubes of different bores to demonstrate capillary action; a slide with a galvanometer to indicate electric current or with a large number of small pivoting magnets to illustrate magnetic fields[27]; with a tuning fork for acoustical demonstrations, or a model steam piston with side valve, in which the action of the steam is demonstrated by cigarette smoke blown into the model during projection (Fig. 6, colour plates).[28] Compendia of some of the most popular experiments could be purchased for 'home improvement', and textbooks and popular articles were written for 'experiments in chemistry, optics, electricity, magnetism, etc. adapted for the magic lantern'.[29]

Closely allied to the encroaching of science into the school curriculum in the last decades of the 19th century was 'recreational science' for the lay public and home amusement. Education and amusement are closely related and it is therefore not surprising how many pieces of demonstration apparatus used in lectures were adapted for public and family entertainment. For instance, many of the demonstration pieces used by itinerant lecturers (such as Abbé Nollet), were the forerunners of toys and scientific amusements made for family entertainment. As the

25 E. Ashby, 'Education for the Age of Technology', in C. Singer et al (eds) *A History of Technology* (Oxford:OUP, 1958), Vol. 5, pp. 779–795.

26 D. Follett, *The Rise of the Science Museum under Henry Lyons* (London: Science Museum, 1978), pp. 1–4; *Science and Art Department of the Committee of Council on Education, Catalogue of the Special Loan Collection of Scientific Apparatus at the South Kensington Museum*, 3rd ed. (London, 1877); Willem Hackmann, 'Nineteenth-Century Scientific Instruments and their Makers, P.R. de Clercq (ed.), *Nineteenth-Century Instruments and their Makers (papers presented at the Fourth Scientific Instrument Symposium, Amsterdam, 23–26 October 1984)* (Leiden: Museum Boerhaave & Amsterdam: Rodopi, 1985), pp. 53–91.

27 The author's large galvanic lantern slide was manufactured for the Pierpont Morgan Trust Fund for science education in schools in the USA. The financier and philanthropist, J.P. Morgan, was keen to promote science and technical education.

28 For a brief note see Willem Hackmann, 'To Smoke or Not to Smoke', *Bull. Scientific Instrument Soc.*, No. 100 (March 2009), p. 35.

29 Wright's important book in this context has already been mentioned in note 19, but a range of publications appeared from the serious to the popular, such as W.J. Chadwick, *The Magic Lantern Manual* (London, 1878, 1884 or 1888); An Expert, *The Art of Projection and Complete Magic Lantern Manual* (London, 1893); J.H. Pepper, *Playbook of Science* (London, 1860, see note 37)); and Walter B. Woodbury, 'Science at Home. A Series of Experiments in Chemistry, Optics, Electricity, Magnetism, &c. adapted for The Magic Lantern, Reprinted from *the English Mechanic', Magic Lantern*, **1** (1874–5), reprinted by the Magic Lantern Society in 1988. There is a great deal of repetition in both the text and engravings between these sources.

historian Huizinga pointed out many years ago, much scientific knowledge is absorbed through play.[30]

The 19th century had an insatiable appetite for self-made and 'self-improving' entertainment, and young and old were catered for in an endless series of books. In England the tone was set by John Henry Pepper's *The Boy's Playbook of Science* (1860) and Sir David Brewster's *Letters on Natural Magic*, first published in 1868. Probably, the most famous in this genre was Larousse's *La Science Amusante* in three volumes published in the 1890s, marvelously illustrated by the renowned engraver Louis Poyet (1846–1913). The book gained such popularity that over 130 editions of the three volumes have been printed.[31] The common theme of these publications was that simple, reasonable scientific explanations could be found for apparently magical or miraculous effects.[32]

In the first decades of the 19th century, as the technology of the magic lantern and lantern slides[33] became more sophisticated and expensive, the magic lantern show moved out of the home and was embraced by a new breed of professional lanternists, who often styled themselves 'Professors of Natural Philosophy'.[34] They gave public shows in educational establishments, music halls, temperance halls, and in establishments more geared towards spectacular demonstrations, such as in London the Royal Polytechnic Institution (founded in 1838),[35] the Egyptian Hall, Crystal Palace in Sydenham, the Colloseum, the Albert Hall and the Zoological Gardens. In this they continued the pioneering lecture demonstrations of Humphrey Davy, William Thomas Brande, and Michael Faraday at the Royal Institution. Similar developments took place in other European capitals,

30 Johan Huizinga, *Homo Ludens* (Boston: Beacon Press, 1955). Dutch edition first published in 1938.

31 Originally printed in a long series of articles published in the French weekly periodical *L'Illustration,* under the title 'La Science Amusante', written by Arthur Good (1853–1928) under the pseudonym 'Tom Tit', and bundled together in three volumes by Larousse (1890–1893). For Pepper's *Playbook*, see note 37.

32 G. L'E. Turner, *Nineteenth-Century Scientific Instruments* (London: Sotheby Publications/University of California Press, 1983), pp. 291–308.

33 Among the improvements in slides were the dissolving views (in which one picture would be dissolved in another by the use of two lanterns), invented by Henry Langdon Childe in 1837, the mechanical slides to be described later, and the photographic slide first shown at the Crystal Palace in 1851, but becoming common in the making of scientific slides such as microscopic images from the 1870s. See for instance, T.C. Hepworth, *The Book of the Lantern* (London, 1888).

34 Herman Hecht, 'The History of Projection 2', *NMLJ*, **6**–2 (1991), pp. 5–6.

35 Brenda Weeden, *The Education of the Eye. History of the Royal Polytechnic Institution 1838–1881* (London: University of Westminster, 2008). During the 1870s a travelling branch of the Polytechnic was set up, and yearly syllabuses arranged which included series of scientific lectures (mostly practical demonstrations), by E.V. Gardener, professor of chemistry and general science, see Lester Smith, 'Education and Amusement, Education and Instruction Lectures at the Royal Polytechnic Institution', in Richard Crangle, Mervyn, Heard, Ine van Dooren, *Realms of Light. Uses and Receptions of the Magic Lantern from the 17th to the 21st Century* (London: The Magic Lantern Society, 2005), pp. 138–145.

where travelling professional showmen interspersed their sophisticated slideshows with demonstrations of the latest scientific discoveries.[36]

Among the most sophisticated and memorable lantern shows were those performed by the flamboyant "Professor" John Henry Pepper at the Royal Polytechnic Institution in mid-century (the title was conferred on him by the Polytechnic management). He was particular proficient in optical experiments, of which the most famous was "Pepper's Ghost", which made its debut at the Polytechnic in a Christmas performance of Charles Dickens' *Haunted Man* in 1862. The effect was created on stage by reflecting the image of an actor, concealed below the stage and brightly lid, from a large inclined glass sheet which was invisible to the audience. Pepper used his 'ghost' not only to explain the underlying principles of optics, but also that his audience should retain a healthy skepticism for the claims made by the spiritualists which was then much in vogue.[37]

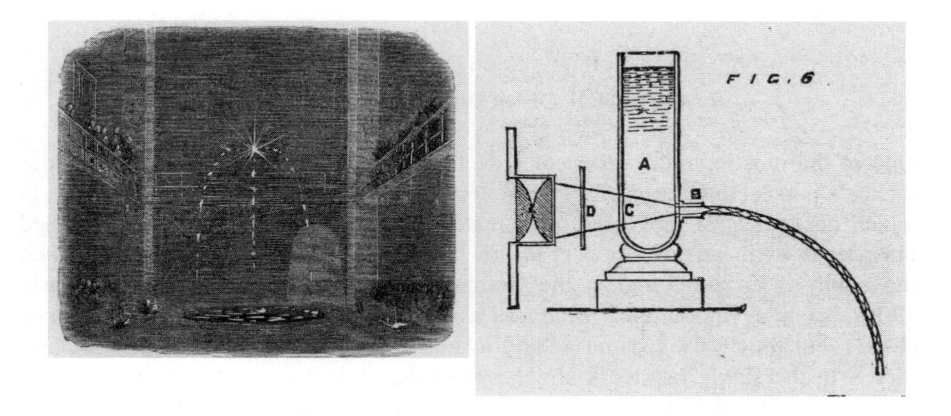

Fig. 7: The 'illuminated cascade' at the Polytechnic and the portable version for home experiment. From Pepper's Playbook and Woodbury 'Science at Home' (note 40).

36 Hecht (note 34), p. 6, cites the case of the brothers Skladanowski who performed such shows all over Europe under their assumed name of Hamilton Brothers.

37 J.R. Secord, 'Portraits of Science: Quick and Magical Shaper of Science', Science, 297, no. 5587 (2002), pp. 1648–1649, revised version of an introduction to Pepper's *Boys Playbook of Science* (reprinted by Thoemmes Press, 2003). See also Verity Hunt, 'Raising a Modern Ghost: The Magic Lantern and the Persistence of Wonder in the Victorian Education of the Senses', *Romanticism and Victorianism on the Net*, no. 52 (November 2008), *http://id.erudit.org/iderudit/01980ar*. On a general history of 'ghostly projections', see Mervyn Heard, *Phantasmagoria. The Secret Life of the Magic Lantern* (Hasting: The Projection Box, 2006). Note that for this paper I have used J.H. Pepper, *The Boy's Playbook of Science* revised by T.C. Hepworth, new edition (London: [1881]).

Nine different lanterns and complicated and beautifully hand-painted slides were used in the Polytechnic shows to produce spectacular superimposed images ('dissolving views'), accompanied by carefully synchronized sound effects.[38] Furthermore, Pepper enthralled his audiences with the sheer number and virtuosity of his scientific lecture-demonstrations that ranged across the gamut of physical phenomena: in mechanics, pneumatics, optics, heat, electricity, magnetism, chemistry and astronomy, and in which demonstrations with the magic lantern was in important component. Scaled-down versions of these demonstrations were incorporated into the science teaching or home experiment repertoire. A case in point is the large illuminated cascade which was built by the distinguished French optician and lantern manufacturer Louis Jules Duboscq (1817–1886)[39] during his visit to the Royal Polytechnic Institution in the summer of 1854, which delighted Queen Victoria and the royal party. Spectacular light effects were created in the falling water by the careful placing of the magic lantern and lenses. Soon thereafter, a small portable version was described so that this effect could be replicated elsewhere (Fig. 7).[40]

RACKWORK DEMONSTRATIONS SLIDES

One of the most exciting group of didactic lantern slides (certainly from the audience's perspective) used in scientific lectures are those operated by rackwork, in which the effects of movement are created by rotating glass discs set in toothed brass rings by means of a ratchet turned by a handle. The first references to such moveable slides are found in the German magic lantern literature of the early 1700s, such as in Samuel *Rhanaeus's Novum et curiosum laternae magicae* (1713), but they only became widely available in the latter half of the 19th century.[41] In the single rackwork slide, a glass disc revolves while another remains stationary; in the double rackwork slide two discs revolve in opposite directions.[42]

38 W.F. Ryan, 'Limelight on Eastern Europe. The Great Dissolving Views at the Royal Polytechnic', *NMLJ*, **4** (1986), pp. 48–55; Weeden (note 35).

39 Two of his important products he introduced in the Polytechnic were his 'Lanterne Photogénique' (a brass scientific lantern with arc light and Foucault regulator introduced in 1850) and the projecting phenakistiscope to be discussed later.

40 Pepper, *Playbook* (note 37), pp. 334–336, Figs 311–313; Woodbury (note 29), pp. 8–9, Fig. 6. This effect is due to the 'total reflection' of light. Pepper, the inveterate teacher, compares this effect to the travelling of the voice in a speaking tube.

41 See 'rackwork slides' in *Encyclopaedia of the Magic Lantern* (note 3), p. 247; Hauke Lange-Fuchs, 'On the Origin of Moving Slides', *NMLJ*, **7**–3 (1995), pp. 10–14; so that Petrus van Musschenbroek (note 21) certainly was not the first but had impact because his textbooks were so widely circulated.

42 The 'windmill' with revolving sails is an example of the single rackwork, while the ever popular 'chromatrope' or firework effect is an example of the double rackwork. For the classification of mechanical rackwork and pulley slides, see John Barnes, 'Classification of Magic Lantern Slides for Cataloguing and Documentation', in *Magic Images. The Art of*

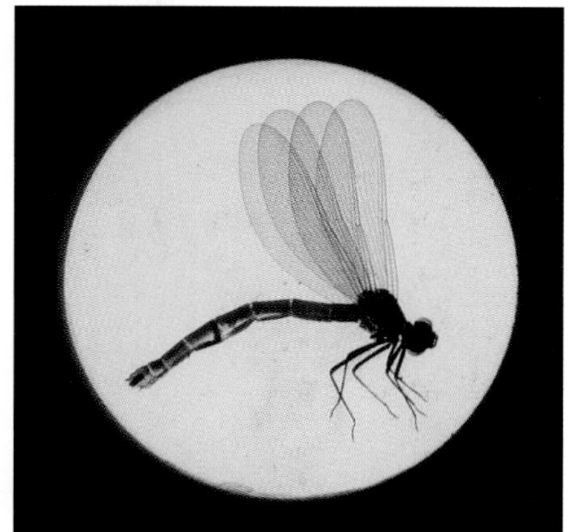

Fig. 4: Prepared slides for lectures: A. hand-painted slide in oils (1860s); B. Dragonfly, real specimen commercial slide (1860s); C. Home-made plant specimen of Anagatis artensis, using same techniques for making a microscope specimen slide (1930s); D. 'Decomposition of sun-light through a prim', hand-tinted diagram in York's 'Astronomy' series, no. 26 (1890s). Author's collection.

Fig. 6: The action of the steam is demonstrated by smoke blown into the model of the steam piston with side valve. This "lantern slide" dates from around 1910, and was probably made by E. Leybold's Nachfolger. Author's collection.

Fig. 8: Two from the set of ten astronomical rackwork slides by Carpenter & Wesley: no. 1 of the "Solar System shewing the Revolution of the Planet with their Satellites round the Sun" and no. 6 showing "the Eccentric Revolution of a Comet round the Sun, and … the appearance of its Tail at different points of its Orbit." Author's collection.

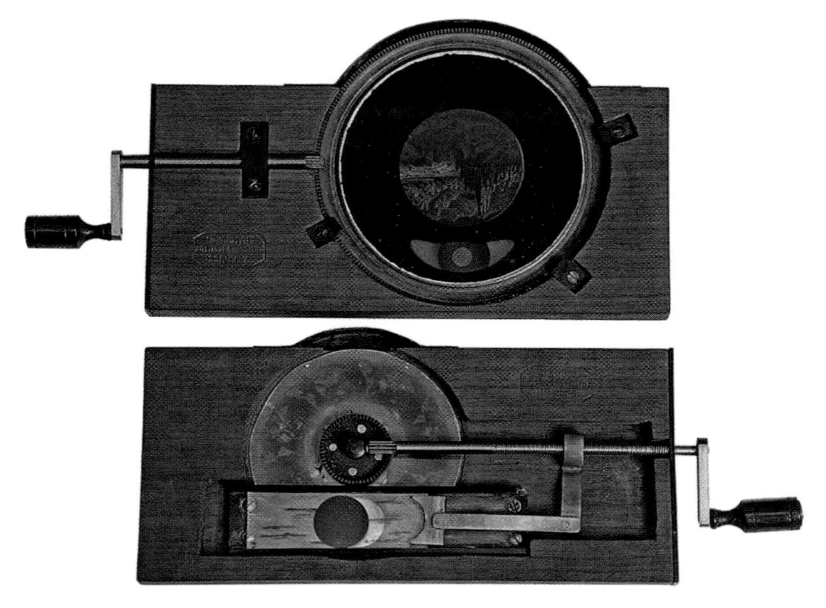

Fig. 9: 'A Day on the Moon': two rackwork slides by A. Krüss, from the 1890s. They are accompanied by a pamphlet 'Populärer Vortrag zu dem mechanisch-beweglichen Bilde Ein Tag auf dem Monde von C. Knoch.' Author's collection.

Fig. 11: Original 'Wheel of Life' design with glass disc shutter of 8 segments (producing 8 slits), and below the Ross's improved design of 1871 of a metal disc shutter with single slit. Also shown their moving images: the donkey engine and the skaters. Author's collection.

The longitudinal vibration of sound waves was either demonstrated by a scaled down version of a wave machine in which the slider moving the pegs traversed across the window of the lantern slide. A more economical method was invented by the physicist André Prosper Crova (1833–1907), a faculty member at the University of Montpellier in France. A glass disc on which are printed non-centric circles is rotated by the rackwork, and the resultant projection through a narrow horizontal slit simulates a travelling longitudinal wave.[43] A version was described by Frederick J. Cheshire of Birckbeck in 1892, which by means of an adjustable slit simulated the sound waves in open and closed organ pipes.[44] Other rackwork slides rotated specimen to demonstrate the startling changes of colour under polarized light, or to rotate fragments of coloured glass to produce pretty patterns when projected through a kaleidoscope.[45]

The most popular scientific rackwork slides were those dealing with astronomical events. Several reasons for this can be suggested. From the end of the 17th century there existed a close connection between optical instrument makers, astronomical instruments and the development of the magic lantern. By the end of the 19th century, manufacturers of astronomical or scientific instruments were still marketing astronomical slides, and this would last until the late 1930s. Interest in astronomy was multifarious. It became fashionable in the 18th century to be knowledgeable about the latest astronomical discoveries, and in the second half of the 19th century, too, rapid advances in astronomy again changed the knowledge of the subject dramatically. Such developments could be captured on lantern slides. Then there was the religious dimension: the study of "The Heavens

Hand-Painted and Photographic Lantern Slides (London: The Magic Lantern Soc. of Great Britain: 1990), pp.75–84.

43 First described by Crova in 1867, which is now quite rare. In the most common version, the 'Dorman's Sound-Wave Slide', four lines of wave motion are shown travelling outwards at right-angles to each other as from a bell.

44 See *Nature,* 44 (1892), pp. 47–348, of which Newton & Co. in London where the sole makers.

45 The cheapest was the 'elbow polariscope', named so because of its shape, and could either be used for direct viewing or for projection with the magic lantern, see 'polariscope' in *Encyclopaedia of the Magic Lantern* (note 3), p. 236; Wright, *Optical Projection* (note 20), pp. 338–348. The peculiar appearance of the elbow polariscope was because of the stacked plates of crown glass which acted as the polarising mirror, but as this arrangement shows it made it an awkward instrument to use. The lantern kaleidoscope was based on the invention by David Brewster in 1816. In his patent of 1817 Brewster suggested that his device could be adapted for the lantern. The first commercially successful lantern kaleidoscope was marketed by J. Darker (Charles?) of London, of which my instrument is an example. More elaborate kaleidoscopes were also made with adjustable inclined mirrors; see 'kaleidoscope, lantern' in *Encyclopaedia of the Magic Lantern* (note 3), p. 147; Wright, *Optical Projection* (note 20), pp. 145–147; J.H. Pepper, *Cyclopædic Science Simplified*, Revise ed. (London, 1872), pp. 31–35; anon., 'The Darker Kaleidoscopic Projection Lens', *NMLJ*, **3**–3(1985), p. 9; for earlier projection kaleidoscopes, see Herman Hecht, 'The Origin of the Kaleidoscope', Magic Lantern Society, *Newsletter*, No. 7 (November 1982).

Declare the Glory of God", as was declared on one of the fine hand-painted astronomical slides by Newton & Co.[46]

In the last decade of the 18th century there was a move to transform the orrery (a mechanical planetarium) into lantern slides that could project the same effects. The earliest known surviving set was manufactured in the 1780s by Mary Dicas, a mathematical instrument maker in Liverpool, who made what was described as a 'Portable Eidouranion or Astronomical Lucernal',[47] which was the precursor of the astronomical rackwork slides that became popular in the 19th century. One of the most influential makers of astronomical slides was Carpenter & Wesley, whose founder, Philip Carpenter, produced the first set 'copper plate slider' mass produced slides in the mid 1820s, with a set of printed lecture notes. By the late 1840s Carpenter & Wesley were advertising their set of ten rackwork slides of which no. 1 of the solar system and no. 6 of the path of a comet were especially complex (Fig. 8, colour plates).[48]

Some quite complex and imaginative astronomical slides were produced. Of particular note is a rare elaborate mechanical slide simulating the transit of Venus, which may have been produced for the 1874 transit.[49] A rare set of two slides was produced by A Krüss of Hamburg in the 1890s. They were made for a popular lecture by C. Knoch entitled 'A Day on the Moon'. The central image of the large rackwork slide shows the mountain ranges of the moon, set in an intense black sky. The slide demonstrates the astronomical events as they unfold during a moon day, which is equivalent to 28 earth days. The continually changing light phenomena are much more intense as there is no atmosphere on the moon. The earth appears as a large black disc surrounded by zodiacal light, which casts a rosy glow over the moonscape. The second slide shows that as the sun rises in the moon's sky, the sickle of the earth continually decreases while the rotation of the earth is observed by the rotation of the continental masses (Fig. 9, colour plates).[50]

46 Psalms 19.1.

47 'Eidouranion' from the Greek 'form of the heavens'. For another description, see Wendy Bird, 'Enlightenment and Entertainment. The Magic Lantern in Late 18th Century and Early 19th Century Madrid', *Realms* (note 35), pp. 90–91.

48 Similar sets were offered by other makers such as Newton & Co. It is not clear who of these two makers pioneered this set, but the ten slides are always in the same sequence. See Mark Butterworth, 'Astronomical Lantern Slides', NMLJ, **10**–4 (2008), pp. 65–68.

49 Only two transits occurred in the 19th century: 1874 and 1882. There were none in the 20th century, and there will be two in this century: 2004 and 2012. See Butterworth, *ibid.*, p. 67, Fig. 5.

50 Andreas Krüss married the daughter of the instrument maker Esmund Gabory who, after apprenticeship with Jesse Ramsden in London, eventually opened his business in Hamburg in 1796. After Gabory's death, Krüss took over the firm, which has evolved into a distinguished instrument company which today is noted for its optical instruments. The firm began to manufacture lantern slides in the early 1860s and by the early 20th century their catalogue listed around 8,000 slides, of which these two astronomical slides are among the most complex.

PERSISTENCE OF VISION

A whole range of optical illusion devices which became popular during the 19th century are based on the physiological phenomenon of the persistence of vision – that is, the ability of the eye to retain the impression of an object for a fraction of a second after its disappearance. This would become a fruitful source of spectacular projections for the 'scientific' lanternist. These 'animation' devices which nearly all attracted elaborate Greek names, were, of course, the forerunner of the modern cinema. The first of these devices was invented by a leading medical man, Dr John Ayrton Paris in 1825 to illustrate his research on persistence of vision. This was the well-known *thaumatrope* ('Wonder-turner'), a card disk with two different figures drawn on the two sides which combine into a single scene when the card is rapidly turned.[51]

In 1831 Michael Faraday demonstrated the stroboscopic effect by means of two cardboard disks with radial slits rotating in the same plane but in opposite directions. Variations of this technique were used by Charles Wheatstone and other physicists to determine the speed of the electric spark and of light. In 1832 the p*henakistoscope* was invented by the Belgian physicist J.A.F. Plateau of Brussels and almost simultaneously by Professor S. Stampfer of Vienna, who called his instrument the s*troboscope*. These instruments consisted of a disk with figures arranged radially in successive positions, and which appear to move when spun. In the case of Plateau's disk the reflections of the figures in a mirror were viewed by looking through radial slits cut into the disk, while in Stampfer's arrangement the figures were viewed through the slits of a rotating second disk, thus more in line with Faraday's original experiments. A similar device was marketed by E.J. Purkinje of Breslau in 1841. These were the first of all the later and more complicated forms of motion picture.[52]

W.G. Horner of Bristol in 1834 invented a variant, the *Zoetrope*, or 'Wheel of Life', but this was not marketed until 1867. In this case a strip of paper with the figures was viewed through the slots of a rotating drum. A sophisticated version of this was the French *praxinoscope*, invented in 1877 by Professor Émile Reynaud, in which case the figures were viewed from their reflections in mirrors secured to the rotating drum. Apart from the 'direct vision' version there was a more expensive one in which the image was projected. In that case the image

51 John Ayton Paris, P*hilosophy in Sport Made Science in Earnest : being an attempt to implant in the young mind the first principles of natural philosophy by the aid of the popular toys and sports of youth* (London, 1827), Vol. 3, p.1. For a general review of optical toys, see Georg Füsslin, *Opticches Spielzeug oder wie die Bilder laufen lernten* (Stuttgart: Verlag Georg Füsslin, 1993), and for the most recent historical analysis on persistence of vision and early cinema, see chapter 5 'Captured by Cinematography', pp. 117–153, in Jimena Canales, *A Tenth of A Second A History* (Chicago and London: The University of Chicago Press, 2009).

52 Maurice Dorikens, Joseph Plateau (1801–1883), *Leven tussen Kunst en Wetenschap* (Provincie Oost-Vlaanderen, 2001). On Farady's disc see the contribution by David Robinson, pp. 254–263 (this work is in Flemish, French and English).

must have been very dim as the light source was an oil lamp. These devices were for domestic use or for lecture-demonstrations.[53]

To animate moving images for large audiences, such as in the Royal Polytechnic, the magic lantern with limelight would have to be used. An expensive 'moving glass slide' usually depicting a dancing skeleton, was the *choreutoscope* invented by Lionel S. Beale, the assistant of the famous physician in Oxford, Sir Henry Acland. In this device, invented in 1866, intermittent action was achieved by rapidly moving six skeletons painted on the glass slide in view by turning a handle, while at the same time a shutter moved across the viewing opening. It was a complicated device and the mechanism was not particularly strong. In a sense its mechanism can be regarded the forerunner of the Maltese cross device used in cinematography.[54]

Perhaps a more promising way forward in these early days was the 'projecting phenakistiscope'. The first to achieve projecting a painted glass 'wheel of life' was T.W. Naylor in 1843.[55] The projected figure was interrupted by a second rotating disk with slits. No such lantern has survived. A similar arrangement was devised independently in 1845 by a German army officer, Franz von Uchatius. In a second version described in 1853, he attempted rather a novel solution in getting a more stable moving picture by having a static disk but a rotating light source.[56]

The scene is now set for Jules Dubosc who began his pioneering experiments on moving images in the early 1850s and by 1853 announced his 'projecting phenakistiscope'.[57] Two parallel disks were projected at different speeds by

53 For a general history on these early persistence of vision devices, see Brian Coe, *The History of Movie Photography* (London: Ash & Grant, 1981) and the old classics Olive Cook, *Movement in Two Dimensions* (London: Hutchinson & Co., 1963), in particular pp. 121–136 and C.W. Ceram, *Archaeology of the Cinema* (London: Thames and Hudson, 1965). For Émile Reynaud's 'Théatre optique' (*c.* 1892), see Mannoni, et al (note 20), pp. 249–253. His set up has been recreated in the Cinèmathèque Française Museum in Paris.

54 Intermittent movement was achieved by means of a circular disc carrying a pin attached to a handle. As the disc revolves the pin engages the notch on the slide, moving it on by one picture and at the same time raising the shutter (Turner, note 32), p. 304. The common version was demonstrated at the Royal Polytechnic in the early 1870s, and consisted of a glass strip with six images (Pepper, *Playbook* (note 37), pp. 355–356). A circular choreutoscope with a six-sectioned Maltese cross was made by T.H. McAllister of New York in the 1890s. For good examples of these scientific slides, see the section 'Science et Enseignement' (pp. 159–171), and pp. 224, 324–325 for various 'Newton discs' in Mannoni, et al. (note 20).

55 The glass phenakistiscope disks were made by Ackermann & Co. on the Strand, London.

56 Coe (note 53), p. 33.

57 Füsslin (note 51), pp. 48–50; Pepper, *Cyclopædic* (note 45), Fig. 80, and pp. 71–86 on persistence of vision; Pepper, *Playbook* (note 37), pp. 372–374, Fig. 354. Duboscq also invented another variation at this time which combined the moving image effect with the 3D effect of the stereoscope just then becoming popular – which he called the *stéréofantascope* or *bioscope*. The observer had to look in a couple of small mirrors through the radial slits of the rotating disk of the phenakistiscope. The mirrors were so orientated that one reflected the upper image to the right eye, while the left eye saw the lower image. Alas, it appears that

means of gearing in front of a magic lantern. A series of images in sequence (between 10 and 16) were painted on the glass disk, which was nearer the light source, while the second one made of wood carried four bulls-eye (condenser) lenses – thus, working both as condenser and as shutter. It became never very popular and an actual demonstration with a restored specimen shows why! It was not found to be possible to replicate the steady image of the jumping boys projected with the Royal Polytechnic instrument as shown in Pepper's *Cyclopædic Science Simplified* (Fig. 10). In fact, the severest limitation is the movement of the image. The speed of rotation is quite critical and there is a great deal of flickering, mostly as the slower rotating mahogany disc with the condensers acts as an inadequate shutter. Obviously it was regarded as an interesting novelty, but it was technically a dead-end and the inventors had to go back to the drawing board.

A more rewarding way forward to achieve 'fluid animation' was the lantern slide phenakistiscope 'Wheel of Life', inspired by the zoetrope, and patented by Thomas Ross in 1869. The original design consisted of a glass disc shutter with 8 segments, rotating in one direction and the glass disc with the images rotating in the other. However, this caused the images to be distorted on projection. In the improved version patented by Ross in 1871, there were 13 silhouettes on the glass disc and a faster contra-rotating shutter with only one aperture. On each rotation of the shutter disc, the picture disc advanced by one image (Fig. 11, colour plates). The 'Wheel of Life' construction, but with a different shutter configurations, was used to demonstrate other optical effects, such as anorthoscopic slides (turning geometrically distorted images into undistorted ones),[58] and chromatic wheels producing colour effects by the visual persistence of complimentary colours. One of the most 'fun' lantern slides was the 'chameleon slide' in which the aperture in the shape of this creature changes colour as the glass disc is rotated.[59]

none have survived, apart from the bioscope disc (see Dorikens, note 52), despite the fact that it was advertised in Duboscq's catalogue until 1885.

58 'Wheel of Life, lantern' in *Encyclopaedia of the Magic Lantern* (note 3), pp. 321–322. The anorthoscopic slide was based on the experiments by Plateau, see Dorikens (note 52), pp. with many examples of these printed discs. The slides, however, are extremely rare.35–53. Edweard Muybridge used the 'Wheel of Life' technique to demonstrate his animated photography of humans and animals in 1880, and called it the *zoopraxiscope*. Georges Demeny used a similar technique in 1891, which he called the *phonoscope*, as he used it in teaching deaf children to lip-read.

59 Wendy Bird: 'Meanwhile, the common people, unwilling or unprepared to reap the benefits of the 'age of reason', clamoured for pseudo-scientific magic and optical entertainment', *Realms* (note 47), p. 91; and her 'Optical Entertainment in Madrid in the time of Goya', *NMLJ*, 9–2 (2002), pp. 19–22.

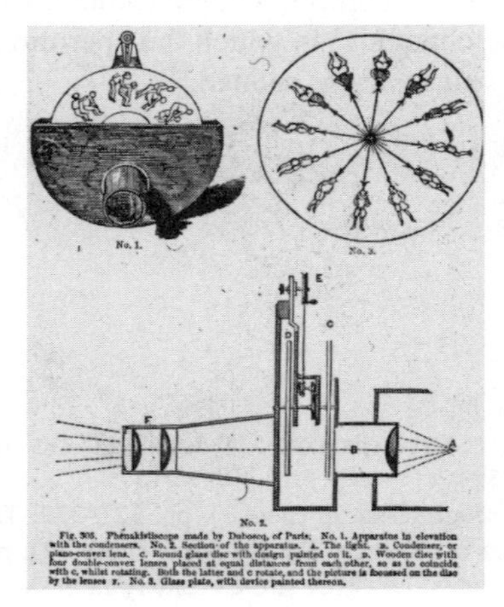

Fig. 10: Duboscq's Projection Phenakistiscope with illustrations from Pepper's Playbook and Cyclopædic Science (note 57). Author's collection.

CONCLUSION

The magic lantern formed the bridge between natural magic and natural philosophy. Giovanni Battista della Porta, Kaspar Schott, Athanasius Kircher and Georg Philip Harsdörffer promoted the teaching of nature by means of fantastic machines and scientific games. The magic lantern, as the name implies, fell very much into this category, but it was exactly because of this use that Christiaan Huygens, was ambivalent towards this optical device. Could the magic lantern have a legitimate function in scientific discourse? Magicians and showmen on the one hand and natural philosophers on the other utilized their knowledge of nature to create effects. The visual scientific culture of the Enlightenment and beyond was both magical and edifying. Instruments, like the magic lantern, could make transparent the workings of nature but could also mystify. How much of the 'science' of the public lectures was regarded as tricks by the audience?

The power of the magic lantern to draw in and to teach audiences was recognized by science lecturers of the stature of Michael Faraday at the Royal Institution, and John Henry Pepper (of 'Pepper's Ghost' fame) at the Royal Polytechnic Institution.

Initially the Magic Lantern accompanied science discourses with projected static images (similar in function to textbook illustrations), but by the mid-19th century considerable improvements in optics and mode of illumination resulted in a boost in the development of the lantern slide, and in the function of the lantern, which could now also be used to create scientific effects and project these onto the screen. As a consequence the Magic Lantern blurred the lines between experiment and performance, between laboratory and theatre, and made the phenomena of science both visible and accessible.[60]

The reconstruction of historical experiments has become an accepted didactic resource in the history of science, either by using original instruments, or by means of facsimiles as pioneered by the Didactic and History of Physics Group at the Carl-von-Ossietzky-University in Oldenburg.[61] There are several websites that demonstrate original historical apparatus. Of note are the recent demonstrations by Elizabeth Cavicchi, utilizing instruments at the Norwegian Museum of Science and Technology (the Norsk Teknisk Museum) in Oslo Norway[62], and of Paolo

60 Weeden (note 35), p. 8, and on the use of the lantern, pp. 43–50.

61 The Oldenburg group has been reconstructing historical instruments for some 25 years. To appreciate the range and variety of these facsimiles, see http://www.histodid.uni-oldenburg.de/22139.html, and for the context of this approach in teaching the history of science, see the publication list at http://www.histodid.uni-oldenburg.de/22102.html, in particular the book published in 2007 based on the experiences of this group: Peter Heering and Daniel Osewold, eds., *Constructing Scientific Understanding through Contextual Teaching* (Berlin: Frank & Timme, 2007).

62 This video made in 2009 of 19th century instruments for the teaching and research of physics held at Oslo University and Norsk Teknisk Museum shows the spectrum cast in sunlight by glass and quartz prisms, made by Soleil in Paris, the sounds from a xylophone, 19th century organ pipes, and tuning forks, which includes the Doppler effect, also Lissajous figures

Brenni using the original instruments in the collection of the Fondazione Scienza e Tecnica in Florence.[63] Using either original historical material or modern-made facsimiles has pros and cons. No matter what level of restoration has been achieved of historical instruments, these will have deteriorated in all sorts of subtle ways that will be almost impossible to rectify fully. Facsimiles, on the other hand, may not have worn out parts or deteriorated in other ways, but it will have been difficult to source all the materials that made up the original instrument being copied, and these subtle differences may have an effect on performance.

Another didactic approach is recreating original experiments "virtually" by means of digital technology. This has the advantage that programmes can be developed so that students can be made to interact with the virtual experiment. A pioneer in this is the History and Didactics Group of the "A.Volta" Department of Physics at Pavia University.[64] This approach can be further developed by recreating the actual experiments with facsimiles or with original instruments allied to computerized graphic displays of what has been demonstrated.[65]

These approaches are the modern extension of the visual culture in which the magic lantern was such an important component pioneered by earlier generations. Such Victorian 'science shows' can, of course, be replicated by using the original historical material, and this indeed has been done, but the equipment is extremely bulky, fragile, and difficult to operate 'in real time'. A way to overcome these problems is to capture the scientific effects produced by the magic lantern and 'scientific' lantern slides on video and to recreate these by means of PowerPoint and digital projector, as was done for the meeting at Regensburg.[66]

created by a mechanical device made by the Norwegian instrument maker P.C. Brantzeg, and a wave demonstration device made by the London maker E. M. Clarke, based on the work of Charles Wheatstone in the late 1840s.
See http://www.youtube.com/watch?v=3DIreqT8Y5bsw.

63 Paolo Brenni has spent many years restoring this almost complete collection of 19th century physics instruments. He has now produced a series of short video clips illustrating the functioning of various instruments (optical pre-cinema apparatus, fire piston, Geissles's tubes, experiments with the pneumatic pump, models of steam engine, Puluj's apparatus for the mechanical equivalent of heat, etc.). These clips are used in their own didactic activities, but will also be used more widely. The clips can be seen on http://www.youtube.com, and click on "fondazione scienza tecnica".

64 In 1996 Fabio Bevilaqua established a course on digital technologies in education at Pavia, and has since then promoted this approach with vigour at a number of institutions, setting up numerous websites, blogs and wikis in the process.

65 An example of this is the STeT ("Science Teacher e-Training Video Gallery") website project which was started in 2008, see http://stet.wetpaint.com, coordinated by the National and Kapodistrian University of Athens, and the Universities of Pavia, Oldenburg, Cyprus and the Aristotle University of Thessaloniki. Three videos in a project on teaching early electromagnetism which can be found here are
http://stet.wetpaint.com/page/Faraday%27s+electrical+motor,
http://stet.wetpaint.com/page/Ampere%27s+current+balance,
http://stet.wetpaint.com/page/The+Jacobi+Engine).

66 This has the advantage that the delays or interruptions caused by changing slides and preparing them for projection which occurs in the making of the lantern show can be reduced,

There is another problem when we attempt to recreate visual culture. We always lose something in our recreations, which can be more or less problematical. To give a simple example: we can only experience a daguerreotype photo fully by seeing an *actual* daguerreotype. Its fleeting reflected image on its mirrored surface that produces the image we see cannot be reproduced in print. All we have in the printed image is the black and white shadow of the original. So in print we cannot distinguish between a conventional black and white photo produced by absorbing light and the daguerreotype which produces its image by scattering of reflected light.

Similar loss has occurred in this paper. The static images that represent the moving images that were shown are a poor substitute and only mere shadows of what the audience experienced. This is what can make the digital technology of the computer such a powerful didactic tool, both for explaining modern science, but also the past. We are only still at the foothills of this development. My attempt to recreate by means of PowerPoint and digital projector some of the scientific effects of these extraordinary Victorian mechanical lantern slides with such exotic names as 'kaleidotrope', 'phenakistiscope', 'wheel of life', 'choreutoscope', and 'chromatic wheel' is only at the beginning of merging the old and the new ways of 'Learning by Doing'.

REFERENCES

'A Mere Phanton', *The Magic Lantern. How to Buy and How to Use It, Also How to Raise a Ghost* (London, 1876 edition).

Ashby, E., 'Education for the Age of Technology', in C. Singer et al (eds) *A History of Technology* (Oxford:OUP, 1958), Vol. 5, pp. 779–795.

Barbaro, D., *La Pratica della Perspettiva* (Venice: Camillo and Rutilio Borgominieri, 2nd ed., 1569).

Barnes, J., 'The Projected Image. A Short History of Magic Lantern Slides', *The New Magic Lantern Journal* (see NMLJ), **3**–3 (1985), pp. 2–7.

Barnes, J., 'Classification of Magic Lantern Slides for Cataloguing and Documentation', in *Magic Images. The Art of Hand-Painted and Photographic Lantern Slides* (London: The Magic Lantern Soc. of Great Britain: 1990).

Canales, J., *A Tenth of A Second. A History* (Chicago and London: The University of Chicago Press, 2009).

Ceram, C.W., *Archaeology of the Cinema* (London: Thames and Hudson, 1965).

Chadwick, W.J., *The Magic Lantern Manual* (London, 1878, 1884, 1888).

Chadwick, W.J., An Expert, *The Art of Projection and Complete Magic Lantern Manual* (London, 1893).

Coe, B., *The History of Movie Photography* (London: Ash & Grant, 1981).

Cook, O., *Movement in Two Dimensions* (London: Hutchinson & Co., 1963).

thereby producing a 'smoother' show for the audience. The two disadvantages are that the audience do not experience the extraordinary vibrant colours of many of the original slides as projected by the magic lantern, nor do they experience the complexities faced by the lanternists during a 'live show'.

Crompton, D., Henry, D., Herbert, S. (eds) *Magic Images: The art of the hand-painted and photographic lantern slides* (London: Magic Lantern Society, 1990).
de Clercq, P., *The Leiden Cabinet of Physics. A Descriptive Catalogue* (Leiden: Museum Boerhaave Communication 271, 1997), pp.108–111.
de Clercq, P., *At the Sign of the Oriental Lamp. The Musschenbroek Workshop in Leiden* (Rotterdam: Erasmus Publishing, 1997).
Dorikens, M., *Joseph Plateau (1801–1883), Leven tussen Kunst en Wetenschap* (Provincie Oost-Vlaanderen, 2001).
Fahie, J.J., *Galileo. His Life and Work. With Portraits and Illustrations* (London: John Murray; Elibron Classics edition, 1903).
Füsslin, G., *Optisches Spielzeug oder wie die Bilder laufen lernten* (Stuttgart: Verlag Georg Füsslin, 1993).
Godwin, J., *Athanasius Kircher's Theatre of the World* (London: Thames and Hudson, 2009).
Garnier, P., *A Manual of Painting on Glass for the Magic Lantern* (London: J. Barnard & Son, n.d., c. 1899).
Hackmann, W.D., 'Natural Philosophy Textbook Illustrations 1600–1800', in Renato G. Mazzolini (ed.), *Non-Verbal Communication in Science Prior to 1900* (Florence: Leo S. Olschiki, 1993), pp. 169–196.
Hammond, J., *The Camera Obscura: A Chronicle* (Bristol: Institute of Physics Publishing, 1981).
Heard, M., *Phantasmagoria. The Secret Life of the Magic Lantern* (Hasting: The Projection Box, 2006).
Hecht, H. 'The History of Projection 1', *NMLJ* (1989), pp. **6**–1, and 'The History of Projection 2', NMLJ, **6**–2 (1991), pp. 5–6.
Hecht, H. (ed. Ann Hecht), *Pre-cinema History: an encyclopaedia and annotated bibliography of the moving image before 1896* (London: Bowker Saur/BFI, 1993).
Heering, P. and Osewold, D. (eds), *Constructing Scientific Understanding through Contextual Teaching* (Berlin: Frank & Timme, 2007).
Hepworth, T.C., *The Book of the Lantern* (London, 1888)
Hockney, D., S*ecret Knowledge: Rediscovering the Lost Techniques of the Old Masters* (London: Thames and Hudson, 2001).
Hunt, V., 'Raising a Modern Ghost: The Magic Lantern and the Persistence of Wonder in the Victorian Education of the Senses', *Romanticism and Victorianism on the Net*, No. 52 (November 2008).
Lange-Fuchs, H., 'On the Origin of Moving Slides', *NMLJ*, **7**–3 (1995), pp. 10–14.
Lefèvre, W., Renn, J., Schoepflin, U. (eds), *The Power of Images in Early Modern Science* (Birkäuser Verlag, 2003).
Mannoni, L., Neke, W. and Warner, M. (eds), *Eyes, Lies and Illusions: Drawn from the Werner Nekes Collection* (London: Hayward Gallery in association with Lund Humphries, 2004).
Mannoni, L., and Campagnoni, D.P., *Lanterne magique et film peint. 400 Ans de Cinéma* (Paris: Éditions de la Martinière and La Cinèmathèque Française, 2009).
Paris, J.A., P*hilosophy in Sport Made Science in Earnest : being an attempt to implant in the young mind the first principles of natural philosophy by the aid of the popular toys and sports of youth* (London, 1827).
Pepper, J.H., *Playbook of Science* (London, 1860), reprinted by Thoemmes Press, 2003.
Pepper, J.H., *Cyclopædic Science Simplified*, Revise ed. (London, 1872).
Rees, T., *Theatre Lighting in the Age of Gas* (*London: The Society for Theatre Research,* 1978).
Robinson, D., *The Lantern Image. Iconography of the Magic Lantern 1420–1880* (The Magic Lantern Society, 1993); and *Supplements* of 1997 and 2009.
Robinson, D., Herbert, S. and Crangle, R. (eds), *Encyclopaedia of the Magic Lantern* (London: The Magic Lantern Society, 2001).
Rossell, D., 'The True Inventor of the Magic Lantern', *NMLJ*, **9**–1 (2001), pp. 7–8
Rossell, D. *Laterna Magica – Magic Lantern Vol. 1* (Füslin Verlag, 2008).

Ryan, W.F., 'Limelight on Eastern Europe. The Great Dissolving Views at the Royal Polytechnic', *NMLJ*, **4** (1986), pp. 48–55.

Smith, L., 'Education and Amusement, Education and Instruction Lectures at the Royal Polytechnic Institution', in Richard Crangle, Mervyn, Heard, Ine van Dooren, *Realms of Light. Uses and Receptions of the Magic Lantern from the 17th to the 21st Century* (London: The Magic Lantern Society, 2005), pp. 138–145.

Talbot, S., ''The Perfect Projectionist': Philip Carpenter, 24 Regent Street, London', in Bull. Scientific Instrument Society, No. 88 (March 2006), pp. 17–20, reprinted in *NMLJ*, **10**–3 (2007), pp. 49–51.

Tebra, W., 'The Magic Lantern of Giovanni da Fontana', *NMLJ*, **2**–2 (1982), pp 10–11.

Turner, G. L'E., *Collecting Microscopes* (London: Studio Vista and Christie's, 1981).

Wagenaar, W. 'The Origins of the Lantern', *NMLJ*, **1**–3 (1980), pp. 10–12.

Weeden, B., *The Education of the Eye. History of the Royal Polytechnic Institution 1838–1881* (London: University of Westminster, 2008).

Wright, L., *Optical Projection. A Treatise on the Use of the Lantern in Exhibition and Scientific Demonstration* (London, 1891); 2nd edition (1906) and the posthumous edition (1920).

Woodbury, W.B., 'Science at Home. A Series of Experiments in Chemistry, Optics, Electricity, Magnetism, &c. adapted for The Magic Lantern, Reprinted from *the English Mechanic', Magic Lantern*, **1** (1874– 5).

Zahn, J., *Oculus artificialis teledioptricus* (Würzburg, 1685).

THE ESTABLISHMENT AND DEVELOPMENT OF PHYSICS AND CHEMISTRY COLLECTIONS IN NINETEENTH-CENTURY SPANISH SECONDARY EDUCATION (1845 – 1861)[1]

Mar Cuenca-Lorente and Josep Simon

ABSTRACT

This paper studies the formation of school cabinets of physics and chemistry in nineteenth-century Spain. It places the initiatives of the Spanish government in connection and comparison with those taken in France decades earlier. Thus, it analyses how the scientific, pedagogical, and commercial relations between France and Spain contributed to the establishment of Spanish school collections, and how these compared to those developed in France. Moreover, this paper evaluates the role of centralised policies in the acquisition of school equipment, and establishes a map of physics and chemistry school collections in Spain. In doing so, we locate the interest in studying the rich record of printed, manuscript, and material heritage in schools which is available to historians, and the importance of international comparison. In this context, we single out the major similarities and differences between Spanish and French collections, as well as within Spain, in schools with various financial, political, and intellectual means of support. Furthermore, we evaluate the capacity of Spanish schools and teachers to update their collections in relation to changes in pedagogical, scientific, and technological knowledge over time.

INTRODUCTION

The creation of school collections in physics and chemistry was central to the endeavour of establishing a national secondary school system in nineteenth-century Spain. The provision of cabinets and laboratories was undertaken almost simultaneously with the organisation of secondary schools, the publication of a national curriculum, the preparation of textbooks, and the establishment of a national institution aimed at training science teachers.

1 This paper is part of a project developed at the University of València by the Catalan Commission of Scientific Instruments (COMIC) and funded by the Institut d'Estudis Catalans. We would like to thank José R. Bertomeu and Antonio García-Belmar who have contributed with foundational work to the preparation of this paper.

The idea of national 'secondary education' was developed and implemented across Europe and the Americas during the nineteenth century. The earliest developments in this field happened at the turn of the eighteenth century in France and the German states (Anderson, 2004; Green, 1990). The Spanish educational reforms of the mid-nineteenth century followed closely the French example. In this period, the French educational system was already mature, and played an important role in the advancement of physics and chemistry as disciplines. This had not always been the case.

The status of the physical sciences in the French school curriculum was low until the late 1830s. Although secondary education had been established in the first decade of the nineteenth century, the French government did not enact any measure to equip the school cabinets and laboratories until 1821, and this provision was not renewed until two decades later. Nonetheless, French instrument makers, textbook authors, and publishers developed their businesses in consonance with the highly profitable establishment of a national network of schools and faculties of sciences and medicine, which constituted a faithful and lasting clientele. Furthermore, the international prestige of French pedagogy, science and technology, the entrepreneurial spirit of French booksellers and instrument makers, and the cultural impact of the Napoleonic imperial expansion in the late eighteenth and early nineteenth centuries contributed to the international expansion of the French trade in scientific instruments and textbooks (Anderson, 1975; Simon, 2009, Chapter 2).

Although the organisation of school science education by the Spanish government relied heavily on the French experience, it also had distinctive features. First, the government provisions tackled the problem of providing schools with appropriate science collections contemporaneously with the establishment of secondary schools. Second, the national curriculum included a new subject 'Física y Química' which coupled the teaching of physics and chemistry, and thus provided a disciplinary space for these subjects distinct from mathematics.[2] Finally, the scheme designed to train teachers for secondary education placed strong emphasis on the preparation of teachers in the physical sciences.

The wide range of educational initiatives implemented by the Spanish government yielded uneven results. Furthermore, they depended considerably on foreign (especially French) production in pedagogy and science. Nonetheless, by the 1860s, physics and chemistry were subjects firmly established in the Spanish secondary school curriculum. They were taught in every school across the Spanish territory, and every school had a physics cabinet and chemistry laboratory. A large amount of these teaching collections have survived, together with associated

2 During the first three decades of the French secondary school system, the physical sciences were in general annexed to the teaching of mathematics as this subject and its teachers had a higher status in the school curriculum and professional system, respectively. Also, there were not many teachers prepared to teach physics and chemistry. This state of affairs had started to change though from the 1830s (Simon, 2009, pp. 34–38).

sources such as manuscript and printed inventories, trade catalogues, equipment invoices, and student notebooks and examinations.

The aim of this paper is to study the making of the Spanish school cabinet of physics, to extract its major characteristics and principal variations, and to reflect on how the study of the characteristics of teaching collections can feed into the study of how they were actually used. While introducing a number of basic historical facts necessary to understand the making of these collections, our historical analysis is based on the combination of comparison with the study of connections. Thus, we will study both how the scientific, pedagogical, and commercial relations between France and Spain contributed to the making of Spanish school cabinets, and how Spanish school collections compared to those developed in France. Furthermore, by comparing different Spanish collections, we intend to show that, in spite of the centralised initiatives of the government, there were different types of collections in different schools, and that this implied an uneven state of the teaching of the physical sciences across the Spanish territory. Finally, we will evaluate the capacity of schools and teachers to update their collections in relation to changes in pedagogical, scientific, and technological knowledge over time.

THE ESTABLISHMENT OF SECONDARY EDUCATION IN SPAIN

The first Spanish secondary schools were created during the 1830s and 1840s through isolated but overall coherent initiatives of municipal political forces, after the disentailment of the property of religious orders by the Spanish Liberal government. Between 1835 and 1844, 24 secondary schools – called *institutos* – were established. In 1845, an educational reform promoted by the Ministro de Gobernación (Home Secretary) Pedro Pidal, and the officer Antonio Gil de Zárate provided secondary education with a legal framework, and gave rise to the establishment of additional schools. In 1868, *institutos* numbered 66, of which 49 were located in the capitals of the Spanish provinces constituting the administrative structure of the country, and 17 were established in other resourceful towns (Gil de Zárate, 1855, II, 61ff; Viñao Frago, 1982, 335 ff).

However, the status of these schools was uneven and related to their capacity to raise funds. There were 11 *institutos* attached to universities (Madrid – with two *institutos* – Barcelona, Granada, Oviedo, Salamanca, Santiago, Sevilla, Valencia, Valladolid, and Zaragoza) which were the only ones offering the whole secondary school curriculum. This consisted of five years of the elementary secondary school curriculum, followed by two additional years which presented two options (literary and scientific) and gave access to university studies.

Provincial *institutos*, located in the province capitals, were funded by the budget of provincial administrations provided by the central government. Most of them (around 30 by mid-century) were able to offer the complete elementary curriculum. Local *institutos* had to rely on funding provided by their town council or foundations established in them, and many of them were unable to offer more than the first four years of secondary education.

In general terms, the *institutos'* funding was based on student fees (around 20 per cent), the contribution of town councils or provincial administrations, and land and property released after the Liberal confiscation of Catholic estates. By the end of the century, many *institutos* were highly profitable institutions, but some of them suffered losses. Some *institutos* had to fight against the fierce competition of private Catholic schools, which were favoured by the Conservative governments who came into power in the 1850s (Delgado Criado, 1994; Díaz de la Guardia, 1988, pp. 461–67; Gil de Zárate, 1855, p. II; Viñao Frago, 1982, 338 ff.).

In this context, physics and chemistry were taught to a minority of students. The 'Física y Química' subject was commonly taught during the fifth year of studies, involving five hours per week. Thus, it was not available in all the *institutos*. The first official syllabus was published during the 1840s, covering the whole spectrum of experimental physics (mechanics, hydrostatics and hydrodynamics, acoustics, heat, optics, electricity and magnetism), together with some lectures on 'notions of chemistry'. The disciplinary coupling of physics and chemistry had its origins in the first Liberal educational reforms of the 1830s, and it was, in principle, a genuinely Spanish characteristic. However, in France this formal coupling had briefly existed after the Revolution and, like in other countries in the nineteenth century, physics and chemistry were often taught together in French schools, within a more generic subject termed 'sciences physiques' (Simon, 2009, Chapter 2; Sisto Edreira, 2007, 183 ff).

The sciences had an important driving agency in the development of modern curricula, but always in tension with the traditional classical curriculum. At the beginning of the century, several Liberal projects allocated a large space to scientific subjects such as chemistry applied to the arts, experimental physics, mathematics, and natural history. But subsequent reforms limited their importance, promoting a more humanistic approach focused on Latin, grammar and literature, and ethics and religion. Javier de Quinto, an influential commentator, as former educational administrator, and editor of the *Boletín Oficial de Instrucción Pública*,[3] remarked that improving the status of the sciences meant diminishing that of the humanities, and that specialisation would have deprived secondary education of its preparatory role for university studies (Moreno, 1988, pp. 252–54).

The debate on the purpose and contents of secondary education – humanities versus science, and specialisation versus *Bildung* – endured even beyond the nineteenth century, and was common to many countries (Delgado Criado, 1994, pp. 159–60; Donnelly, 2002; Fournier-Balpe, 1994). Scientific subjects were then restricted to an early training in elementary mathematics, together with courses in 'physics and chemistry' and natural history, only provided in the last years of the secondary school curriculum (López Martínez, 1999).

In fact, by the mid 1840s, it was also difficult to find in Spain teachers able to teach the physical sciences. The first initiatives to provide secondary schools with physics and chemistry teachers were taken independently by some *institutos*, be-

3 The Spanish Parliamentary Papers.

fore the introduction of Pidal's legal framework. The call for positions was designed as a public competition. The qualifications included the preparation of a syllabus, details on its pedagogical implementation, and references to the textbook or textbooks in Spanish which would be used. There was an oral examination, which in certain cases included the performance of some basic demonstrations and experiments.[4]

After 1845, the Spanish government published an official syllabus for the subject and established a national legal framework for the hiring of secondary school teachers. However, the professional profiles of physics and chemistry teachers were uneven, since in the faculties of philosophy the presence of the physical sciences was heterogeneous, and in general poor. In his retrospective analysis of Spanish education, the Education officer Gil de Zárate noted that most teachers were not experts or lacked the appropriate pedagogical training, thus communicating knowledge and using a teaching style often inadequate for the education of secondary school students (1855, pp. 44–45, p.66–67).[5]

The educational and professional profiles of the first 'physics and chemistry' teachers were thus heterogeneous. Most of them had been trained as apothecaries or medical doctors. Around a third of them had obtained a diploma in the sciences after concluding their elementary secondary education, and a similar number had previously taught physical sciences in the faculties of philosophy. Only a few of these teachers had a doctorate in sciences or in pharmacy. Others had previously been educated in law and pursued informally their interest in science.[6]

In order to confront this lack of preparation and professionalisation, the Spanish government followed the French example in attempting to establish a special institution aimed at training science teachers for the secondary school network (Fournier-Balpe, 1994, pp. 119–42). The preparation of science teachers was considered a priority, for they were more scarce than other teaching staff. However, in contrast to the French case, the existence of such an institution was too ephemeral and did not have a major impact.

After an early trial experiment in the late 1840s, in 1850 the Escuela Normal de Filosofía was established with the exclusive aim of training science teachers (Gil de Zárate, 1855, II, pp. 66–67; Moreno, 1988, pp. 310–11). To gain admission to this school, candidates should have successfully completed secondary education and were offered a scholarship to attend lectures for four years. The curriculum was organised in three different sections: literature, 'physico-mathematical sciences' and 'natural sciences', and the lectures were given by university pro-

4 These positions were advertised by the *institutos* of Cáceres, Tudela, and Sanlúcar de Barrameda in the *Gaceta de Madrid* (1 September 1841 and 16 May 1843).

5 In fact, during the first half of the nineteenth century, similar remarks can be found in the reports of French education inspectors concerning the performance of physics teachers in the French colleges (Balpe, 1997).

6 Lists of all the Spanish secondary school teachers, specifying their educational and professional backgrounds and the history of their educational contracts, appeared from 1860 periodically in publications such as the *Gaceta de Madrid* and the *Revista de Instrucción Pública*.

fessors. On graduating, students were automatically conferred a university degree in science and were given preference in the awarding of secondary school teaching positions. Subsequently, they were expected to serve as teachers for a decade, and breaking their contract was penalised by the loss of their teaching and academic status (Seijas Lozano, 1850, pp. 33–34). The Escuela Normal produced a selected number of science teachers who found positions in the Spanish *institutos* and universities. However, it was suppressed for political reasons in 1852 and was thus unable to have a significant effect on the teaching of the sciences (Delgado y Vargas, 1860; Yanes Cabrera, 2006). From 1857, the newly established faculties of sciences assumed the role of providing teachers for the secondary school system (Moya Cárcel, 1991).

The first secondary school teachers were influential actors in their local context. Their high and broad educational qualifications and intellectual authority allowed them to participate in public health committees; perform chemical analyses; be part of literary and scientific societies; collect meteorological data for the government; perform mineralogical surveys; collect minerals, botanical, and zoological specimens; and actively participate in the public sphere through contributions in general and specialised periodicals. A small but not negligible group of teachers published textbooks and sometimes scientific papers, and also gave lectures at industrial schools, faculties of science, and other institutions engaged in furthering the country's scientific, industrial, and agricultural improvement (López Martínez, 1999; Moreno, 1988).

As the first initiatives to provide the *institutos* with 'physics and chemistry' teachers show, the production of science textbooks in the national language was a major priority. Three physics textbooks had a major circulation in the Spanish *institutos*: *Manual de física y nociones de química* (1847) by Manuel Rico y Sinobas and Mariano Santisteban; *Programa de un curso elemental de física y nociones de química* (1848) by Venancio González Valledor and Juan Chavarri; and *Tratado elemental de física esperimental y aplicada* (1856) by Adolphe Ganot. All three saw many editions.

Juan Chavarri and González Valledor were professors of physics at the Central University in Madrid, Santisteban held the chair of 'physics and chemistry' at the *instituto* of San Isidro (the largest in Spain), and Rico y Sinobas was professor of physics at the University of Valladolid. These authors took the responsibility of preparing the first 'physics and chemistry' textbooks for the Spanish secondary schools (Egido et al., 2000; Guijarro Mora, 2002). Some of them had already been involved in this task through translation work. For instance, González Valledor had translated a major physics textbook by the French secondary school teacher Nicolas Déguin. Translations from the French had been usual in previous decades. Hence, from the 1850s Adolphe Ganot's textbook found a place in the Spanish market, in spite of important competitors. It was translated by José Monlau, a former student of the Escuela Normal.

While these three books dominated the market, there were many other physics textbooks prepared by Spanish secondary school teachers. The contribution of Spanish authors rose from the late 1860s, attaining a high level of independence from foreign production, and making the introduction of French textbooks in the Spanish market very difficult – with the exception of Ganot's translation which survived the competition. Another fundamental tool for the implementation of the teaching of the physical sciences was the provision of school collections. As we will see, the French model was also influential in this case.

THE CREATION AND DEVELOPMENT OF PHYSICS AND CHEMISTRY COLLECTIONS

The creation of physics cabinets and chemistry laboratories was a high priority for the Spanish government, and administrative measures were taken to this end shortly after the official establishment of secondary education in 1845. Centralised purchases took place, aimed at establishing collections which would contribute to the development and consolidation of the teaching of the physical sciences in the *institutos*. The first purchases were organised thanks to the initiative of the Education secretary, Antonio Gil de Zárate.[7]

Before joining the state education administration, Gil de Zárate had been educated in the first two decades of the eighteenth century in France, later attending lectures on experimental physics at the Reales Estudios de San Isidro in Madrid. He subsequently returned to Paris to pursue his preparation, but did not succeed in becoming a physics teacher in Spain due to adverse political upheavals (Gil de Zárate, 1850, pp. iv–xvi). His educational background was crucial in his contribution to the reform of university and secondary school curricula in the 1840s. After surveying the Spanish university collections, a reference catalogue of physics and chemistry collections was compiled by a commission of university professors appointed by the Spanish government (Pidal, 1846).

At the end of 1846, Gil de Zárate – accompanied by Juan Chavarri – travelled to Paris to organise the purchase of physics and chemistry instruments for the universities. In November, they met Mateu Orfila (1787–1853), a Spaniard who had made a successful career in France as a professor of medical chemistry and dean of the Paris Medical Faculty (Bertomeu-Sánchez and Nieto-Galan, 2006). Thanks to Orfila's advice, Gil de Zárate obtained the services of four Parisian instrument makers: Messrs Pixii and Deleuil for the physics instruments; Lizé & Clech for the glassware and porcelain; and the brothers Rousseau for chemistry products and instruments. The importance of the purchase allowed him to negotiate a deal and acquire more instruments than initially expected, thus expanding the range of recipient institutions. The purchase included "physics instruments for eleven cabi-

7 Some *institutos* already had collections, though. The *institutos* located in university towns initially used the university cabinets. Others, such as that of San Isidro in Madrid, inherited important collections held by previous eighteenth-century and early nineteenth-century institutions.

nets", "precision scales", steam machine models, chemical apparatuses and substances, a mineralogical collection and a large number of medical charts.

By the same token, in September 1846 a reference catalogue was established to equip the physics and chemistry cabinets of secondary schools (Gil de Zárate, 1846). The list of instruments was based on the catalogues of the French makers Lerebours and Pixii, including 152 physical instruments (valued at 9,531 francs) and 133 chemical items (valued at 6,448 francs) (Gil de Zárate, 1847 and 1855, III, pp. 255–57; Simon Castel, García Belmar, and Bertomeu Sánchez, 2005). However, the collection was subsequently reduced, particularly in the case of chemistry, for which the funds were reduced to just 10 per cent of the initial amount (that is, 600 francs). The collection of physics instruments was reduced to 116 items only, although for almost half the price of the original amount (5,000 francs). The fields better represented in the physics list were electricity and magnetism (39), mechanics (15) and pneumatics (19) (Pastor Díaz, 1847a).

In the making of the reference catalogue, Gil de Zárate had considered that the teaching of the physical sciences in secondary education should not be based only on oral presentations of the subject. It should especially include the examination of instruments and the performance of experiments and manipulations. Furthermore, teachers should not limit themselves to teaching. It was also their mission to get involved in research and thus contribute to the patriotic advancement of the country through the production of original science and its applications to industry. His dealings in Paris in relation to university collections saw a larger purchase than originally expected, and thus a certain number of additional instruments might have been distributed to some *institutos*. For this reason, perhaps, the reference catalogue for the latter was subsequently reduced. Furthermore, Gil de Zárate had already established commercial relations with the instrument maker Pixii who provided instruments both for the universities and *institutos* and, therefore, he managed to get discounts which surely reduced the price of the *institutos'* purchase. In addition, by 1847, the subject matter of chemistry in the Spanish secondary education curriculum had been considerably reduced and, as a consequence, it was deemed appropriate to limit its associated collections (Pastor Díaz, 1847a).

Gil de Zárate had followed procedures similar to those promoted a few years earlier in France by Louis-Jacques Thenard. In 1842, Thenard had initially sent copies of the catalogues of instrument makers Deleuil and Pixii to all French schools, followed by a centralised survey of their collections and the publication of a reference catalogue. Pixii had also been one of the major instrument makers recommended by the French government in its previous reference catalogue published in 1821. Both the 1821 and 1842 French reference catalogues suggested that, although the major aim in the development of collections was pedagogical, when possible schools would also purchase instruments intended for research work by their physical sciences teachers (Belhoste, Balpe, and Laporte, 1995; Conseil royal de l'instruction publique, 1842, pp. 181–91).

By the mid nineteenth century, according to a Spanish governmental report, 19 secondary schools had a complete cabinet of physics, 11 other cabinets were almost complete, and only five secondary schools were ill equipped.[8] A few years later, Gil de Zárate remarked proudly that many *institutos* (such as Palma de Mallorca, Girona, Lleida, and Orense) had larger collections than those prescribed by the model catalogue (Gil de Zárate, 1855, II, pp. 80–161). So, how important was the impact of the Spanish government's reference catalogue in the quantitative and qualitative constitution of the *institutos'* physics collections? And what were their main characteristics across the Spanish territory?

In the early 1860s, most *institutos* published complete catalogues of their teaching collections. They were included in the yearly reports submitted by every *instituto* to the government and published in their *Memorias*. The publication of the *institutos' Memorias* had been established by the Spanish government in the late 1840s. They documented quantitatively and qualitatively life at the *instituto*, including information about its collections, libraries, premises, staff and students, preceded by a presentation by the school principal. This presentation was based on the annual speech made by every school principal to inaugurate the academic year. In this ceremony, attended by the major social representatives in every town, copies of the previous year *Memorias* were presented. Furthermore, the *institutos* often exchanged their *Memorias* between themselves.[9] Thus, the *Memorias* had a combined administrative and social function. They obliged the schools to keep a regular record of their activities and they helped the government to control them; they were also used as a tool of social and institutional prestige at the local and national level, since they publicly displayed the relative affluence and capabilities of each *instituto* (Simon Castel, 2008).

These catalogues display clearly the heterogeneity of the *institutos'* collections around 1860.[10] In general, the school collections followed the foundational pattern provided by the 1847 government reference catalogue, but they had introduced some upgrades, replacements, and additions. Most of the *institutos* located in university towns had collections which trebled the number of items recommended in the 1847 secondary school reference catalogue, and doubled those in the reference catalogues for Spanish universities (1846) and French collèges (1842). A considerable number of provincial *institutos* had also managed to increase their collections beyond the recommendations of the Spanish university

8 The report appeared in the *Gaceta de Madrid*, 7 September 1850, pp. 1–3. It did not include the university secondary schools and the secondary schools in which physics and chemistry were not taught.

9 Thus, for instance, the *instituto* of Valencia has preserved in its library a large set of *Memorias* of almost all the schools in Spain, which allowed us to compare the instrument catalogues of a large number of schools.

10 In this paper we have worked through comparisons with a set of printed collection catalogues published in 1861–62 in the *institutos' Memorias*. The set includes four university *institutos* (Granada, Oviedo, Salamanca, and Valencia), 21 provincial *institutos* (Alicante, Badajoz, Baleares, Burgos, Cáceres, Castellón, Ciudad Real, Cuenca, Gerona, Huelva, Huesca, Jaén, León, Lérida, Logroño, Málaga, Orense, Palencia, Pamplona, Pontevedra, and Soria), and two local *institutos* (Figueras and Monforte de Lemos). The complete bibliographical references of these catalogues are available in the bibliography at the end of this paper.

and school catalogues. But many others could only match or approach the recommendations made more than a decade earlier. This was also the case for most local *institutos*.

The reference catalogue published by the Spanish government in 1847 for the *institutos* was a reduced version of that published the year earlier for the universities. The two catalogues were roughly similar both in quantitative and qualitative terms, but the university catalogue contained almost a third more physics instruments and triple the number of chemistry items. The university collection allowed for the exposition of a wider range of physical and chemical phenomena. Furthermore, the school reference collection was cheaper, indicating that university instruments were probably of a greater quality and sophistication which could be used not only in teaching but, in certain cases, also in research. Examples of this can be found in the range of thermometers, barometers, telescopes, and electric machines included in the two catalogues. Moreover, the university catalogue included items such as a polariscope and an apparatus to demonstrate the development of magnetism by rotation – both devised by François Arago – which were more closely connected to contemporary research.

On the other hand, comparing the 1846 Spanish university and the 1842 French collège catalogues shows that the French school reference model contained a fifth more physics instruments, but a similar number of chemistry items.[11] The French reference collection contained a larger number of barometers and thermometers, more advanced instruments for the study of heat, and electricity and magnetism, and recent industrial applications such as magneto-electric apparatuses. Many of these instruments were related to research work conducted in Paris by physicists such as François Arago, Alexandre-Edmond Becquerel, and Macedonio Melloni. The Spanish catalogue was very poor in acoustics instruments in comparison to the French school reference collection. In contrast, the latter had fewer instruments to illustrate the mechanics of solids, surely because mechanics was considered a subject independent from general physics in France. In many respects, the 1847 Spanish *instituto* catalogue was similar to the 1821 French collège catalogue in listing a limited number of instruments aimed at illustrating simple physical phenomena (Conseil royal de l'instruction publique, 1821, 1842, 1843; Pastor Díaz, 1847b; Pidal, 1846).

By the early 1860s, the state of the physics collections in the Spanish *institutos* was diverse. The analysis of the collection catalogues of a sample of 27 *institutos* shows that, quantitatively, those located in university towns had trebled the number of instruments recommended in 1847, and doubled that of the reference collection for the universities suggested a year earlier. Their collections were larger now than the physics collection of the 1842 model catalogue for the French colleges.[12] Some provincial *institutos* were also in this range (Orense, Lérida, and

11 But chemical substances – in large numbers in the French catalogue – were not included in the Spanish reference collection.

12 This was the case of Valencia, Salamanca, and Oviedo. Granada was an exception, which still needs an explanation. In 1861, the number of physics instruments of the Granada *instituto*

Baleares). A similar number of provincial schools were above the Spanish university catalogue but below the French collège model (Pontevedra, Gerona, Burgos, and Pamplona). A larger number of these schools were above the Spanish school reference catalogue but below the university catalogue model (Logroño, Figueras, Soria, Ciudad Real, Alicante, Málaga, and Castellón). Thus, almost three quarters of the *institutos* had increased their collections beyond the 1847 government recommendations. However, only half of these had increased considerably their collections, surpassing even the recommendations made in 1846 for Spanish universities. Only a select number of *institutos* located in university and provincial towns excelled in the update of their physics cabinets, by surpassing considerably the size of the collections recommended in previous Spanish and French government reference catalogues.[13]

There were also differences in the ways in which the Spanish *institutos* updated their collections in the years between their establishment and the general survey of 1861–62. In the early 1860s, the pattern of the 1847 reference catalogue for the *institutos* could be clearly seen in all the physics collections of the Spanish secondary schools. The largest collections had built on this pattern, and increased and updated it. The smaller collections were similar to the 1847 reference collection or still tried to match it a decade-and-a-half later. Thus, it is without doubt that the initiative of Gil de Zárate had a major impact in the making of the Spanish school collections. Besides the government purchases, the exemplary value of the government model collection, designed by Gil de Zárate and collaborators such as Juan Chavarri (physics professor in Madrid and a major textbook author), had a huge impact on Spanish collections.[14] However, there were differences which affected not only the largest schools, but most of the *institutos*.

The largest *institutos*, in university towns and some provincial capitals, were able to increase their collections with a large number of instruments, especially related to electricity, but also in optics and heat. These additions allowed diversi-

was lower than that recommended in the 1847 reference catalogue for Spanish schools. This might have been that the *instituto* was using the university collections, or because it had financial and professional problems in relation to its professorship of 'physics and chemistry' which we have not been able to elucidate yet.

13 In chemistry, however, only a few *institutos* (Valencia, Oviedo, Salamanca, Lérida, Soria) appear to have matched or surpassed the recommendations published in the 1840s for the Spanish universities and French colleges, and many of them only matched those for the *institutos*. However, these results are more preliminary, since chemistry collections are more difficult to count. The chemistry catalogues are more heterogeneous and less systematised. They could, for instance, only contain apparatuses or also include glassware or chemical substances. In our analysis we have not taken into account the latter, which were not recorded in many catalogues.

14 It is a question for further research to determine how representative the 1847 Spanish model collection was in the international context of physics teaching. A first attempt has been made in this paper to compare it with the French government school pattern, showing differences in the number of instruments in mechanics and acoustics in both models and a lower capacity in Spanish collections to update in relation to contemporary research. However, further comparative work needs to be done.

fying the range of natural phenomena that teachers could demonstrate in the classroom. This included, for instance, phenomena of light polarisation, thermo-electricity, and electromagnetic induction which had been investigated for the first time in the previous decades. By the 1870s, a large university *instituto* such as Valencia had introduced in its collection instruments of most recent invention such as a Crookes radiometer, a telephone, and a microphone.

Many smaller *institutos*, which did not have great purchasing capacity to cope with recent advances in physics, showed nonetheless an interest in updating their collections. Thus, for instance, many schools introduced polarimetry instruments such as Arago's polariscope (Oviedo Salamanca, Burgos, Baleares, and Gerona, but Ciudad Real, and Cuenca, as well) and Norremberg's polarimeter (Salamanca, Baleares, Burgos, and Gerona, but also Málaga, Cuenca, and Monforte de Lemos, a local school), induction apparatus such as Ruhmkorff's coil (Valencia, Salamanca, Lérida, Baleares, Gerona, Alicante, and Pontevedra), and instruments for the study and illustration of discharges in gases such as Geissler tubes (Lérida, Baleares, Gerona – provincial but not university *institutos*). The smaller *institutos* had to be selective in their purchases, and thus they often focused only on one branch of physics, typically electricity. Most schools expanded their collection range of batteries and included some illustrations of industrial or commercial applications of electricity such as a telegraph model, an electro-medical apparatus or electromagnetic apparatus. Other typical additions, which had an explicit pedagogical purpose, were stereoscopes, magic lanterns, and photographic cameras.

But size, political location, and economic affluence were not the only parameters which explain collection composition. Thus, for instance, the only galvanometers available in Spanish schools were to be found not in the largest university or provincial *institutos* (with the exception of Salamanca), but in small schools such as that of Logroño and especially Huesca and Monforte de Lemos. Analogously, many *institutos* increased considerably their collections of thermometers and barometers, and acquired high precision instruments in relation to their meteorological stations. Indeed, secondary schools and their science teachers contributed to the establishment of a national meteorological network, coordinated by the Astronomical Observatory in Madrid. Meteorological data were processed in Madrid, but were also published in local newspapers and in the school Memorias. Some of these observations were also compiled and analysed by the physics professor and major textbook author Manuel Rico y Sinobas (Anduaga, 2003; Sisto Edreira, 2007).[15]

During the 1850s and 1860s, many schools were equipped with meteorological stations. Small schools such as Cuenca, Huesca, and León could thus increase their instrument collections. In 1856, for instance, the 'physics and chemistry' teacher Francesc Bonet i Bonfill established a meteorological station in Lérida and two years later started to send data to Madrid (Casals Berges, 2006, pp. 243–46). In 1861, the director of the Cáceres *instituto* established another meteorologi-

15 We would like to thank Aitor Anduaga for providing us with some data and advice on the Spanish meteorological network.

cal station in its premises. The instruments, made in Paris, were calibrated in the Madrid Observatory before being sent to the school (Sergio Sánchez, 1861, p. 9). However, in general the meteorological instruments of the *institutos* made an exception to the general pattern of dependence on the French instrument trade, which was not challenged by the timid emergence of the Spanish instrument industry. In schools with large meteorological collections such as Oviedo, Salamanca, and Burgos, many of the instruments were made by leading makers in London of Italian origin such as Casella, and Negretti & Zambra (Ruiz Castell, Simon, and Bertomeu Sánchez, 2002; Williams, 1994).

CONCLUSION

About fifteen years after the official establishment of physics and chemistry school collections in Spain, secondary schools still depended considerably on foreign instrument makers, and many of them had not managed to considerably update their collections. However, almost all the schools updated giving due priority – and according to their budgets – to the most topical areas of research at the time, and all of them made significant additions to their library of experimental demonstrations. The centralised initiatives of the Spanish government had a major impact on the configuration of physics and chemistry collections. But their subsequent development depended in great measure on the financial capacity of the different *institutos*, which was related to the number of students, their connection with the local political administrations, and their geopolitical location. The largest schools, located in university and provincial towns, were able to update their collections, increasing the pedagogical repertoire of science teachers and incorporating some contemporary advances in the physical sciences and instrument making. The smaller schools, located in some provincial towns and in some small towns with educational traditions, faced major problems in increasing and updating their collections. However, in many cases they were able to change with the times by including new items for pedagogical illustration of the most topical areas of physics.

The analysis of the Spanish school collections in physics and chemistry offers valuable approximations to the study of science in the classroom. However, a more in-depth study is still needed in order to elucidate how the collections were used. This research will have to integrate other sources as well, such as scientific instrument collections, teacher and student notebooks, textbooks, and trade catalogues. The recovery of these sources is currently being developed through an online project supported by the Catalan Scientific Instrument Commission (COMIC) which is currently creating a repository integrating these sources and offering updated information tools on Spanish projects in this field.[16]

16 The details of this project and the current state of its repository can be checked at www.instrumentscientifics.com.

REFERENCES

Ametller y Marill, M. 1861. 'Inventario de los instrumentos, aparatos y demas objetos que posee el Instituto de Gerona para la enseñanza de las diversas asignaturas', in *Memoria leída en la apertura del curso académico de 1861 a 1862 en el Instituto Provincial de 2ª Enseñanza de Gerona.* Imprenta de F. Dorca, sucesor de I. Grases, pp. 33–135.

Anon. post-1873. 'Material Científico' [manuscript inventory of the scientific collections of the Instituto of Valencia, before and after 1873]. Archive of the Instituto Luis Vives, Valencia.

Anderson, A. 2004. The Idea of the Secondary School in Nineteenth-century Europe', *Paedagogica Historica*, 40 (1 and 2), pp. 93–106.

Anderson, R. D. 1975. *Education in France, 1848–1870.* Oxford: Clarendon Press.

Anduaga, A. 2003. 'Ciencia, ideología y política en España: Augusto Arcimís (1844–1910) y la creación del Instituto Central Meteorológico', *BILE*, 53, Diciembre), pp. 1–15.

Balpe, C. 1997. 'L'enseignement des sciences physiques: naissance d'un corps professoral (fin XVIIie – fin XIXe siècle)', *Histoire de l'éducation*, 73, janvier, pp. 49–85.

Belhoste, B., C. Balpe, and T. Laporte. 1995. *Les sciences dans l'enseignement secondaire français. Textes officiels.* Paris: INRP-Éditions Economica.

Bertomeu-Sánchez, J. R., and A. Nieto-Galan (eds). 2006. *Chemistry, Medicine, and Crime: Mateu J. B. Orfila (1787–1851) and His Times.* Sagamore Beach: Science History Publications.

Boix y Monrós, J. 1861. 'Inventario del material científico existente en las clases y gabinetes de este Instituto', in *Memoria leída en el Acto de apertura del curso académico de 1861 a 1862 en el Instituto Local de Segunda Enseñanza de la Villa de Figueras.* Figueras: Imprenta de Lorenzo Mirgeville, Rambla, pp. 1–16.

Casals Berges, Q. 2006. *Tots a l'escola? El sistema educatiu liberal en la Lleida del XIX.* València: PUV.

Conseil royal de l'instruction publique. 1821a. 'Circulaire relative au catalogue des instruments de physique qui doivent composer les cabinets des collèges royaux, 16 janvier 1821', in B. Belhoste, C. Balpe, and T. Laporte (eds), *Les sciences dans l'enseignement secondaire français. Textes officiels.* Paris: INRP, pp. 98–101.

———. 1821b). 'Circulaire annonçant l'envoi aux recteurs de l'état des objets dont les collèges royaux peuvent proposer l'acquisition pour leurs laboratoires de chimie, 29 août 1821', in B. Belhoste, C. Balpe, and T. Laporte (eds), *Les sciences dans l'enseignement secondaire français. Textes officiels.* Paris: INRP, pp. 101–103.

———. 1842. 'Arrêté déterminant la liste des instruments de physique nécessaires dans chaque collège royal, 27 décembre 1842', in B. Belhoste, C. Balpe, and T. Laporte (eds), *Les sciences dans l'enseignement secondaire français. Textes officiels.* Paris: INRP, pp. 181–85.

———. 1843. 'Arrêté déterminant la liste des instruments de chimie et produits chimiques nécéssaires dans chaque collège royal, 27 janvier 1843', in B. Belhoste, C. Balpe, and T. Laporte (eds), *Les sciences dans l'enseignement secondaire français. Textes officiels.* Paris: INRP, pp. 186–91.

De los Herreros, F. M. 1861. 'Inventario y coste de los instrumentos, aparatos, modelos y demás objetos que posee este Instituto, para las enseñanzas de Física y Química y sus aplicaciones', in *Memoria leída por el Sr. D. Francisco Manuel de los Herreros, Director y Catedrático del Instituto Provincial de Segunda Enseñanza de las Baleares, en la solemne apertura del curso de 1861 á 1862.* Palma: Imprenta de D. Felipe Guasp.

Delgado Criado, B. 1994. *Historia de la educación en España y América: La educación en la España contemporánea (1789–1975).* Madrid: Ediciones Morata.

Delgado y Vargas, A. 1860. 'La Escuela Normal de Filosofía y algunas consideraciones sobre el profesorado de segunda enseñanza', *Revista de instrucción pública*, 9 (4) Octubre, pp. 149–52.

Díaz de la Guardia, E. 1988. *Evolución y desarrollo de la Enseñanza Media en España de 1875 a 1930. Un conflicto político-pedagógico.* Madrid: C.I.D.E.

Diaz y Monasterio, R. 1861. 'Inventario de los instrumentos, aparatos y demas objetos que posee el Instituto de Oviedo para la enseñanza de las diversas asignaturas', in *Memoria leída el día 16 de setiembre de 1861 en el Instituto de Segunda Enseñanza de Oviedo, en el acto solemne de la apertura del curso académico de 1861 a 1862, por el doctor D. Rafael Diaz y Monasterio.* Oviedo: Imp. y Lit. de Brid, Regadera y Compañia, pp. 23–121.

Dominguez, I. 1861. 'Inventario y coste de los aparatos é instrumentos existentes en el Gabinete de Física y Química', in *Memoria leída el día 16 de Setiembre en la inauguración del curso de 1861 á 1862 en el Instituto Provincial de Segunda Enseñanza de Palencia por el Doctor D. Inocencio Dominguez.* Palencia: Imprenta y librería de José Maria Herran.

Donnelly, J. F. 2002. 'The 'humanist' critique of the place of science in the curriculum in the nineteenth century, and its continuing legacy', *History of Education*, 31(6), pp. 535–55.

Egido Rodríguez, Á. del, L. González de Lastra, V. Guijarro Mora, E. Hidalgo Cámara, and R. M. Martín Latorre. (2000). *Instrumentos científicos para la enseñanza de la física. Estudio realizado por el Museo Nacional de Ciencia y Tecnología de la colección histórica de instrumentos científicos de la facultad de ciencias físicas de la Universidad Complutense de Madrid.* Madrid: Ministerio de Educación, Cultura y Deporte.

Fariña, F. 1861. 'Inventario de los instrumentos aparatos y demas objetos que posee el Instituto de Monforte in *Memoria leída en la apertura del curso académico de 1861 á 1862 en el Instituto Provincial de Monforte por el Lic. Don Francisco Fariña, Srio. Honorario de S. M. y Catedrático Director Propietario en el mismo, el 16 de Setiembre de 1861.* Monforte: Imprenta de Ramon Cortiñas.

Fournier-Balpe, C. 1994. *Histoire de l'enseignement de la physique dans l'enseignement secondaire en France au XIXe siècle.* unpublished PhD thesis,Université Paris XI..

Gil de Zárate, A. 1846. 'Circular previniendo a los Institutos se provean de los instrumentos necesarios para la explicación de las ciencias físicas y naturales, 15 de Setiembre de 1846', in *Colección Legislativa de España. Tercer trimestre de 1846. Tomo XXXVIII.* Madrid: Imprenta Nacional, pp. 354–65.

———. 1847. 'Real órden por la cual se manifiesta al señor director general de instrucción pública que S. M. ha visto con agrado el buen éxito de la comisión que llevó al estrangero para adquirir máquinas y útiles necesarios en las universidades', *Boletín Oficial de Instrucción Pública*, 5, pp. 129–37.

———. 1850. *Obras dramáticas de D. A. Gil de Zárate. Edición precedida de una noticia biográfica y dada a luz por D. Eugenio de Ochoa.* Paris: Baudry.

———. 1855. *De la instrucción pública en España.* Madrid: Imp. del Colegio de Sordo-Mudos.

Green, A. 1990. *Education and State Formation. The Rise of Education Systems in England, France and the USA.* Basingstoke and London: Macmillan.

Guijarro Mora, V. 2002. *Los instrumentos de la ciencia ilustrada: física experimental en los Reales Estudios de San Isidro de Madrid (1770–1835).* Madrid: UNED.

Herrero, D. 1861. 'Inventario y coste de los aparatos de Física, Química y sus aplicaciones', in *Memoria del estado del Instituto de 2ª Enseñanza de Castellón de la Plana durante el curso de 1860 a 1861, leída en la solemne apertura del de 1861 a 1862 por su Co Director interino D. Domingo Herrero.* Castellón: Imprenta y Librería de Vicente Perales y Navarro, pp. 9–15.

La-Rosa, M. 1861. 'Inventario y coste de los instrumentos y demás objetos de los gabinetes de Física y Química', in *Memoria leída por D. Manuel La-Rosa, Catedrático de Psicología Lógica y ética y Director del Instituto de Segunda Enseñanza de Lérida en la solemne apertura del curso de 1861 a 1862.* Lérida: Establecimiento topográfico de D. José Sol, pp.25–42.

Lopez de Cerain, D. 1861. 'Gabinete de Física y Química. Aparatos para la enseñanza de la Física y Química', in *Memoria leída en la apertura del curso académico de 1861 á 1862 en el Instituto Provincial de 2ª Enseñanza de Soria, por D. Dionisio Lopez de Cerain, Catedrático y Director de dicho Establecimiento.* Soria: Imprenta de D. Manuel Peña, pp. 23–30.

Lopez, G. 1861. 'Inventario de los instrumentos y aparatos de Química y Física', in *Memoria leída el día 16 de Septiembre de 1861, en la solemne apertura del Instituto de Segunda Enseñanza de Ciudad Real. Por el Director D. Genaro Lopez*. Ciudad Real: Imp. de C. Clemente Rubisco, pp. 18–24.

López Martínez, D. 1999. *La enseñanza de la física y la química en España en la educación secundaria en el primer tercio del siglo XX en España*. Unpublished PhD thesis, Universidad de Murcia.

Martinez Rives, J. 1861. 'Gabinete de Física-Química. Inventario de todas las máquinas, aparatos y efectos existentes en dicho gabinete en 1° de Octubre de 1861', in *Memoria del Instituto de Segunda Enseñanza de Burgos*. Burgos, pp. 1–11.

Moreno, A. 1988. *Una ciencia en cuarentena. La física académica en España (1750–1900)*. Madrid: CSIC.

Moreno González, J. M. 1861. 'Inventario del Gabinete de Física', in *Memoria del curso académico de 1861 a 1862 del Real Colegio de San Bartolomé y Santiago. Instituto de Segunda Enseñanza de la Provincia de Granada*. Granada: Imprenta de D. Francisco Ventura y Sabatel, pp. 20–24.

Moya Cárcel, T. 1991. *La enseñanza de las ciencias. Los orígenes de las facultades de ciencias en la universidad española*. Unpublished PhD thesis, Valencia: Universidad de Valencia.

Muñoz y Garnica, M. 1861. 'Inventario de los instrumentos y aparatos de Física y Química', in *Memoria leída en la solemne apertura del curso académico de 1861 a 1862, y en la distribución de premios ordinarios y extraordinarios del curso 1860 a 1861, por el Director del Instituto provincial de Jaén, D. Manuel Muñoz y Garnica*. Jaén: Imprenta de D. Francisco Lopez Vizcaino.

Muntada y Andrade, J. 1861. 'Gabinete de Física. Inventario y costo de los instrumentos, aparatos, modelos y demás objetos para el estudio de dicha asignatura', in *Memoria acerca del estado del Instituto de Badajoz, leída en la apertura del curso de 1861 á 62 por su director Don José Muntada y Andrade*. Badajoz: Imprenta de Arteaga y Compañía, pp. 27–31.

Orodea y Urdaneta, J. 1861. 'Inventario de los instrumentos y aparatos que contiene el Gabinete de Física del Instituto provincial de Logroño', in *Memoria acerca del estado del Instituto de 2ª Enseñanza de Logroño, leída el día 16 de setiembre de 1861, por el Director, Don Julian Orodea y Urdaneta, en el acto solemne de la apertura del curso de 1861 á 62*. Logroño: Imprenta y Litografía de Ruiz, pp. 1–8.

Pastor Díaz, N. 1847a. 'Circular para que los Institutos de segunda enseñanza se provean de los instrumentos que les faltan para la enseñanza de física (10 de Abril de 1847)', in Ministerio de instrucción pública, *Colección de Órdenes generales y especiales relativas a los diferentes ramos de la instrucción pública secundaria y superior*. Madrid: Imprenta Nacional, pp. 330–38.

———.1847b. 'Catálogo Modelo de los instrumentos de física-química necesarios para las demostraciones en las cátedras de los institutos provinciales de segunda enseñanza (10 de Abril de 1847)', in Ministerio de instrucción pública, *Colección de Órdenes generales y especiales relativas a los diferentes ramos de la instrucción pública secundaria y superior*. Madrid: Imprenta Nacional, pp. 332–38.

Paula de Sola, F. 1861. 'Gabinete de Física y Laboratorio de Química', in *Memoria acerca del estado del Instituto de 2ª Enseñanza. Provincia de Málaga*. Málaga: Imp. de D F. Carreras é hijos, pp. 1–8.

Perejon y Campoamor, L. G. 1861. 'Catálogo de los instrumentos y aparatos de Física y Química existentes en el día de la fecha en el gabinete de dicha asignatura', in *Memoria del Instituto de Segunda Enseñanza de Orense, leída en la Apertura del curso de 1861 á 1862 por su Director y Catedrático, Lic. D. Leoncio G. Perejon y Campoamor*. Orense: Imprenta de Don Agustín Moldes.

Pidal, P. J. 1846. 'Real Orden mandando se adquieran instrumentos y aparatos de física y química para proveer a las Universidades, Madrid, 28 de Octubre de 1846', in *Colección legislativa de España. Cuarto trimestre de 1846, Tomo XXXIX*. Madrid, pp. 76–77.

Rodríguez Garcia, V. 1861. Inventario y coste de los Gabinetes de Física y Química. In *Memoria acerca del estado del Instituto de 2ª Enseñanza de Huelva leída el día 16 de septiembre de 1861 por el Director del mismo Don Vicente Rodríguez Garcia, en el acto solemne de la apertura del curso de 1861 a 1862.* Huelva: Imprenta de D. J. Reyes y Moreno.

Rueda, A. 1861. 'Inventario y coste de los instrumentos y aparatos de Física y Química con sus aplicaciones', in *Memoria acerca del estado del Instituto de 2ª Enseñanza de León.* León: Imprenta, litografía y librería de Manuel Gonzalez Redondo.

Ruiz Castell, P., J. Simon, and J. R. Bertomeu Sánchez. 2002. 'Los fabricantes de instrumentos de la Universitat de València', in J. R. Bertomeu Sánchez and A. García Belmar (eds), *Abriendo las Cajas Negras: Instrumentos Científicos de la Universidad de Valencia.* València: Universitat de València, pp. 367–80.

Ruiz, S. 1861). 'Catálogo de las máquinas y aparatos existentes en el Gabinete de Física en el día 31 de Diciembre de 1860', in *Memoria que en la solemne apertura de los estudios del Instituto de 2ª Enseñanza de Salamanca, leyó el 16 de Setiembre de 1861, según lo dispuesto en el artículo 96 del Reglamento de 22 de Mayo de 1859, el Dr. D. Salustiano Ruiz, su Director y Catedrático de Matemáticas.* Salamanca: Imprenta de Diego Vazquez, pp. 22–43.

Sanchez, L. S. 1861. 'Instrumentos y aparatos destinados al estudio de la Física y Química y sus aplicaciones', in *Memoria leída el día 16 de Septiembre en la inauguración del curso de 1861 a 1862 en el Instituto Provincial de Segunda Enseñanza de Cáceres, por Don Luis Sergio Sanchez, Director del mismo Establecimiento.* Cáceres: Imprenta y librería de D. Antonio Concha.

Seijas Lozano, M. de. 1850. *Plan de estudios.* Madrid: Imprenta de la V. de Perinat y Compañía, á cargo de D. S. Compagni.

Senante, M. 1862. 'Gabinete de Física y Química. Catálogo de los aparatos é instrumentos existentes en el mismo', in (Ed.), *Memoria leída el día 16 de Septiembre de 1862 en el Instituto de Segunda Enseñanza de Alicante, en el acto solemne de la Apertura del curso académico de 1862 a 1863, por el Señor D. Manuel Senante, Doctor en Jurisprudencia, Catedrático de Retórica y Poética y Director del mismo.* Alicante: Imprenta y Litografía de Pedro Ibarra, pp. 19–27.

Sergio Sánchez, L. 1861. *Memoria leída [...] en el instituto provincial de segunda enseñanza de Cáceres.* Cáceres: Antonio Concha.

Simon Castel, J., García Belmar, A., and Bertomeu Sánchez, J.R. 2005. 'Instrumentos y prácticas de enseñanza de las ciencias físicas y químicas en la Universidad de Valencia, durante el siglo XIX', *Endoxa*, 19, pp. 59–124.

Simon Castel, J. 2008. 'Les col·leccions de física i química dels instituts de secundària: Catalogació, estudi i metodologies', *Actes d'Història de la Ciència i de la Tècnica*, 1(1), pp. 85–94.

Simon, J. 2009. *Communicating Physics in Nineteenth-Century France and England: The Production, Distribution and Use of Ganot's Textbooks.* Unpublished PhD thesis, University of Leeds.

Sisto Edreira, R. 2007. *A Disciplina de física e química na educación secundaria do século XIX, modelos, recursos e produción do coñecemento, o modelo español á luz das ideas vixentes en Europa e da súa concreción en Galicia.* Unpublished PhD thesis, Universidade de Santiago de Compostela.

Sobrino, L. M. 1861. 'Inventario y coste de los instrumentos, aparatos, modelos y demás objetos para el estudio de la Física y Química', in *Memoria acerca del estado del Instituto de Segunda Enseñanza de Pontevedra, leída el día 16 de Setiembre de 1861 por el Doctor D. Luis María Sobrino, Catedrático de Psicología, Lógica y Ética, y Director del mismo, en el acto solemne de la apertura del curso de 1861 á 1862.* Pontevedra: Imprenta de D. José Vilas, pp. 39–46.

Uriarte, J. d. M. 1861. 'Física-Química. Inventario de los aparatos, instrumentos y demás objetos que para la enseñanza de esta asignatura posee el Instituto', in *Memoria leída el día 16 de Setiembre de 1861 en la solemne apertura de los estudios del Instituto de 2ª Enseñanza de*

Pamplona por el Dr. D. Juan de Mata Uriarte, Catedrático de Física-Química y Director de dicho establecimiento. Pamplona: Imprenta de Francisco Erasun, pp. 3–7.

Ventura y Solana, V. 1861. 'Inventario y coste de los instrumentos, aparatos, modelos y demás objetos existentes en el Gabinete de Física y Química y en el Observatorio meteorológico', in *Memoria del Instituto Provincial de Segunda Enseñanza de Huesca*. Huesca: Imprenta de Mariano Castaneda.

Viñao Frago, A. 1982. *Política y educación en los orígenes de la España contemporánea. Examen especial de sus relaciones en la enseñanza secundaria*. Madrid: Siglo XXI.

Williams, M. E. W. 1994. *The Precision Makers: A history of the instruments industry in Britain and France, 1870–1939*. London: Routledge.

THE DEATH AND LIFE OF THE PLANT SPECIMEN

Dawn Sanders

INTRODUCTION

Botany was a contested subject in Anglo-American educational spheres for much of the 19th and 20th centuries. In the latter part of the 20th century it became absorbed into a generic biological education, in which it struggled to remain visible. Botanical education in the early 19th century was strongly evident, although rife with specific notions of femininity and class, through a proliferation of popularisations. As the 20th century unfolded, the identity of botany, as a distinct curriculum subject, became a frequently contested pedagogical space. By the late 20th century botany had become an indistinct feature of a generic biological education. Drawing on Philip Pauly's work on the development of high school biology in New York City between 1900 and 1925 this chapter examines the role of plant specimens, both living and dead, throughout this period and highlights key pedagogical debates surrounding their use.

BOTANICAL EDUCATION

Pauly's paper "The development of High School Biology: New York City 1900–1925"[1] is a pivotal discussion of the growth of American biology education. For historians of botanical education he reveals the competitive tensions between a triad of life sciences: physiology, botany and zoology in late 19th century New York and documents an increasing emphasis on social biology, informed by an expanding urban population into the early twentieth century. Although Pauly's study is site specific, it attends to issues that are equally pertinent to the emergence of British biological education. This chapter draws on Pauly's work to consider the death and life of the botanical specimen in these Anglo-American educational contexts.

The development of botanical education, in both formal and informal contexts, has not been a smooth journey.[2] Indeed, for much of its history the subject

1 Pauly, P. J. (1991) The Development of High School Biology: New York City, 1900–1925. *Isis*, 82, 662–688.

2 Hershey, D.R. (1996) A Historical Perspective on problems in botany teaching *The American biology Teacher*, 58, 341–347.

content of botany has either been stridently debated[3] or visibly demoted within school biology curricula.[4] In Victorian and Edwardian England, botany was culturally rooted in everyday life, as contemporary commentators have observed.[5]

In addition, it was explicit in the literature of the time, see for example "Mary Barton – a tale of Manchester life."[6] However, this historical period was also a time when avid plant collectors searching for ferns (an obsession known as Pteridomania) and orchids, decimated large swathes of land. In the 21st century comparable paradoxes exist; gardening is one of the most subscribed to pastimes in England[7] and yet knowledge of the native flora continues to decrease, particularly among children and young people.[8]

TEACHING WITH PLANTS

Living, dead and dried plant specimens occupy a key position in the evolution of botanical education, particularly when discussed in conjunction with the environments in which they were studied. Writing in 1894 J.G. Rooper, a British school inspector supportive of the nature study movement stated,

> "plants and animals should not be abstracted from their surroundings, as when studied in laboratories, but in the closest connection with them. The living organism should be studied as part of the panorama which surrounds the child in which he himself is included, and not as a specimen in a museum, dried and ticketed."[9]

Concern had been expressed by *"vitalists"*, such as Patrick Geddes and J. Arthur Thomson, that "object lessons" in schools were dry exercises in observing and rendering likenesses of a mausoleum of dead specimens without any biological contextualization. Rooper's call for a living experience was an important turning point for biological education, although one that was not to be resolved in the 19th century, as this quote from a twentieth century American biology educator reveals:

3 Boney, A.D. (1991). The "Tansley manifesto" affair. *New Phytologist*, 118, 3–21.

4 Honey, J. (1987). Where have all the flowers gone? – The place of plants in school science, *Journal of Biological Education*, 21, 3, 185–9.

5 See Shteir, A.B. (1996). *Cultivating Women, Cultivating Science*. Baltimore: John Hopkins and Secord, A. (1996) Artisan botany. In: Jardine, J., Secord, A. & Spary, E.C.(Eds) *"Cultures of Natural History"* pp.378–393. Cambridge: Cambridge University Press.

6 Gaskell, E. (1848). *Mary Barton – a tale of Manchester life*. London: Chapman and Hall.

7 See for example Hoyles, M. (1994). *Lost Connections and New Directions: The private garden and the public park*. London: Comedia and Evans, P. (2002). Flower power (Guardian Society), The Guardian, 3 April, 9.

8 Bebbington, A. (2005) The ability of A-level students to name plants *Journal of Biological Education* 39, 2 62–67.

9 Rooper, J.G. (1894). New Methods in the Lower Standards. *Teacher's Aid* **XVII**, 441, 530–533.

> "too much of our biology, the science of living organisms, is drawn from dead examples or text book pages."[10]

Daglish[11] considered the learning of a long list of plant names a, "dull and unsatisfying affair", when botany lessons emphasised this process, as a large part of their content, he felt the subject remained "a dreary science", especially when using "dry and often dusty" pressed specimens.[12]

He advocated, as many other botanical educators had,[13] and still do,[14] observing fresh, living plant specimens and through exploring these, ascertains that the learner will remember the name by becoming familiar with the plant's shape and colour, and other characteristics such as smell. He stated that this personal association with the plant would have far greater meaningfulness than the rote learning of a list of disassociated names. During the period (1930's England) Daglish was writing, the predominate botanical teaching and learning culture was a didactic one, which emphasised rote learning utilising preserved plants, described by some as "botanical cadavers."[15] As discussed earlier, the challenge of teaching a living science with dead specimens was not an isolated concern.

BOTANICAL MODELS: THE BRENDELS AND BLASCHKAS

From the mid-nineteenth century to the early twentieth century Germany was at the heart of a botanical modelling industry led by the Brendels – Robert and Reinhold – based in Breslau and then Berlin, and the Blaschkas – Leopold and Rudolf – based in Hosterwitz near Dresden.

PAPIER-MÂCHÉ PLANTS

The Brendels created botanical models made from papier-mâché, wood, cotton, rattan, pulp cane, glass beads, feathers and gelatine. Many of their constructions had articulated parts to aid explanations in teaching and were primarily used in universities rather than schools. In contemporary Britain several museums still hold extensive collections of Brendel plant models – for example Liverpool Museum (circa 200) and the University of Aberdeen Zoology Museum (circa 150), some remain on view as remnants of a botanical pedagogy that reached its heyday

10 Youngpeter, J. M., (1961) Field Trips *The American Biology Teacher* **23**,5, p.273–275.

11 Daglish, F.E. (1932). *How to see plants*. London: Dent.

12 Daglish, F.E. *op. cit. supra* (1932) p.2.

13 See Lindley, J. (1858). *School botany and vegetable physiology or the Rudiments of botanical science.* London: Bradbury and Evans and Stopes, M. (1906). *The study of plant life for young people*. London: Alexander Moring.

14 Walker, T. & Allen, L. (1999). "Seeking partnerships to teach biodiversity. *Roots*, 19, 30–2.

15 Pool, R.J. (1919) About High School and College Botany. School *Science and Mathematics*, 19, 487–500 cited in: Hershey, D.R. (1996) A Historical Perspective on problems in botany teaching *The American Biology Teacher*, **58**, 341–347.

at the end of the 19th century. The models themselves have captured the attention of individual collectors, and thus their identities have shifted from articulated, functional teaching models to static, aesthetic objects retained for artistic display. Modern commentators, such as Reiling, consider these models to "represent a reality that is more related to museums and education than with life in nature."[16]

GLASS FLOWERS

The Blaschkas originally from North Bohemia, specialised in the creation of fine glassware. Their first botanical commission was for Prince Camille de Rohan[17] to model his orchid collection, their work then focused on marine biology; specifically invertebrates inspired by the work of naturalists such as Phillip Henry Gosse and their own childhood memories.[18]

Professor George Lincoln Goodale, founder of the botanical museum at Harvard University, commissioned "life-like representations of the plant kingdom for teaching botany"[19] from the Blaschkas. Over 3,000 glass models of 847 different plant species were made from 1887–1936.[20]

Today, the "Glass Flowers" exhibit at Harvard Museum of Natural History continues to attract extensive visitor attention. It is not just the detailed character of these models that created their original educational use, but their everlasting nature "since the glass flowers are always in bloom, tropical and temperate species may be studied all year round"[21] – a stark contrast to the constraints of seasonality presented by the use of actual plant specimens. However, glass has its own vulnerabilities and the models need cleaning and conserving lest their "blooms" crack, and fade with the passing of time.

INFLUENTIAL TEXTBOOKS

Recognising the substantial influence of New York's science educators as authors of 'general biology' textbooks, Pauly notes that between 1900 and 1925 their publications catalysed the teaching of general biology as a "standard subject", one which became the 'most widely taught scientific discipline in high schools' across the country.[22]

Similarly, in 19th century England, botanical textbooks had flourished, indeed Henslow, Darwin's botanical mentor at Cambridge, authored his own "Botany for

16 Reiling, H. (2003) Beter dan de natuur available in English translation at http://members.ziggo.nl/here/neo.html accessed June 2010.
17 http://designmuseum.org/design/leopold-rudolf-blascka accessed June 2010.
18 http://designmuseum.org/design/leopold-rudolf-blascka accessed June 2010.
19 http://www.hmnh.harvard.edu/on_exhibit/the_glass_flowers.html accessed June 2010.
20 http://www.hmnh.harvard.edu/on_exhibit/the_glass_flowers.html accessed June 2010.
21 http://www.hmnh.harvard.edu/on_exhibit/the_glass_flowers.html accessed June 2010.
22 See Pauly, P. J. *op. cit.* (1991) p.671.

Children".[23] Botany textbooks for both artisans and school/university students were equally popular. These texts celebrated the examination of living botanical specimens through dissection, microscopy and drawing. Henslow recommended 'the critical examination' of the "minute differences" of flowers and "the comparative study of kindred forms" while recording such arrangements through drawing.[24]

Thome, in his structural and physiological botany textbook,[25] urged the reader to:

> "Make himself practically acquainted, with the aid of a microscope, and, if possible under the guidance of a competent teacher, with the minute structure of plants and with the life history of various forms."

Significantly, in the context of plant specimens, he considered the study of botany through – "mere book-knowledge"– to be "valueless". Balfour makes an eloquent plea to go to 'the book of nature', in his "Class Book of Botany" published in 1864:[26]

> "In prosecuting the science of botany, the student must ever bear in mind that it is only by examination of plants in the garden and in the fields, by careful dissections, and by microscopic investigations of living and dead tissues, that he can acquire a correct knowledge of the subject. No book can make up for want of this; no description nor illustrations can supply its place. All that a teacher can do by his lectures and textbook is to direct the pupil in his researches, and to refer him to the book of nature as his guide in his investigations. The student must not be led away by human authority, however distinguished."[27]

Each of these writers appears to embrace Gosse's view that a "true knowledge of the material world" comes from "protracted aquaintance" with actual plants and animals.[28] However, in English and Welsh elementary schools of the 19th century, engagement with the living world, as noted earlier, was limited by the facilities available to teachers and their own pedagogical outlooks, thus object lessons could be extremely variable in their content. Children might be confronted by a disparate array of inanimate objects, including the occasional, "stuffed dogs and alligators",[29] experience "incoherent" sequences of object study led by a teacher using regimented texts to harness their attention.[30] One such text, taken from

23 Henslow, J.S. (1880) *Botany for Children: An illustrated elementary textbook for junior classes and young children.* London: Edward Stanford.
24 Henslow, J.S. *op cit* (1880) preface.
25 Thome, O.W. (1885) *Thome's Structural and Physiological Botany : Adapted for the use of artisans and students in public and science schools*. London Longmans,Green & Co.
26 Balfour, J.H. (1864) *Class Book of Botany: An introduction to the study of the vegetable kingdom.* Edinburgh: Adam and Charles Black.
27 Balfour, J.H. (1864) *op cit* p. viii–ix.
28 Twaite, A. (2002). *Glimpses of the Wonderful: The Life of Philip Henry Gosse 1810–1888* p. 125 London: Faber & Faber.
29 Jenkins, E.W. and Swinnerton, B. J. (1998). *Junior School Science Education in England and Wales since 1900: From steps to stages.* P. 14 London and Portland, Oregon: Woburn Press.
30 Jenkins, E.W. and Swinnerton, B. J. (1998) *op cit supra* p.14.

Louise Walker's book for teachers (1895),[31] deals with an object lesson on Cork from the cork oak tree *(Quercus suber)*:

> "Shew a large piece of cork. Ask what it is (a piece of cork). Where do we find bark? (On trees). On what part of trees? (On the stem). Teacher to show the stem of a tree on which the bark still remains. Tell the children the bark before them is from the cork oak and is called 'cork bark'. Repeat together 'cork is the bark of a small oak tree'. Write on blackboard...."[32]

These lessons are a far cry from explorations in "the book of nature" beloved by Henslow, Thome, Balfour and the American nature-study writer/educators-Anna Comstock and Liberty Hyde-Bailey. By 1908 The British Association announced that the object lesson had become, in many cases, "a laborious elucidation of the obvious."[33]

An attempt to transform nature-based pedagogy in the classroom came in 1910 with the publication of a three-volume book for teachers – "Nature Teaching on the Blackboard".[34] Here was a publication that "encouraged interactive blackboard sessions and demonstrations"[35] alongside object-based study and field-trips. In the context of botanical study this text book stresses a unification between drawn representations of plant life and actual specimens:

> "It is important that stress should be laid upon the fact that the drawings can only be helpful in such lessons when they are used along with, and not apart from, actual specimens of the plants illustrated. Each pupil should be provided with a specimen which can be handled, examined and dissected. Only when such provision of actual specimens is made can Nature Study, as applied to plants, be successfully taught, and the habit of accurate observation, which such study promotes, inculcated."[36]

As the twentieth century unfolded, experimental science came to the fore, influencing both school curricula and the physical spaces in which teaching and learning took place. Textbooks for secondary (high) schools contained, the by now familiar, line drawings of plant form and structure, but in addition offered experiments informed by scientific studies on processes such as photosynthesis and osmosis. Student investigations of plant tissues became a pedagogical coalition of chemistry and botany. Moreover, the powerful emergence of genetics, in the latter half of the century, dramatically altered the profile of botanical speci-

31 Walker, L. (1895) *Cusack's Object Lessons, Mineral and Vegetable World and Common Objects.* London: City of London book depository cited in . Jenkins, E.W. and Swinnerton, B. J. (1998).

32 Walker, L. (1895) *op cit* Part II p.169.

33 Jenkins, E.W. and Swinnerton, B. J. (1998). *Junior School Science Education in England and Wales since 1900: From steps to stages.* P. 14 London and Portland, Oregon: Woburn Press.

34 Pycraft, W.P. and Harvey-Kelman, J. (1910). *Nature Teaching on the Blackboard Vols 1–3.* London : Caxton Publishing Company.

35 http://blogs.nature.com/scottkeir/2010/04/03/the-booms-of-popular-science accessed May 18 2010. See also doctoral research of Caitlin Wylie, Department of History and Philosophy of Science, University of Cambridge.

36 Pycraft, W.P. and Harvey- Kelman, J. (1910). *Op Cit*, vol 1 p. V.

mens in science education and *Arabidopsis thaliana*, described by Endersby[37] as "a fruit-fly for the botanist", entered the school laboratory, and a new era for botanical education.

URBAN NATURE

Pauly highlights issues of engaging with nature in urban contexts noting that Manhattan "seemed a particularly unpromising site for the study of living nature."[38] He goes on to comment on the "insuperable problems"[39] field trips posed for urban educators:

> "The logistics of transporting hundreds of urban teenagers to a place where they might see something significant were daunting. Students whom were hard to control in classrooms for one-hour periods could be impossible outdoors or destructive in their enthusiasm."[40]

In response to this issue educators such as George Hunter at DeWitt Clinton High School in New York City decided on two urban opportunities for fieldwork; vacant lots to "study the interactions between insects and weeds"[41] and visits to the American Museum of Natural History. However, Pauly argues that although the museum dioramas provided opportunities to view "habitat groups" it gave

> "little opportunity for active observation for the study of living organisms. Moreover in their perfection the museum's framed tableaux aestheticized nature and gave students the impression that the best place to experience it was indoors on a rainy day."[42]

He notes that although promoters of the museum saw it as a central educational resource, biology educators in the main considered the school laboratory the primary site for tying the "pedagogical ideal of experiential learning to the physiological content of the biology course."[43] In this context pupils could "cut and poison plants without opposition."[44] During the latter stages of the twentieth century the American botanist Hershey, proposed that as plants did not "bite, run away, or produce odours"[45] they were ideal subjects for use in the classroom environment. Taylor, writing fifteen years earlier in the journal *American Biology Teacher* suggested that plants are the perfect teaching organism as they can be

37 Endersby, J. (2007) *A Guinea Pig's History of Biology: The plants and animals who taught us the facts of life*. London: Heinemann.
38 See Pauly, P. J. *op. cit.* (1991) p.679.
39 See Pauly, P. J. *op. cit. supra* (1991) p.679.
40 See Pauly, P. J. *op. cit. supra* (1991) p.679.
41 See Pauly, P. J. *op. cit. supra* (1991) p.679.
42 See Pauly, P. J. *op. cit. supra* (1991) p.680.
43 See Pauly, P. J. *op. cit. supra* (1991) p.680.
44 See Pauly, P. J. *op. cit. supra* (1991) p.680.
45 Hershey, D.R. (1990). "Plants for education" (letter), *Journal of Biological Education* 24, 2, p. 68.

> "inverted, bent, pinned and regionally subjected to chemical analysis, acid, heat, or knife without torture as they are nerveless"[46]

and even in death are no problem as

> "their corpses, which are more likely to desiccate then putrefy, may be discarded with paper refuse or kept indefinitely as inexpensively mounted demonstrations of the effects of certain treatments."[47]

Fig. 1: Girls Gardening – Botany Gardens James Allen's Girls' School 1910

Clearly there are ethical benefits to using plants, unlike the moral minefield of animal dissection, which Pauly discusses at length. For the teachers of young men in New York during the late nineteenth and early twentieth century there was a difficult line to tread between "vulgar boyish cruelty towards animals and effeminate sentimentality about nature."[48] Much of this dilemma concerned the fate of domestic cats, as one New York City School inspector commented:

> "There is no great fuss about pulling bugs apart, but if you touch a cat you create a commotion at once in a community."[49]

The quiet demise of plants passed unnoticed.

46 Taylor, M.C. (1965). Live Specimens. *The American Biology Teacher*, 27, 116–117. P. 117.
47 See Taylor, M.C. *op. cit. supra* (1965) p.117.
48 Pauly, P. J. (1991) The Development of High School Biology: New York City, 1900–1925. *Isis*, 82, 662–688.
49 See Pauly, P. J. *op. cit.* (1991) p.681.

Fig. 2: Dune Garden – Botany Gardens James Allen's Girls' School.

LEARNING CONTEXTS

The contexts in which botanical studies, by both teachers and students, are undertaken have received a great deal of attention over an extensive period of time.[50] In the 21st century it continues to be a contested pedagogical space. Much of this attention has focused on using living rather than preserved plant material within a discovery-based pedagogy. Some of these past commentaries have highlighted the role that botanic gardens can have in the teaching and learning of botany.

The British paleobotanist Marie Stopes in her 1906 publication *"Young People and Plant Life"* extolled the virtues of taking young people "to the plants themselves and asking them to teach us."[51] Lilian Clarke, working in a South Lon-

50 See for example: Lindley, J. (1858). *School botany and vegetable physiology or the Rudiments of botanical science*. London: Bradbury and Evans; Stopes, M. (1906). *The study of plant life for young people*. London: Alexander Moring; Brightwen, E. (1913). *Glimpses into plant life: an easy guide to the study of botany*. London: Fisher Unwin; Clarke, L. (1922). *The botany gardens of James Allen's Girls' School, Dulwich: their history and organisation*. London: London Board of Education; Clarke, L., 1935. *Botany as an experimental science in laboratory and garden*. London: Oxford University Press; Daglish, F.E. (1932). *How to see plants*. London: Dent; Shaw, E.E. (1930). "Nature study for teachers and children", *School Nature Study*, 25, 100, 80–3. Hutchinson, M. (1947). *Children as Naturalists*. London: Allen and Unwin; Montessori, M. (1962). *The discovery of the child*. Oxford: Clio Press. And Tranter, J. (2004). Biology: dull, lifeless and boring? *Journal of Biological Education*, 38 (3): 104–105.

51 Stopes, M. (1906). *The study of plant life for young people*. London: Alexander Moring p. 196.

don girls' school, also prioritised the use of living specimens and encouraged her girls to create their own books from their observations in the school botanic garden. In her book *Botany as an Experimental Science in Laboratory and Garden,*[52] which was published posthumously, Clarke highlighted two key elements of her philosophy on botany teaching, elements that are particularly pertinent to the challenges of botanical education today. She observes that

> "since the end of the last century more importance has been paid at the James Allen's Girls' School to the plant as a living organism than to any other branch of botany."[53]

Significantly, in the context of botanic gardens, she considered that:

> "The experimental method of studying botany has been greatly helped by the development of botany gardens. The gardens have been made gradually in response to the needs of the work. They have become, in many cases, out-of-door laboratories, and the work indoors and out of doors is one."[54]

For further discussion of Clarke's methods see Sanders, 2008.[55] Montessori, in her work *"The Discovery of the Child"* also advocated a dynamic engagement with plants:

> "Children indeed love flowers, but they need to do something more than remain among them and contemplate their coloured blossoms. They find their greatest pleasure in acting, in knowing, in exploring."[56]

During the late 19th century, and for much of the earlier twentieth century British botanic gardens such as the Royal Botanic Gardens Kew and Chelsea Physic Garden provided a free plant specimen supply service to schools.

PROVISION OF SPECIMENS

The archive of school letters and photographs stored at the Royal Botanic Gardens in Kew offers a rich collection of texts and images describing the type of plants and plant parts that teachers were requesting to borrow from the garden between 1877 and 1914, and how they were using them in their classrooms. Some of these archived letters provide useful evidence of the plant groups that teachers were most commonly using in their teaching. They also demonstrate that the problem of using pictures to explain the structure of plants to pupils was an issue in the past, as indeed it can be now. This was a free service offered by the garden to schools and advertised in a magazine called *"Teacher's Aid."*

52 Clarke, L., (1935). *Botany as an experimental science in laboratory and garden.* London: Oxford University Press.

53 See Clarke, L. *op. cit. supra* (1935) p.vi.

54 See Clarke, L. *op. cit. supra* (1935) p.vi.

55 Sanders, D. (2008). Balancing the interplay between botanical gardens and schools: the work of William Hales and Lilian Clarke in *Studies in The History of Gardens and Designed Landscapes*, 28, 3 and 4, p.1460–1176.

56 Montessori, M. (1962). *The discovery of the child.* Oxford: Clio Press. P. 74.

Teachers were keen to impress upon the staff at Kew that their specimens would be put to good educational use, as one headmaster wrote,

> "I assure you sir that any specimens you may kindly send will not be stored away in a forgotten corner, but will be well lectured upon to the best of my ability."[57]

Specimens were not always used for botany lessons, as a teacher at Gloucester Road Boy's School in Peckham, South London demonstrated in 1914. His focus, as illustrated by this extract from his letter to Kew, was,

> "to teach a certain amount of geography by means of an exhibition of products of the empire."[58]

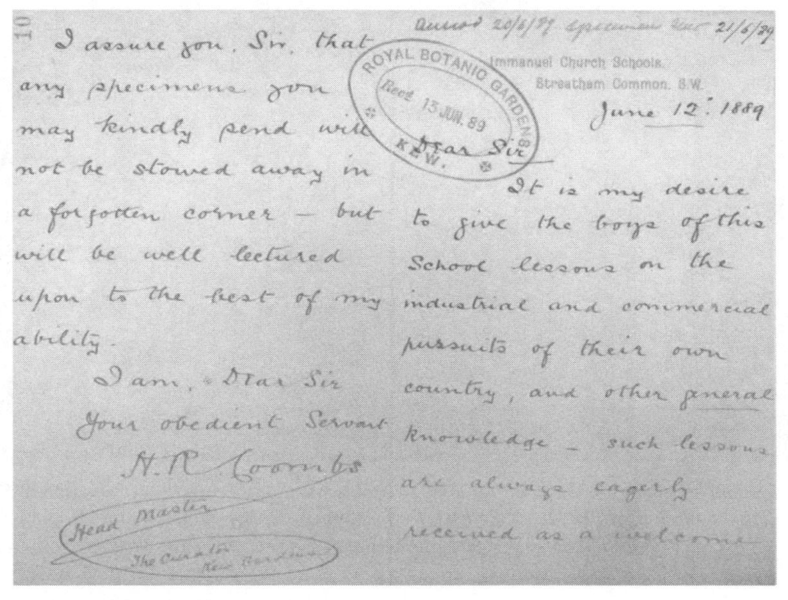

Answd 20/6/89 specimens [illegible] 21/6/89

ROYAL BOTANIC GARDENS Recd 13 JUN. 89 KEW.

Immanuel Church Schools.
Streatham Common. S.W.
June 12th 1889

Dear Sir

It is my desire to give the boys of this School lessons on the industrial and commercial pursuits of their own country, and other general knowledge – such lessons are always eagerly received as a welcome

I assure you, Sir, that any specimens you may kindly send will not be stowed away in a forgotten corner – but will be well lectured upon to the best of my ability.

I am, Dear Sir
Your obedient Servant
H.R. Coombs
Head Master

The Curator
Kew Gardens

Fig. 3: Example of teachers letter from the Royal Botanic Gardens, Kew archives – ***Source*** *Royal Botanic Gardens, Kew Economic Botany Collection Reproduced by permission of the trustees of The Royal Botanic Gardens, Kew*

This type of "products of empire" use of plant specimens is mentioned regularly in this historical collection of letters, although on this occasion the request was refused due to a dearth of duplicate objects. It reflects the type of geographical

57 H.R. Coombs, Headmaster of Immanuel church Schools, Streatham Common, London on June 12th 1889 sourced from three volumes of school letters from 1877–1914 Royal Botanic Gardens, Kew archival collection.

58 Letter from teacher at Gloucestor Road boys school,l Peckham, South London 1914 sourced from three volumes of school letters from 1877–1914 Royal Botanic Gardens, Kew archival collection.

education practised in English schools during the late nineteenth century. Indeed, in England during this period,

> "much geography for schools and training colleges was influenced by the prevailing imperial outlook of the time. Few books resisted the temptation to glorify British achievements overseas and to laud Britain's civilizing mission in Africa and elsewhere. There were stated to be religious, economic and cultural aspects to this mission, thus not stressing the exploitative nature of much activity in the colonial territories."[59]

Fig. 4: Plant specimens and objects on display Gloucester Road Boy's School, Peckham, South London – Source Royal Botanic Gardens, Kew Economic Botany Collection Reproduced by permission of the trustees of The Royal Botanic Gardens, Kew.

TEACHER TRAINING IN BOTANICAL GARDENS

Although both the Royal Botanic Gardens Kew and Chelsea Physic Garden in London regularly supplied specimens during the late nineteenth and early twentieth centuries, I have found no historical record of teacher training programmes located in these botanic gardens during this time. However, there are records that demonstrate how teacher training in the early part of the twentieth century, was embedded into the work of Brooklyn Botanical Garden in New York. In a 1930 article in *School Nature Study*, Ellen Eddy Shaw, an educator

59 Graves, N. (2001). *School Textbook Research: the Case of Geography 1800–2000* (Bedford Way Paper 17). London: University of London, Institute of Education.

from Brooklyn Botanical Garden described the garden's functions and their educational role:

> "The Brooklyn Botanic Garden, an institution similar to Kew Gardens in London, situated in the City of New York, not only carries on departments of scientific research and plans its grounds so that they themselves may be a lesson in botany for students and the public, but it also supports educational departments which offer to the children, the teachers-in training, and the teachers of New York City opportunities to pursue at first hand their studies in nature."[60]

Students from the Maxwell Training School for teachers would study a botanical and horticultural programme at the garden, which included; greenhouse work, fieldwork, and laboratory work. In the same article Shaw also describes groups of teachers *"threading their way"* to the botanic garden, for after school *"extension"* courses in botany, field botany, school gardening and greenhouse work. These courses lasted over thirty weeks and teachers received accreditation for them. Encapsulated in Shaw's article, is an important record of how one botanic garden structured their teacher training programme from the 1920's onward, a training that provided many opportunities for New York teachers to work with living plant specimens.

The contradictions in teaching a living science with dead specimens alongside text-book drawings were the subject of a special edition of *The American Biology Teacher* in 1961. The following comment by one of the contributors, H.G. Liebherr exemplifies the main focus of this edition:

> "How many times have you emphasised to your classes that biology is the science of living things and then proceeded to teach the course using dead specimens?"[61]

THE RELEGATION OF BOTANY AND THE RISE OF HUMAN BIOLOGY

The subtext of Pauly's article is the emergence of a biological curriculum that was centred on human biology, often with social undertones such as the impact of narcotics on the body or the lifecycle of disease spreading insects. British statistics from the early to mid twentieth century also show a sharp decline in botanical studies and a preference towards a biology curriculum dominated by zoology.

Kelly's *International Review of Trends in Biological Education*, published in 1967, highlights the steep drop in students taking exams in botany in contrast to the rise in biology students; he notes that in 1928

> "some 26% of students entering for the school certificate exam took botany, less than 1% took biology. Come 1950 and 31% were taking biology and about 1% botany."[62]

60 Shaw, E.E. (1930). "Nature study for teachers and children", *School Nature Study*, 25, 100, 80–3.

61 Liebherr, H.G., (1961). Field biology in the High School Program. *The American biology Teacher*, 23,5 p.285–287.

In nineteen eighties Britain separate botanical studies at ordinary and advanced level were discontinued and botany became absorbed, as had zoology, into a generic biology curriculum.

PLANT SPECIMENS TODAY

What impact has this move towards general biology had on the use of plant specimens in the UK classroom today? Several commentators have suggested that the range of living plants both studied in schools and presented in textbooks are limited. Tranter has observed that

> "in too many schools, the wealth of living or once living organisms which pupils are required to study is often reduced to little more than the geranium and the potato."[63]

Besides this dearth of living specimens, experimental plant material in biology textbooks is repeatedly

> "drawn from a relatively restricted number of species-geranium, Canadian pond weed, broad bean seeds, tomatoes and mustard and cress."[64]

Slingsby, suggests that plants "get a bad deal in science curricula" and are, "often just victims in a series of photosynthesis experiments which do not always work."[65] Moreover, research,[66] both in North American and UK schools, has demonstrated that teaching with and about plants is considered to be a pedagogical challenge by many biology educators, and students tend to prefer animals. Current use of the genetically well-mapped, but morphologically uninspiring rapid-cycling brassica *(Arabidopsis thaliana)* appears to have done little to change this state of affairs.

Pauly's paper ends with a quote from the evolutionary entomologist Alfred Kinsey who, during the Great Depression, wrote a comprehensive guide to high school biology in which he argued the goal of the course was to, "teach the typical student that it is an endlessly interesting world in which he is living."[67] Kinsey

62 Kelly, P. (1967). Trends in Biological Education – An International Review. *Journal of Biological Education*, 1,1, 1–12.

63 Tranter, J. (2004). Biology: dull, lifeless and boring? *Journal of Biological Education*, 38 (3): 104–105. P.41.

64 Collins, N. & Price, R. (1996). Plants. In: Reiss, M. (Ed) *Living biology* (pp.29–45). London: Institute of Biology. p.29.

65 Slingsby, D. (2006) Biological Education: Has it gone anywhere since 1875? *The Biologist* 53,6 283–284.

66 See for example Kinchin, I. (1999). Investigating Secondary – School Girls' Preferences For Animals Or Plants: A simple "head to head" comparison using two unfamiliar organisms. *Journal of Biological Education*, 33, 2, 95–9; and Wandersee, J. (1986). "Plants or animals? Which do junior high school students prefer to study?" *Journal of Research in Science Teaching*, **23**, 5, 415–26.

67 Pauly, P. J. (1991) The Development of High School Biology: New York City, 1900–1925. *Isis*, 82, 662–688.

then went on, as Pauly notes, to focus on the "aspect of biology that most stimulated young Americans."[68] Plants, dead or alive, do not feature in this mammalian reproductive landscape. Perhaps botanical gardens, through charismatic mega flora, such as *Amorphophallus titanum*, can once again draw urban student attention towards living plant specimens[69] and the underpinning importance of botanical science in both private and public contemporary life.

REFERENCES

Balfour, J.H. (1864) *Class Book of Botany: An introduction to the study of the vegetable kingdom.* Edinburgh: Adam and Charles Black.

Bebbington, A. (2005) The ability of A–level students to name plants *Journal of Biological Education* 39, 2 62–67.

Boney, A.D. (1991). The "Tansley manifesto" affair. *New Phytologist*, 118, 3–21.

Brightwen, E. (1913). *Glimpses into plant life: an easy guide to the study of botany*. London: Fisher Unwin.

Clarke, L. (1922). *The botany gardens of James Allen's Girls' School, Dulwich: their history and organisation*. London: London Board of Education.

Clarke, L., (1935). *Botany as an experimental science in laboratory and garden*. London: Oxford University Press.

H.R. Coombs, Headmaster of Immanuel church Schools, Streatham Common, London on June 12th 1889 sourced from three volumes of school letters from 1877–1914 Royal Botanic Gardens, Kew archival collection.

Collins, N. & Price, R. (1996). Plants. In: Reiss, M. (Ed) *Living biology* (pp.29–45). London: Institute of Biology.

Daglish, F.E. (1932). *How to see plants*. London: Dent.

Endersby, J. (2007) *A Guinea Pig's History of Biology: The plants and animals who taught us the facts of life*. London: Heinemann.

Gaskell, E. (1848). *Mary Barton–a tale of Manchester life*. London: Chapman and Hall.

Graves, N. (2001). *School Textbook Research: the Case of Geography 1800–2000* (Bedford Way Paper 17). London: University of London, Institute of Education.

Henslow, J.S. (1880) *Botany for Children: An illustrated elementary textbook for junior classes and young children.* London: Edward Stanford.

Hershey, D.R. (1990). "Plants for education" (letter), *Journal of Biological Education* 24, 2, p. 68.

Hershey, D.R. (1996) A Historical Perspective on problems in botany teaching. *The American biology Teacher*, 58, 341–347.

Honey, J. (1987). Where have all the flowers gone? –The place of plants in school science, *Journal of Biological Education*, 21, 3, 185–9.

Hoyles, M. (1994). Lost Connections and New Directions: The private garden and the public park. London: Comedia and Evans, P. (2002). Flower power (Guardian Society), The Guardian, 3 April, 9.

Hutchinson, M. (1947). *Children as Naturalists*. London: Allen and Unwin.

http://blogs.nature.com/scottkeir/2010/04/03/the-booms-of-popular-science accessed May 18 2010. See also doctoral research of Caitlin Wylie, Department of History and Philosophy of Science, University of Cambridge.

68 See Pauly, P. J. *op. cit.* (1991) p.681.

69 Sanders, D. (2005). Boring Botany? Rethinking teaching about plants in schools. *Topic* 33 April 2005 4–8. Slough: NFER.

http://designmuseum.org/design/leopold-rudolf-blascka accessed June 2010.
http://designmuseum.org/design/leopold-rudolf-blascka accessed June 2010.
http://www.hmnh.harvard.edu/on_exhibit/the_glass_flowers.html accessed June 2010.
http://www.hmnh.harvard.edu/on_exhibit/the_glass_flowers.html accessed June 2010.
http://www.hmnh.harvard.edu/on_exhibit/the_glass_flowers.html accessed June 2010.

Jenkins, E.W. and Swinnerton, B. J. (1998). *Junior School Science Education in England and Wales since 1900: From steps to stages*. P. 14 London and Portland,Oregon: Woburn Press.

Kelly, P. (1967). Trends in Biological Education–An International Review. *Journal of Biological Education*, 1,1, 1–12.

Kinchin, I. (1999). Investigating Secondary–School Girls' Preferences For Animals Or Plants: A simple "head to head" comparison using two unfamiliar organisms. *Journal of Biological Education*, 33, 2, 95–9.

Letter from teacher at Gloucestor Road boys school, l Peckham, South London 1914 sourced from three volumes of school letters from 1877–1914 Royal Botanic Gardens, Kew archival collection.

Liebherr, H.G., (1961). Field biology in the High School Program. *The American biology Teacher*, 23,5 p.285–287.

Lindley, J. (1858). *School botany and vegetable physiology or the Rudiments of botanical science*. London: Bradbury and Evans

Montessori, M. (1962). *The discovery of the child*. Oxford:Clio Press.

Pauly, P. J. (1991) The Development of High School Biology: New York City, 1900–1925. *Isis*, 82, 662–688.

Pool, R.J. (1919) About High School and College Botany. *School Science and Mathematics*, 19, 487–500. Cited in: Hershey, D.R. (1996) A Historical Perspective on problems in botany teaching *The American Biology Teacher*, **58**, 341–347.

Pycraft, W.P. and Harvey- Kelman, J. (1910). *Nature Teaching on the Blackboard Vols 1–3*. London: Caxton Publishing Company.

Reiling, H. (2003) Beter dan de natuur. Available in English translation at http://members.ziggo.nl/here/neo.html accessed June 2010.

Rooper, J.G. (1894). New Methods in the Lower Standards. *Teacher's Aid* **XVII**, 441, 530–533.

Sanders, D. (2005). Boring Botany? Rethinking teaching about plants in schools. *Topic* 33 April 2005 4–8. Slough: NFER.

Sanders, D. (2008). Balancing the interplay between botanical gardens and schools: the work of William Hales and Lilian Clarke. In *Studies in The History of Gardens and Designed Landscapes*, 28, 3 and 4, p.1160–1176.

Secord, A. (1996) Artisan botany. In: Jardine, J., Secord, A. & Spary, E.C.(Eds) *Cultures of Natural History,* pp. 378–393. Cambridge: Cambridge University Press.

Shaw, E.E. (1930). "Nature study for teachers and children", *School Nature Study*, 25, 100, 80–3.

Shteir, A.B. (1996). *Cultivating Women, Cultivating Science*. Baltimore: John Hopkins

Slingsby, D. (2006) Biological Education: Has it gone anywhere since 1875? *The Biologist* 53,6 283–284.

Stopes, M. (1906). *The study of plant life for young people*. London: Alexander Moring.

Taylor, M.C. (1965). Live Specimens. *The American Biology Teacher*, 27, 116–117.

Thome, O.W. (1885) *Thome's Structural and Physiological Botany: Adapted for the use of artisans and students in public and science schools*. London Longmans, Green & Co.

Tranter, J. (2004). Biology: dull, lifeless and boring? *Journal of Biological Education*, 38 (3): 104–105.

Twaite, A. (2002). *Glimpses of the Wonderful: The Life of Philip Henry Gosse 1810–1888*. London: Faber & Faber.

Walker, L. (1895) *Cusack's Object Lessons, Mineral and Vegetable World and Common Objects*. London: City of London book depository cited in Jenkins, E.W. and Swinnerton, B. J. (1998). *Junior School Science Education in England and Wales since 1900: From steps to stages*.

Walker, T. & Allen, L. (1999). "Seeking partnerships to teach biodiversity. *Roots*, 19, 30–2.
Wandersee, J. (1986). "Plants or animals? Which do junior high school students prefer to study?" *Journal of Research in Science Teaching*, **23**, 5, 415–26.
Youngpeter, J. M., (1961) Field Trips. *The American Biology Teacher* **23**, 5, p.273–275.

LEARNING IN THE LABORATORY: THE INTRODUCTION OF "PRACTICAL" SCIENCE TEACHING IN ONTARIO'S HIGH SCHOOLS IN THE 1880S[1]

Michelle Hoffman

In 1885, the Department of Education of the Province of Ontario initiated a concerted effort to introduce practical, laboratory-based science teaching into its high schools. Though a few natural science subjects had been prescribed on the curriculum since 1872, the reform effort of the late 1880s was truly a systematic overhaul, prompted by enthusiasm for the German model of scientific research that was quickly gaining currency in the United States and Canada. Within five years, many of the features that are today considered part and parcel of science classes were instated in Ontario's high schools. At the helm of the project was newly-appointed high school inspector John Seath, who applied himself zealously to the task and quickly became a gadfly to many school trustees. The top priorities were equipping schools with apparatus and dispatching inspectors to ensure that students and teachers were indeed using it. "In every High School and Collegiate Institute, . . . Chemistry and Physics should be taught experimentally, and Botany practically; and it shall be the duty of the High School Inspectors to report specially those schools in which this recommendation is not observed," announced new regulations issued in August of 1885.[2] In physics and chemistry, the importance of experiments – performed not by the teacher, but by the students themselves – was paramount. In botany and zoology, the hands-on collection and examination of specimens were considered the natural analogues of the laboratory experiment, and nature itself was the laboratory. Grants were instituted to induce reluctant school boards to invest in apparatus and specimens, and a round of new textbooks was commissioned to canonize the modern methods of school science.

These sweeping changes were not unique to Ontario. In fact, they were representative of a wider North American trend in secondary education. Within Canada, Ontario's reforms provided a template for other provinces to follow. At Confederation in 1867, the British North America Act had placed education squarely within the jurisdiction of the individual provinces. Even so, throughout the nineteenth century, Ontario's textbooks were widely used in other provinces

1 The author would like to thank Chen-Pang Yeang, Elizabeth Smyth, and Roland Wittje for their helpful suggestions. This research was supported by the Social Sciences and Humanities Research Council of Canada.

2 Ontario Department of Education, *Report of the Minister of Education for the year 1885*, 29.

and arguably served as a de facto national curriculum.[3] Moreover, even as Inspector Seath officiated over the introduction of apparatus and labs into Ontario's schools, similar reforms were being carried out in American high schools in response to the influential "Harvard lists" of experiments developed by physicist Edwin Hall and chemist Josiah Cooke.[4]

At the crux of these reforms was a growing allegiance to a novel pedagogical principle: that of independent, hands-on learning. Like many of their colleagues abroad, Ontario educators increasingly embraced the conviction that real learning required students to reason things through autonomously rather than be spoon-fed information for memorization. Critics took aim at the outdated, authoritarian ways of teaching repudiated by Herbert Spencer in his influential essays: "To *tell* a child this and to *show* it the other, is not to teach it how to observe, but to make it a mere recipient of another's observations: a proceeding which weakens rather than strengthens its powers of self-instruction."[5] The schools' longstanding reliance on book work, rote and drill was sharply criticized. In this effort, science, more than any other subject, held particular promise. Properly taught, it was seen as an ideal field for putting into practice the pedagogical ideals of practical, self-directed learning, encompassing as it did an array of skills in observation, interpretation, and handiwork.

In one respect, this episode of science teaching reforms is a story about the identity of the high school itself. In the 1880s, Ontario's high schools had only just managed to carve out a distinct niche within the school system. For many decades, they had been plagued by persistent overlap between their work and that of the universities. Universities, starved for students, kept entrance standards low enough to compete directly with the high schools for enrolments. At mid-century it was not uncommon for rural students, especially, to bypass high school entirely and head to university directly from a common (public) school, doing any necessary remedial work at university. By the 1870s, most of this redundancy was eliminated, and high schools had secured their role as gatekeepers to the universities. In the 1880s, however, a different problem persisted: each university had its own distinct entrance requirements. This was a major logistical challenge for high school teachers and headmasters, who struggled to accommodate the various

3 G. Tomkins, *A Common Countenance: Stability and Change in the Canadian Curriculum* (Vancouver: Pacific, 2008), 64.

4 On the rise of teaching laboratories in American high schools, see for example K. Sheppard & G. Horowitz, "From Justus von Liebig to Charles W. Eliot: The Establishment of Laboratory Work in U.S. High Schools and Colleges," *Journal of Chemistry Education* 83.4 (2006), 566–570; A.E. Moyer, "Edwin Hall and Emergence of the Laboratory in Teaching Physics," *Physics Teacher* 14 (1976), 96–103; J.L. Rudolph, "Epistemology for the Masses: the Origins of 'The Scientific Method' in American Schools," *History of Education Quarterly* 45.3 (2005): 349–354; K.M. Olesko, "German Models, American Ways: The 'New Movement' Among American Physics Teachers, 1905–1909," in *German* Influences *on Education in the United States to 1917* (Washington: German Historical Institute, 1995), 129–153.

5 Herbert Spencer, *On Education: Intellectual, Moral, Physical* (New York: Appleton & Co., 1866, c1861), 135.

kinds of preparation that their matriculating students required.[6] (American high schools at this time were similarly beleaguered by the multiplicity of college entrance requirements, a problem that prompted the Committee of Ten to convene in 1892 in an effort to standardize the curriculum.[7]) In Ontario as in the United States, overhauling the science curriculum was unquestionably part of the project to harmonize high school work with university entrance standards and impose order on the system as a whole. High schools fell in line with the science requirements established at the University of Toronto, just as high schools across the United States gradually geared their programs to the entrance standards set by Harvard.

It would be a mistake, however, to see the reform effort simply as a top-down process dictated by the universities. The changes that were imposed on facilities, equipment, course content, teaching methods and textbooks also provided an opportunity for the Department of Education to assert control over its far-flung network of high schools and to uphold the authority of a standardized curriculum over the autonomy of individual teachers in the classrooms of Ontario. Furthermore, the high schools were seen as serving a very different social function than the universities. Though they unquestionably catered to an educational elite, they aspired to provide a general education. Accordingly, although the laboratory model was inherited from the universities, the pedagogical rationales that accompanied it had to be filtered and transformed to match the stated mission of the high schools. It was not uniformly obvious to teachers and school trustees that science was an important element of a general education, nor that the laboratory was the best forum in which to teach it. Indeed, the varied response to the Department's interventions illustrates how top-down decisions about the investment of time and money in the teaching of science became a catalyst for debate about how learning takes place, what a secondary education should offer to students, and what the proper role of the teacher should be.

In view of such debates, this account draws attention to the distinct pedagogical benefits that were ascribed to the study of science. Here, Ontario educators and officials drew on well-entrenched justifications popularized by Herbert Spencer, Thomas Henry Huxley, and particularly James M. Wilson, a science master who had been instrumental in bringing a laboratory to Rugby College, a pioneering move for the teaching of science in Britain. Following the lead of Wilson and others, Inspector Seath and Minister of Education George Ross anchored science subjects to the pedagogical dogma of mental discipline. The mental discipline attributed to the study of science was thoroughly conflated with the hands-on, inductive learning processes that science demanded. School officials could therefore argue that science not only deserved time on an already overburdened program of studies, but also required significant fiscal investments in renovations, furniture, apparatus, and specimen collections. As Goodson has shown, curricu-

6 R.D. Gidney and W.P.J. Millar, *Inventing Secondary Education: The Rise of the High School in Nineteenth-Century Ontario* (Montreal: McGill-Queen's, 1990), 254–272.

7 Sheppard & Horowitz, 568; Moyer, 97–100.

lum reform often triggers competition among school subjects for status, territory, and resources.[8] This was indeed the case for science reform in the 1880s, as demonstrated by the often hostile letters that flew between local school boards, the Department, and its inspectors, wrangling over the allocation of time and money to laboratory science. Moreover, science teachers also wrote to the *Canada Educational Monthly,* the leading educational journal of English-speaking Canada, emphasizing the uniquely powerful rewards of learning science, both moral and intellectual, and measuring its benefits against those of other subjects.

In 1889, in the wake of his unyielding campaign to compel change in the province's high schools, Inspector Seath surveyed the work done over the previous five years and felt justified in declaring it largely complete. "The most gratifying increase," he noted, "has taken place in the value of the scientific apparatus."[9] Yet in spite of the accolades for science teaching publicized by Seath and others, criticisms mounted. Significantly, these criticisms came not from those who opposed the expanded role of science in the schools, but from science teachers themselves. As the optimism of the 1880s gave way to a measure of disillusionment about the successes of science teaching, several Ontario educators expressed strong doubts about the effectiveness of practical science teaching and the much-acclaimed inductive method. Finally, challenges to faculty psychology in the 1890s meant that the rationale of mental discipline that had given impetus to the teaching of science lost much of its authoritativeness. As we will see, however, thanks to the strongly centralized authority of the curriculum, the malleability of the notion of "practical" teaching, the rising cultural influence of the natural sciences, and mounting pressures on the traditional high school program, laboratory science retained its place on the curriculum and withstood the social and pedagogical changes of the 1890s.

1. UNIVERSITIES PAVE THE WAY: THE RESEARCH LABORATORY COMES TO ONTARIO

During the 1870s and 1880s, several factors combined to make the teaching of high school science a clear priority for reform. Most notably, significant changes were afoot at the province's universities, particularly the University of Toronto. From the 1870s onward, university science departments in Canada were gradually transformed as their traditional teaching mandate was expanded to include research. Gingras has traced the changing tenor of teaching and publications among Canadian physicists as a new generation of European-trained researchers gradually took over the scientific professoriate. This transformation entailed a major change in purpose for university science courses. Science had previously been taught as part of a broad-based arts education for the upper classes. Under

8 I. Goodson, "Subjects for Study: Aspects of a Social History of Curriculum," *Journal of Curriculum Studies* 15.4 (1983), 39.

9 *Report of the Minister, 1888/89*, 187–188.

the new model, university students were trained in specialized research with a view to pursuing careers as professional scientists or skilled practitioners of medicine or engineering.[10]

At the University of Toronto, physics and mathematics chair James Loudon emerged as a vigorous promoter of the German research model and lobbied for the introduction of laboratory research and teaching. When an engineering school was proposed by the provincial government in 1873, Loudon fought successfully for it to be affiliated with the University, thereby assuring the survival of the University's science departments – which, he was convinced, would have atrophied in the face of competition with an independent engineering school. In 1878, the University of Toronto's first instructional laboratories for chemistry, physics, mineralogy, geology, and biology were opened within the new School of Practical Science. That same year, laboratory work was introduced as an option for physics students; by 1885, it was mandatory for anyone seeking an honours degree in physics.[11] Loudon was appointed president of the University in 1892 and in this capacity redoubled his advocacy for scientific research in the province's universities, often pointing to Johns Hopkins University as an institution that had successful grafted the German university model onto the traditional higher education system inherited from Britain.[12]

The laboratory-based instruction that defined the transformation of university science departments simultaneously became the pedagogical standard for high school science teaching endorsed by the Department of Education. This was a pattern also exhibited in American high schools, which like Canadian schools took their cues from the program requirements of colleges and universities.[13] Further reinforcing the universities' influence on Ontario high schools was the fact that the universities trained the province's high school teachers. Indeed, earning a university degree was the standard route to the first-class certificate required for high school teaching.[14] By 1896, an aspiring teacher could acquire "specialist"

10 Yves Gingras, *Physics and the Rise of Scientific Research in Canada* (Montreal: McGill-Queen's University Press, 1991), 26–31.

11 Gingras, *Physics*, 17.

12 J. Loudon, "The Universities in Relation to Research," *Science* 15.391 (1902), 1001–1009.

13 Rudolph, 348.

14 The institutional structures for teacher training underwent frequent changes in the last decades of the twentieth century. As of 1885, a university degree was no longer deemed sufficient for a high school teaching certificate: graduates now had to do a four-month "professional" training course in one of the province's five new Training Institutes (which were in fact specially designated collegiate institutes). This new requirement was resented within university circles, where professors perceived it as an implicit reproof of their teaching methods. In 1890, the Training Institutes were abolished and professional teacher training was centralized at the Toronto School of Pedagogy, which itself was relocated to Hamilton in 1897. In 1906, faculties of education were established at Queen's University and the University of Toronto and high school teacher training was henceforth undertaken within the universities, despite longstanding opposition from many university professors. See J.G. Althouse, *The Ontario Teacher* (Toronto, Ontario Teachers' Federation: 1967), 137–156;

status by earning an honours degree in science, history, math, classics or a modern language. A specialist certificate qualified its holder to work (at a higher salary) in one of the province's collegiate institutes, an elite tier of high schools. As science graduates trained in the new university laboratories permeated the high schools, they no doubt brought their laboratory experience to bear on their own teaching.

In 1885, the year that laboratory work became compulsory for honours degrees at Toronto, Inspector Seath went on the offensive in his inaugural high school inspection report. Running at more than 14,000 words – nearly five times the length of his fellow inspector's – Seath offered a sweeping critical appraisal of the high school system, prompting one commentator to describe it as "exhaustive in more senses than one."[15] Seath's assessment of science teaching was particularly grim. "Probably no other subjects have been, confessedly, so badly taught as chemistry, physics, and botany," he charged. He nevertheless expressed optimism that high school science would henceforth receive more attention, thanks to "the recent science additions to the matriculation curriculum of Toronto University."[16] By this time, science had a stable, though not uncontested, position on the secondary curriculum. Natural philosophy, chemistry and natural history had made intermittent appearances in Ontario's grammar schools since mid-century, but had been definitively introduced in 1871 thanks to a pivotal high school reform that restructured the entire curriculum, granting physics, chemistry and botany a lasting place on the program. The landmark 1871 act was also a key step towards a centralized school system: it renamed the grammar schools, which henceforth were called high schools, made them legally co-ed, and brought them under the authority of the province's Department of Education (then known as the Department of Public Instruction), securing its control over the curriculum, the authorization of textbooks, the setting of exams, and the training and certification of teachers.[17]

Seath was an ambitious and influential figure in Ontario's educational system. The son of a Scottish engineer, he had emigrated from Scotland with his family in 1844, a BA from the University of Glasgow in hand. He immediately embarked on a successful career in teaching that saw him rise steadily through the educational ranks, earning a second BA from the University of Toronto along the way. After a prestigious headmastership at a collegiate institute that attracted students from all over the province, he was promoted to high school inspector, and eventually to Superintendent (effectively third-in-command of the Department of Education), a position he held until his death in 1919.[18] Seath's policies were reform-minded and comprehensive: most notably, he relentlessly championed the cause

Robert M. Stamp, *The Schools of Ontario, 1876–1976* (Toronto: University of Toronto Press, 1982), 44.

15 "Editorial: Report of the Minister of Education for the Year 1885," *Canada Educational Monthly* 8 (1886), 190.

16 *Report of the Minister, 1885,* 161.

17 On the wide-ranging changes introduced by the 1871 Act, see Gidney and Millar, 231–253.

18 Stamp, *Schools of Ontario*, 41.

of technical education throughout his career.[19] In his role as inspector, Seath rarely minced words and was clearly not intimidated by the grumblings of indignant trustees. Convinced that a strongly centralized school system was the best way to improve education, he was committed to bringing recalcitrant schools into line with Department policies. This conviction extended to his approach to science education reform. In his inaugural report he asserted: "No plea should be necessary for the study of science. Its claims are now admitted by all, except, perhaps, the few whose liberality is bounded by the horizon of their own attainments or their own selfishness."[20] This kind of polemical statement did not help to heal the increasingly tense relationship developing between the inspectorate and local trustees.

2. ENTICEMENTS AND THREATS: GETTING APPARATUS INTO ONTARIO'S CLASSROOMS

The inspectors saw schools' lack of apparatus as both a symptom and a cause of ineffective science teaching. "Owing to the want of suitable apparatus and, in some cases, of the application of proper methods, of real science teaching there is very little," Seath remarked.[21] Even veteran high school inspector John E. Hodgson, whose criticisms were nearly always much milder than Seath's, noted pointedly that students had very few opportunities to perform experiments.[22] Science also stood out as having a particularly high failure rate in the grade 13 examinations (the notoriously difficult "Departmentals"). Hodgson, reporting the figures, attributed the failure rate to a dearth of apparatus for practical science teaching. (The failure rate for mathematics, by contrast, he attributed to the difficulty of the exam.)[23]

By this point, the Department of Education was on a clear campaign to change the situation. To encourage schools to invest in scientific apparatus, it announced in 1884 that it would provide annual grants based on the value of their apparatus collections. The grant scheme was somewhat complicated and slow to catch on: three years after it had been introduced, Seath complained that many boards had yet to realize how good a deal it was and tried to persuade them that it was even worth borrowing money if they needed to.[24] Properly understood, the new grants meant that schools could recoup the costs of their equipment within as

19 R.M. Stamp, "John Seath," in *Profiles of Canadian Educators*, ed. R.S. Patterson et al (Toronto: D.C. Heath, 1974), 235–247.

20 *Report of the Minister, 1885*, 161.

21 *Idem.*

22 *Report of the Minister, 1884*, 188.

23 *Report of the Minister, 1885*, 152. Only forty-two percent of the 4500 candidates who took the Departmental exam passed it, Hodgson reports. Of those who failed, twenty-five percent failed in science (chemistry, botany, physics and statics). The next-highest failure rate (eighteen percent) was in mathematics.

24 *Report of the Minister, 1887*, 166.

little as five years.[25] For Ontario's collegiate institutes, which were larger, higher-status high schools generally seen as providing a superior education, the apparatus regulations were not merely an incentive; they had teeth. The regulations stipulated that for a high school to qualify for or retain collegiate institute status, it had to own a minimum of $450 worth of science apparatus. By 1887, however, only four collegiate institutes of the ten inspected had met this requirement, and Seath was threatening to demote non-compliant institutes to high schools.[26] (Among other requirements, collegiate institutes were required to have a well-equipped laboratory and a science specialist on their staff of at least four teachers.)[27]

Significantly, the provision of science apparatus had become a point of demarcation between regular high schools and the more prestigious collegiate institutes. The collegiate institutes had been created in 1871 under the direction of then Chief Superintendent Egerton Ryerson, who had envisioned them as the last bastions of a classical education. Latin, he had anticipated, would gradually be phased out of smaller high schools. But Ryerson's plan failed: the enduring cultural prestige of Latin meant that parental demand kept it in schools throughout the province. By 1883, the distinction between high schools and collegiate institutes was based on staff qualifications and physical facilities – including, notably, science laboratories – rather than course offerings.[28] In a sense, then, science facilities and apparatus had replaced Latin as a status-conferring distinction of Ontario's collegiate institutes. As Seath remarked, the new equipment would mean that the inspectors would have "less difficulty hereafter in recognizing the difference between a Collegiate Institute and several of the high schools."[29]

The Department's new rules, particularly the apparatus requirement for collegiate institutes, did not pass uncontested. Much ink was spilled in letters between the inspectors and local boards indignant at the low grading of their facilities and the expense of meeting the new requirements. The board of St. Mary's Collegiate Institute reacted with outrage to Seath's 1887 inspection, venting its anger in an open letter to the Minister of Education. The board complained that it had only recently spent $3000 on a new wing, and was now being asked to set up a miniature laboratory for each pupil. It inveighed against the "autocratical" inspector whose "dictum" they had to obey lest they lose their standing, who would have students fritter away their time in "profitless experiments with expensive scientific

25 The grants amounted to ten percent of the total value of the apparatus owned by a school, up to a designated maximum ($275 or $450, depending on the number of teachers on staff). Because the province's grant was often matched by the county, schools stood to get a 20% annual return on their investment. These annual grants also applied to other forms of capital investment like library books, charts, maps, globes, and gymnasium equipment.

26 *Report of the Minister, 1887*, 165, 177.

27 "Act respecting high schools and collegiate institutes," in *Revised Statutes of Ontario for 1887*, 2453. A similar strategy of enforcement would be adopted in New York State, where high schools in the rapidly expanding school system were required to build laboratories in order to qualify for state funding (Sheppard & Horowitz, 569).

28 Stamp, *Schools of Ontario*, 7.

29 *Report of the Minister, 1885*, 154.

toys, while that which will be of importance to them throughout their whole life work is almost wholly neglected."[30] Similarly, the board of another high school voiced typical annoyance at Seath when it complained of the all-too-frequent changes demanded by the educational department, "which entail such heavy expense upon both school boards and parents."[31]

To build their apparatus collections, schools turned to Toronto-based dealers like Charles Potter, an enterprising optician and instrument-maker who quickly recognized the business opportunity presented by both the School of Practical Science and the high schools laboratories. Founding the Map and School Supply Company in the late 1880s, Potter began tailoring his apparatus to the high school textbooks and successfully expanded his instrument shop into a national school supply firm.[32] But even as instrument dealers capitalized on the new requirements, the Department responded to the chorus of cost complaints from trustees by trying to persuade teachers to build their own apparatus. Seath, in particular, advocated this solution, insisting that "at least as satisfactory experiments can be performed with home-made as with bought apparatus." This notion was pervasive at the time: Seath was drawing on a trope of the educational literature, the image of the scientist as an inventive experimenter tinkering away in a homespun lab.[33] "It is a well-known fact," he wrote, "that some of the most important results in science have been obtained by means of very simple apparatus," adding that pupils, too, should be encouraged to build their own equipment.[34]

Predictably, Seath's suggestion was not universally welcomed by science teachers. One teacher bitterly dismissed it in the *Educational Monthly*, writing that "Costly apparatus is necessary. It is useless to tell the teacher that with a little ingenuity he can prepare what is necessary. . . . It is hardly fair to expect spare time to be devoted to instrument-making, when so many other claims imperatively demand the teacher's attention."[35] Despite such complaints, the practice of

30 St. Mary's Collegiate Institute, *Report of a Committee Appointed to Consider the Inspector's Report on Buildings, Equipment, etc.*, 1888, Department of Education select subject files, RG 2–42–0–4649, Microfilm reel 5656, Archives of Ontario.

31 Paris High School, *Board of Education Responds to Inspector Seath's Negative Report*, 1885, Department of Education select subject files, RG 2–42–0–4256, Microform reel 5653, Archives of Ontario.

32 Julian A. Smith, "Charles Potter, Optician and Instrument Maker," *Journal of the Royal Astronomical Society of Canada* 87.1 (1993): 25–27.

33 Alfred Gage, for example, asked rhetorically in the preface of *The High School Physics:* "Have lecture-room displays proved very effectual in awakening thought and in kindling fires of enthusiasm in the young? Or would a majority of our practical scientists date their first inspiration from more humble beginnings, with such rude utensils, for instance, as the kitchen affords? Is the efficiency of instruction in the natural sciences to be estimated by the amount of costly apparatus kept on show in glass cases, labelled 'hands off,' or by its rude pine tables and crude apparatus bearing the scars, scratches, and other marks of use?" See A.P. Gage and C. Fessenden, *The high School Physics* (Toronto: W.J. Gage, 1887), iv.

34 *Report of the Minister, 1887*, 167.

35 D.F.H. Wilkins, "Observations Regarding the Teaching of Science in Our High Schools," *Canada Educational Monthly* 9 (1887), 89. This complaint was not confined to the high

building apparatus by hand persisted well into the twentieth century. Thirty years later, a teachers' guide issued by the Department of Education indicated the host of workshop skills that students were expected to master: "In both Physics and Chemistry, practice in the preparation and manipulation of apparatus should form part of the Course. Where practicable, the Course should also include simple operations in glass-blowing and lathe work, and in hard and soft soldering."[36]

3. SCIENCE'S CONTESTED CLAIM ON A CRAMMED CURRICULUM

Another major obstacle to the Department's reform effort was the already crowded curriculum. With the increased class time required not only by science courses but also by new subjects like bookkeeping and physical education, headmasters felt their resources being stretched to the limit.[37] Reacting to yet another of Seath's critical reports, the headmaster of the Stratford Collegiate Institute protested that if teachers were spending too little time on newly-prioritized subjects, it was the Department's fault for "putting on *too many* subjects" and demanding that pupils get "more lessons than there are spaces on the Time Table."[38] St. Mary's, too, added the overloaded program to its litany of grievances, targeting science classes for particular criticism:

> [T]he tendency of our educational system is to sacrifice thoroughness for brilliancy – to run over a great number of subjects, and to master none – to cram for examinations and to neglect thorough, honest work. It is of little importance that our children can blow hydrogen bubbles or demonstrate the binomial theorem, if they can not check over their washer-woman's account or write an ordinary letter without murdering the Queen's English, and mis-spelling Saxon words of four letters.[39]

schools. Physicist James Loudon likewise recalled the labour of building apparatus in his early days in the University of Toronto physics laboratory. By 1907, he could rejoice that "the old drudgery" was largely a thing of the past: modern scientific equipment, he claimed, was delicate and expensive, requiring highly specialized skills. See J. Loudon, "The Evolution of the Physical Laboratory," *University of Toronto Monthly* 8 (1907), 44.

36 Ontario Dept of Education, *Manual of Suggestions for Teachers of Science* (Toronto: L.K. Cameron, 1910), 226.

37 The amount of time devoted to science remained officially at the discretion of the headmaster, but in practice inspectors strongly recommended minimum time allotments: Seath advised 45 minutes, 3–4 times a week for Form II chemistry, and 45 minutes, 4 times a week for Forms I and II botany and physics. (Kincardine High School, *Complaint from Science Teacher (AH Smith) That His Subjects are Given a Disproportionately Small Share of Time,* 1889, Department of Education select subject files, RG 2–42–0–4543, Microfilm reel 5656, Archives of Ontario.)

38 Stratford Collegiate Institute, *Improvements in Response to Inspector Seath's Report, Particularly as Regards Laboratory,* 1888, Department of Education select subject files, RG 2–42–0–4609, Microfilm reel 5656, Archives of Ontario.

39 St. Mary's Collegiate Institute. *Report, 1888.* Department of Education select subject files, RG 2–42–0–4649, Microfilm reel 5656, Archives of Ontario.

Overextended school boards and headmasters were not the only ones complaining. Parents wrote too, complaining about the "cramming system" and the harm it caused to students' health – especially, one letter-writer alleged, for girls.[40] Inspector Seath's reply to such complaints was characteristically uncompromising. The wisdom of knowing "a few things well" he dismissed as nonsense, saying that there were many subjects that an educated person could not afford to be ignorant about. "That man is best educated who knows something of a good many subjects and knows at least one subject well," he declared.[41]

These complaints and the inspector's response suggest that the teaching of science had become a flash point for changing ideas about the proper function of a high school education. The board of St. Mary's challenged the value of science not on its academic or pedagogical merit, but on its alleged irrelevance to the more pressing skills of everyday life. But as Seath's comments indicate, the high schools were defined by their broad-based, non-specialized program of studies. Despite calls from some quarters for applied coursework that was better attuned to developing life skills, the Department of Education resolutely adhered to the tenet of mental discipline as its guiding priority.

Mental discipline was the pedagogical application of the doctrine of faculty psychology, which held that the various powers or faculties of the mind required different kinds of mental training, just as the various muscles of the body required specific exercises.[42] In North America as in Britain, this notion of formal discipline was the principal purpose of a secondary school education throughout most of the nineteenth century. Any practical skills or useful knowledge that pupils might glean from their studies were merely a secondary benefit, as the Minister of Education's comments about the grammar course reveal. In an 1886 memo to headmasters, he announced that the grammar exam would be "constructed in accordance with the view that, while the subject is a science which is capable of important practical applications, it has a distinct value as a means of mental training, to which the practical applications are subordinate in a high school course of study."[43]

In view of the overarching mission of the high school curriculum, the Minister's instructions concerning science teaching would have come as no surprise:

> A general literary acquaintance with scientific facts is undoubtedly of practical value . . . but the main reason for the introduction of the study of Science into our schools is the mental discipline to be obtained therefrom. The training of the reasoning powers and the acquisition of the scientific habit of mind are the objects with special reference to which the method of in-

40 *Letter from G.H. Fawcett, Ottawa, Deploring the "Cramming System," 1887*, Department of Education select subject files, RG 2–42–0–4462, Microfilm reel 5655, Archives of Ontario.

41 *Report of the Minister, 1885*, 165.

42 Seath, for example, distinguished between subjects like geography and history, which "involve largely the exercise of the 'portative' memory," and those such as languages, mathematics, and the inductive sciences, which "are intended to promote thought" in *Report of the Minister, 1885*, 166.

43 G. Ross, "Memo to Head Masters" in *Report of the Minister, 1885/1886*, 21.

struction should be chosen, and these also will be the main objects of the examination papers.[44]

Ross seems to have been drawing on the arguments of James M. Wilson, a science master at Rugby School in England, who in 1859 had set up a chemistry teaching laboratory at the school under the guidance of Lyon Playfair. Though Ross does not directly invoke Wilson, the latter's 1867 essay "On the Teaching of the Natural Sciences in Schools" laid out in strikingly similar words the advantages of studying science.[45] Wilson's essay originally appeared in F.W. Farrar's influential *Essays on a Liberal Education*, which challenged the monopoly of the traditional classical curriculum in late nineteenth-century Britain and set the stage for related developments in Ontario schools.

As Donnelly has noted, the doctrine of mental discipline, traditionally the preserve of the classics, was fairly easily co-opted by scientists and science boosters in Britain.[46] Similar tensions surrounded the curriculum in Ontario. Gidney and Millar have shown that by the 1860s, the traditional Georgian distinction between a liberal and an "ordinary" education was breaking down. As science and other new subjects gradually made inroads into the high schools, reformers challenged the exclusive claim of the classics on the development of mental culture. Furthermore, advocates of science argued that subjects like botany, chemistry, and physics were particularly well-suited to the dual mission of conferring mental culture and meeting the "utilitarian demands of a new age."[47]

Arguments founded on mental discipline were most often fuelled by the conviction that science was the best possible mind trainer, because it employed the range of skills required in everyday reasoning. Thomas Huxley, of course, had famously claimed that science was "nothing but trained and organised common sense."[48] In Ontario as elsewhere, this justification was still widely cited in the 1880s. In view of the crammed curriculum and its perceived excess of new subjects, such arguments defending the value of science took on added significance. One science teacher by the name of A.P. Coleman published a "Plea for More Science" in the *Educational Monthly* that systematically compared the educational advantages of natural science with those of other subjects. Mathematics,

44 *Idem.*

45 Cf. J.M. Wilson, "On Teaching Ntural Science in Schools," in *Essays on a Liberal Education*, ed. F.W. Farrar (London: Macmillan & co., 1867), 262, 264. Wilson's essay was reprinted in Henry Barnard's *American Journal of Education* 23 (1872). On Wilson's pioneering role at Rugby as well as his advocacy, alongside T.H. Huxley and John Tyndall, for the teaching of botany in England, see D. Layton, *Science for the People* (London: George Allen & Unwin, 1973), 73–74.

46 J. Donnelly, "The 'Hhumanist' Critique of the Place of Science in the Curriculum in the Nineteenth Century, and its Continuing Legacy," *History of Education* 31.6 (2002), 540–541.

47 Gidney and Millar, 231–253.

48 T.H. Huxley, "On the Educational Value of the Natural History Sciences [1854]," in *Science and Education* (New York: Collier, 1902), 45. Huxley continues: "The man of science… simply uses with scrupulous exactness the methods which we all, habitually and at every moment, use carelessly."

Coleman reasoned, did not reflect the complexities of everyday problem-solving because it built up from a few intuitive premises; the languages, in all their complexity, exercised the judgement more than mathematics, but were taught too superficially to provide real proficiency; history and geography, meanwhile, taught useful facts, but mainly trained the memory, which was "far from being man's most lordly part." The physical sciences, by contrast, excelled at providing mental discipline because they followed "the methods of actual life, only with more exactitude, gathering facts by observation, arranging them, and generalizing and theorizing from them."[49]

It was this careful, inductive process of scientific reasoning that made science so promising in the eyes of many educators. Seath likewise emphasized the kinship between scientific reasoning and ordinary, everyday thinking in his monumental report of 1885. Like the Minister of Education, Seath drew on James Wilson's 1867 essay, which he quoted at length:

> The investigations and reasoning of science, advancing as it does from the study of simple phenomena to the analysis of complicated actions, form a model of precisely the kind of mental work which is the business of every man, from his cradle to his grave. . . . Science teaches what the power and what the weakness of the senses is; what evidence is, and what proof is. There is no characteristic of an educated man so marked as his power of judging of evidence and proof.

For Seath, it was clear that these benefits were contingent on the implementation of proper methods in the classroom. In chemistry and physics, this meant imposing a four-step inductive method: *experiment*, *observation*, *inference*, and when possible, *generalization* to broader principles. Pupils were to carry out experiments themselves, under the teacher's guidance, and to write a careful account of their procedures and findings under each of these headings. In botany, the inductive approach entailed manipulation and examination of specimens with minimal guidance from the teacher, who would provide technical terms only after the observational work was completed.[50]

Although the stated benefits of a practical and experimental approach to teaching science were primarily linked to mental discipline, they were not confined to it. Laboratory work, it was held, developed not only the mind but also the body and the character. "Science," Coleman contended, "trains us to observe with unprejudiced eyes before forming a judgment. The chemist, making an important analysis, if he spills a drop or two from his [beaker], patiently goes back and begins anew, even though he has spent days on his work, and is on the point of finishing."[51] Coleman's image of the patient and meticulous chemist exemplifies another frequently asserted outcome of the study of science: the cultivation of a "scientific" temperament – cautious, methodical and conscientious. Working with delicate apparatus was also said to help curb unruly behaviour. In the opinion of one contributor to the *Educational Monthly*, science combatted the inherent ten-

49 A.P. Coleman, "A Plea for More Science," *Canada Educational Monthly* 2 (1880), 147.
50 *Report of the Minister, 1885*, 161–162.
51 Coleman, "A Plea for More Science," 147.

dency of boys to "muck about." Working with a fine balance, for example, required patience and carefulness; with practice, the instrument commanded the student's respect. The natural sciences, their proponents argued, offered lessons that literature and mathematics could not: they refined the skills of the eyes and the hands.[52] Moreover, while rote learning of definitions might train the memory, proper scientific method instilled sound judgement and wisdom. Just as science obliged restless school boys to settle down and behave in the classroom, so it helped to promote a broader social order: "Scientific men, in general, it will be observed, are not revolutionary in their opinions; they work on patiently, and hate nothing so much as premature production of results," observed one editorialist.[53]

Arguments citing the moral and humanistic rewards of studying science were presented alongside pragmatic appeals to economic development and national prosperity. Donnelly notes that "as natural science gained ground, supported by its growing economic significance, the question of its wider curricular claims receded."[54] It is not surprising that in Canada, a recently established nation, calls for improved science education were often underpinned by concern that the dominion would lag behind the pace of industrial growth abroad, particularly in Germany. "What a list of discoveries and inventions, profitable to us and honourable to us as a nation would result from this attention to science during the next half-century!" effused Coleman, the Cobourg science master. "We are justly proud of our system of education in Ontario, but the world is moving and if we would keep our rank, we must move too."[55] Science teacher (and later textbook author) Archibald P. Knight similarly pointed to the urgency of science teaching for harnessing the resources of a new country. Drawing on a passage from Herbert Spencer's *Essays on Education,* Knight argued that the urgency of national development dictated a clear hierarchy among the school subjects. "Until the forces of nature in this land are conquered to man's use, the study of science in its various branches is an indispensable necessity," he wrote. "History, poetry, music, logic, moral philosophy, classical literature, are excellent as ornament; but as they must, in the present stage of our country's development, occupy the leisure part of life, so they should occupy the leisure part of education."[56] Even Seath, despite his thorough support for the primacy of mental discipline, acknowledged that the economic aspects of science should receive "more emphatic recognition."[57]

52 "The Teaching of Scientific Method," *Canada Educational Monthly* 13 (1891), 263–300.

53 "The Relation of Science to Culture," *Canada Educational Monthly* 7 (1885), 139.

54 Donnelly, 550.

55 Coleman, "A Plea for More Science," 148.

56 A.P. Knight, "The High School Curriculum," *Canada Education Monthly* 6 (1884), 286. (Cf. Spencer, 72–75.) Queen's University chemistry professor W.L. Goodwin also correlated industrial progress with science education, arguing that "the arts and manufactures flourish most vigorously in countries where liberal provision is made for diffusing a knowledge of the principles and applications of science." See W.L. Goodwin, "A School of Science," *Canada Education Monthly* 9 (1887, 85).

57 *Report of the Minister, 1887*, 179.

4. PHYSICS AND CHEMISTRY

How did this wide range of educational ideals translate into pedagogical practice in Ontario's classrooms? This is inevitably difficult to assess, but inspection reports provide clues. Throughout the 1880s, physics came in for special criticism by Seath. Though schools were reasonably well equipped for chemistry experiments, he reported that the situation was dire for physics: no inspected high school had a decent collection of physics apparatus, and in many places an "antiquated air pump" was the only item of note. Physics lagged behind chemistry in terms of facilities and support largely because natural philosophy had been primarily a mathematical, theoretical subject throughout the nineteenth century, in contrast to the more empirical nature of chemistry.[58] As Seath noted, "The instruction in physics has been, so far, chiefly of a mathematical character. The truth is, both physics and chemistry, the former especially, have run to mathematics."[59] This reality extended to teacher training: an ability to teach mathematics had long been assumed sufficient to teach science.[60]

Steering teachers toward an experimental approach meant de-emphasizing the mathematical aspects of the course. As of 1886, high school physics examinations – particularly exams for public school teaching certificates – were revised to reduce the traditional emphasis on quantitative topics like statics.[61] Math teachers felt their territory shrinking. "English and Science are thrust forward, almost offensively—Mathematics pushed into the back ground," protested one disgruntled mathematics master. Physics had been better taught in the past, he complained, "in spite of our present elaborate apparatus, and pretence of experimental and inductive study."[62] The wedge driven between physics and mathematics in the high schools reflected institutional changes in the teaching of physics at the University of Toronto. In 1887, physics gained a distinct institutional identity within the University when the professorship in natural philosophy and mathematics was officially separated into distinct chairs in physics and mathematics.[63]

These curricular changes were reinforced by the introduction of new textbooks in the late 1880s. This undertaking was partly motivated by efforts to modernize the curriculum, but also reflected the Minister's ambition to replace all the province's textbooks, many of which had been imported from the United States, with books by Canadian authors. "Our text books should reflect Canadian sentiment," he stated. "I believe that there is no better, and consequently I desire to see it pervade without obtrusiveness, all the literature placed in the hands of our

58 M. Phillips, "Early History of Physics Laboratories for Students at the College Level," *American Journal of Physics* 49 (1981): 525–526.

59 *Report of the Minister, 1885*, 161.

60 *Report of the Minister, 1887*, 175.

61 "Memorandum on the Examination in Physics," in *Report of the Minister, 1885/1886*, 7.

62 W.J. Robertson, "The Mathematical Condition of Our High Schools," *Canada Educational Monthly* 15 (1893), 168, 164.

63 Gingras, 15.

school children."[64] The Minister was also convinced that teachers should prepare the new textbooks. The "practical teacher," by virtue of having grappled with the difficulties of how to present a subject to students, was best equipped for the task. These textbooks were to be considered coextensive with the curriculum: "In every subject, the text-book prescribed contains the whole course, and as a rule the text-book follows the order in which the subjects should be taken up by the teachers."[65]

The new physics book, however, was an apparent exemption to the Minister's mission of Canadianization. The task of preparing this textbook had been assigned to Napanee science teacher Cortez Fessenden. After surveying existing textbooks, Fessenden opted to adapt Alfred Gage's widely-used American textbook, *Elements of Physics*, rather than start from scratch. In 1887, Gage's textbook was introduced in Ontario schools under the title *The High School Physics.* Gage's preface, retained in the Canadian edition, articulated the pedagogical principles that underpinned his approach. Gage reiterated the widely-held principle that proper learning required direct contact with nature. Noting that many textbooks opened with abstract definitions, Gage asked, "Why should the pupil so frequently, to his great discouragement, be called upon to break through a wall of such difficulties before coming in contact with Nature?" In his own foreword, Fessenden added that he had introduced changes in the Canadian edition, having removed excessive description in an effort to ensure that pupils would perform the experiments themselves rather than "simply taking them for granted as is too often the case."[66]

The 1887 physics curriculum was a far cry from the natural philosophy taught a decade earlier. In 1876, natural philosophy had been grouped with arithmetic, algebra and geometry as a branch of mathematics and had focused on problems like the composition and resolution of forces, moments, centre of gravity, and hydraulic pressure. In 1887, the physics course was applied and thematic. The first-year program included units on heat, electricity, sound and light, as well as brief discussions of topics such as the constitution of matter, physical and chemical changes, force, and the states of matter. The second year program introduced topics in dynamics (velocity, acceleration, momentum, energy, work, etc) and hydrostatics, as well as applications of the latter in pumps, siphons and the barometer.

A textbook of "practical chemistry" was also introduced in 1887, written by Archibald P. Knight, then Principal of Kingston Collegiate Institute. Knight noted in the introduction that he had long believed that a beginners' textbook in chemistry "should consist mainly of directions for performing a series of experiments." These experiments should be designed such that "all the prominent facts

64 "Text-books," *Report of the Minister, 1887*, xliii. Ross noted the Department's progress in this regard: whereas in 1883, only five of 21 subjects had textbooks with Canadian authors, by 1887 all but five did. (None of the five exceptions was a science textbook.)

65 "Text-books," *Report of the Minister, 1887*, xxxviii.

66 Gage and Fessenden, vii–viii.

and principles of the science could be re-discovered, as it were, by any intelligent pupil." He was sharply critical of textbooks that spoon-fed the details of experiments to students, insisting that observations be taken by pupils with no help from the teacher or the textbook. Knight also described how the subject was taught in his own classroom: all the simpler experiments (about two-thirds of them) were performed by students individually, working at separate tables, while the more difficult experiments were performed by him before the class. Knight believed that even beginners should attempt to engage in original research. To that end, he included "a few simple problems which are intended to stimulate in the pupil a desire for original research, as well as to test his power of applying the knowledge he has already acquired to the solution of new problems."[67]

In 1887, Seath presented a brief account of the extent to which the laboratory method had permeated Ontario schools. His observations suggest that instructional methods were slow to adapt to the ideals promoted by the Department. The preferred method, whereby each student performed experiments individually, was taken up in only three schools for chemistry and nowhere for physics. In many schools, a few students performed experiments while the class collectively made observations and drew conclusions. This method, Seath conceded, was likely to remain common in physics. A third approach, whereby the teacher performed experiments and students interpreted them, remained common in many schools. Teachers pled lack of time in defence of this approach, but Seath disapproved of it nonetheless: "Though possessing value, it is defective [in that] the pupil is not brought into direct contact with nature, and, under the circumstances, cannot make satisfactory observations." Finally, Seath noted the persistence of the "lecture method," whereby the teacher both performed the experiment and led the class through its interpretation. He remarked with satisfaction that instances of this were now rare and could be regarded as "anachronistic survivals of an almost extinct species."[68]

5. BOTANY AND ZOOLOGY

It was generally assumed that hands-on analysis of specimens in botany and zoology was the methodological counterpart to laboratory experimentation in physics and chemistry. As with physics and chemistry, direct contact with nature was stressed: students would examine and sketch fresh specimens in spring and fall and study seed and fruit samples and dried leaves in winter. Use of the microscope, if one was available, was encouraged. This direct contact propelled the natural unfolding of the inductive learning process. Seath believed that by examining and drawing a sufficient diversity of specimens, students would uncon-

67 A. P. Knight, *High School Chemistry* (Toronto: Copp Clark, 1887), v–vii. The research questions he mentions might include the following, drawn from the chapter on chlorine: "If the waste pipe of a kitchen sink were foul smelling, devise a method of deodorizing it" (160).

68 *Report of the Minister, 1885*, 174.

sciously absorb the principles of classification. If, for example, they examined the sweet pea in one lesson, and, without any hints from the teacher, were introduced to a clover plant in the next, "there [would] not be one whose face [would] not light up" as they recognized the resemblance, Seath promised.

In some instances, nature and the laboratory were explicitly conflated: nature itself became the student's makeshift laboratory. "The study of the natural sciences furnishes the very thing wanted [to cultivate the spirit of inquiry] – that is, *the study by direct reference to and questioning of nature herself*," wrote A. McGill, the science master at Ottawa Collegiate Institute. "No text-book work here. A laboratory is wanted, to be sure, but every roadside, every ditch, every day and night of the year furnish you a laboratory." McGill's remarks highlight the ambivalence and criticism that sometimes surrounded the use of textbooks in the botany course. Educators frequently suggested that overreliance on the textbook undermined the central tenet of direct contact with nature. It was foolish to assume that botany "or any other science taught from a text-book bristling with technical terms" could interest pupils, McGill argued. "We must bring them face to face with nature herself; make them observers; help them skilfully, but not obtrusively; taking the greatest care to avoid any attempt to *see for* them, or to *think for* them."[69]

The premise that plants held particular interest to children was one of the main reasons that botany was taught in the lower years of high school. "Of all natural subjects there can be no doubt whatsoever that plants are the most suitable for the young observer to begin with. They are naturally attractive to the young, and they can be had everywhere without cost, and in sufficient abundance to enable every pupil of a class to handle and examine a specimen for himself," wrote Henry B. Spotton, headmaster of Barrie Collegiate Institute. Like many others, Spotton advocated a natural progression through the sciences that began with botany and proceeded to natural history, physics, and chemistry in turn. Botany fit neatly into the rationale of mental discipline, being regarded as the ideal purveyor of skills in observation. Observation was regarded by many as the fundamental skill that advanced science courses would subsequently call upon. Furthermore, botany helped students learn to describe things with accuracy – Spotton referred to this as "translation out of nature into one's own speech" – a skill that, like observation, was considered foundational for the meticulous reporting that would be required in chemistry and physics laboratory work.[70] Many secondary benefits were attributed to the study of botany, including the cultivation of an aesthetic

69 A. McGill, "The Natural Sciences in Relation to the Work of Higher Schools," *Canada Educational Monthly* 4 (1882), 202, 204. Emphasis in the original.

70 H.B. Spotton, "The Inductive Method," *Canada Educational Monthly* 2 (1884), 176–177. V.M. Spaulding, professor at the University of Michigan, similarly emphasized the importance of clarity of expression in the study of botany: "It is certainly [the Botany teacher's] business to impress the cardinal truth that whoever has occasion to write a scientific description has no right to leave it in any other than accurate, clear, and concise form." See V.M. Spaulding, "Botany in the High School," *Canada Educational Monthly* 12 (1890), 295.

sensibility, a practical knowledge of principles of scientific agriculture, and the development of moral virtues. All these positive outcomes were contingent on the pupil's interaction with the natural world. Time spent in nature, much like time spent in a garden, "tends to the eradication of certain moral defects and the development of various excellences," observed a teacher from the town of Arthur.[71]

Many changes to the program were necessary if hands-on learning was to become the norm. For Seath, the main culprit was the examinations. Public school teachers' exams in particular needed to be more practical, requiring candidates to examine and compare actual plants. Seath also recommended that the Department provide summer classes for teachers. Two years later, a week-long summer course was offered by Spotton, who also prepared Ontario's high school botany textbook. The summer course included lessons on common plants and discussions about how to introduce the subject to young students. Mornings were spent in lectures illustrated with microscope sections, and afternoons in outdoor fieldwork.[72] By 1887, Seath was praising the improvements he saw in the teaching of botany. He noted with pleasure that the practice of testing students on a few definitions or descriptions, which had been nearly universal in 1885, was disappearing.[73]

Fall of 1887 also saw the introduction of Spotton's newly authorized textbook, *High School Botany*. The new manual came as a long-awaited replacement for Asa Gray's *How Plants Grow*, an American textbook that had been in use in Ontario since 1867. Spotton's manual featured common Canadian plants, many of which had not appeared in imported books. Indeed, within the Ontario Teachers' Association, Gray's textbook was criticized as being unfit for Canadian schools.[74] But by the late 1880s it was falling out of favour even among some American educators: "Whoever had tried to teach the subject by making the learning of Gray's Lessons the main part of the work . . . has found out the dreary dissatisfaction of it and does not need to be told how very dry such botanizing is," wrote a University of Michigan professor.[75] Spotton, who in the preface acknowledges his debt to Joseph Hooker, Asa Gray, Robert Bentley and Daniel Oliver, sought to introduce what he believed was a "more rational" method of studying botany by designing a course that progressed from systematic botany to morphology and finally to vegetable histology. Previously, students had been required to memorize a host of technical terms before handling any plants. Spotton's approach deliberately reversed this order, guiding the student through the examination of various plants, then "lead[ing] him, by his own examination of these, to a knowledge of their various organs – to cultivate, in short, not merely his memory, but also, and

71 A. Stevenson, "A Wider Botany for High Schools," *Canada Educational Monthly* 17 (1895), 205–209. "Next to fine literature botany is the subject on our school programme along with which can best be given some measure of esthetic [*sic*] cultivation," Stevenson claimed (206).

72 "Circulars from the Minister: Botany Class." *Report of the Minister, 1887*, 57.

73 *Report of the Minister 1887*, 174.

74 Albert G. Croal, *The Hstory of the Teaching of Science in Ontario, 1800–1900* (D.Paed, University of Toronto, 1940), 165.

75 Spaulding, "Botany in the High School," 297.

chiefly, his powers of observation."[76] Spotton emphasized that the book's many woodcuts were taken from living specimens – implying, it would seem, that they were just one remove from nature itself – yet cautioned that they should not replace the study of living plants: "It is strongly urged upon teachers and students not to be satisfied with them as long as the plants themselves are available."[77]

Zoology, considered the natural adjunct of botany, was introduced into the course of studies in 1887.[78] Robert Ramsay Wright, the University of Toronto's first professor of biology, prepared the province's authorized zoology textbook, which was introduced alongside Spotton's botany textbook in 1887.[79] Zoology was to be taught in Form II, once students had already received a year of botany. Wright conceded that zoology was less well-suited than botany to teaching basic observation skills because animal forms were more difficult to draw and specimens more difficult to collect, but argued that it was especially successful at cultivating a love of nature, by "awakening an interest in the habits of animals."[80] As he noted in the textbook's preface, zoology offered students an "equally valuable discipline [to training in observation] – the tracing of the modifications of form throughout less nearly allied groups."[81] Accordingly, zoology could introduce students to the broader principles of biology, which Wright considered to be a primary objective of the course.[82] A zoology summer course for teachers was held in 1889, following on the success of Spotton's summer course in botany. Perhaps because it did not correspond as well as botany did to the ideal of practical, hands-on science teaching, but more likely because it was never placed on the Departmental exams, zoology was rarely taught in nineteenth-century high schools.[83]

76 H.B. Spotton, *Elements of Structural Botany: With Special Reference to Canadian Plants* [aka *High School Botany*] (Toronto: Gage & Co, 1888, c1879), 1.

77 Spotton, "Preface to the First Edition," in *Elements of Structural Botany,* [iv].

78 Seath indicates that its inclusion was controversial but does not indicate why this was so: it may have been simply because the program of studies was already thought to be overfull.

79 R.R. Wright, *An Introduction to Zoology: For the Use of High Schools* [aka *High School Zoology*] (Toronto: Copp Clark, 1889). Like many of Ontario's textbooks of this period, it was used widely in other provinces. Wright introduces an explicitly Darwinist understanding of evolution, making reference to both Darwin and A.R. Wallace.

80 R.R. Wright, "Introduction of Zoology into High Schools," *Canada Educational Monthly* 10 (1888), 254–255.

81 Wright, "Preface," in *An Introduction to Zoology*, v–vi.

82 Wright, "Introduction of Zoology," 254.

83 Croal, 177, 180–181.

6. CRITIQUES AND DISAPPOINTED HOPES

In 1889, Inspector Seath took stock of the situation in science teaching and reported significant progress:

> The most gratifying increase has taken place in the value of the scientific apparatus. Five years ago, what is now known as science was taught in only two or three of the high schools in the Province. Then, botany was a matter of "getting up" definitions and memorizing characteristics, physics was purely mathematical, and chemistry was taught practically in but few of the schools. Now botany is taught practically in every High School in the Province; thanks to the amended examination requirements and to the fact that it is now to the financial interests of boards to equip their schools with scientific apparatus, physics and chemistry – the latter more particularly – are taught in most cases as elementary science should be taught; and zoology, the necessary biological complement of botany . . . has been successfully introduced into most of our leading schools.[84]

Despite the increased availability of apparatus and Seath's positive appraisal, however, the optimism that marked the reform efforts of the 1880s was giving way to a measure of skepticism and disillusionment. In 1890, William Lawton Goodwin, professor of chemistry at Queen's University, signalled disappointed hopes about the effectiveness of high school science: "It seems to be a pretty general opinion that, as a school subject, science has not fulfilled the expectations of her friends," he reflected. "The study of science in the schools has too often developed neither accuracy of observation nor clearness of thought." Goodwin alleged that the curriculum still relied far too much on abstruse definitions. It was misleading to call the physics and chemistry course experimental – after all, he noted, the experimental treatment of abstract subjects like the constitution of matter, attraction, sound waves, refraction of sound, and electric polarization "must tax rather severely, not only the ingenuity of the master, but the capacity of the pupils."[85] His reproaches compromise Seath's tidy picture of pedagogical progress in the late 1880s, and cast doubt on the assumption that "practical" science teaching had largely swept aside the rote learning of the past. The province's science masters did not mutely convert to the new pedagogical methods, as evidenced by a letter to Seath from one high school principal who reported that "only the other day my science teacher told me he thinks little of *experimental* work" and worried that the teacher's recalcitrance would ruin the whole science department.[86]

Outside the official spheres that prescribed the new teaching methods, some educators had expressed doubt about the efficacy of experiment-based pedagogy from the outset. In a markedly critical 1887 *Educational Monthly* essay, science teacher D.F.H. Wilkins of Beamsville High School took aim at the teaching of

84 *Report of the Minister, 1888/89*, 187–188.

85 W. L. Goodwin, "The high School Curriculum in Science," *Canada Educational Monthly* 13 (1891), 88.

86 Listowel High School, *Report Re: Improvements, 1890*, Department of Education select subject files, RG 2–42–0–4427, Microfilm reel 5655, Archives of Ontario. Emphasis in the original.

chemistry in particular. Chemistry's equipment was expensive, its experiments time-consuming and prone to failure, and its preparation and clean-up requirements a tax on the overburdened science teacher, Wilkins alleged. But his chief criticism struck at the core of the pedagogical method advocated by the Department of Education. Do "brilliant and showy" experiments really lead the student to the underlying law? he asked. Wilkins argued that reliance on experiment frequently led to one of two evils. Most often, teachers stopped at the experiment and went no further. The experiment was assumed to speak for itself, and students never drew a connection to the law it was supposed to illustrate. And then there was the opposite danger – that students would jump to sweeping conclusions based on the scanty evidence of one or two experiments. Wilkins concluded (speaking specifically about chemistry):

> It is entirely out of the question to expect the pupil to perform an elaborately detailed set of costly experiments, in order to verify a "law of Nature"; and yet, in order that the student may acquire the correct 'scientific method' of reasoning, nothing else ought to be done. . . . To expect aught else than crude guesses, hasty generalizations, imperfect abstractions, confusion of analogy with induction, etc., is to put altogether too sweetly serene a faith in the embryonic student nature, besides losing the value of chemistry as a 'mind-trainer.'[87]

One prominent university professor, Nathan Fellowes Dupuis, likewise emerged as at outspoken challenger to the gospel of learning by experiment. Despite his appointment as Dean of the new Faculty of Applied Science at Queen's University in 1894, Dupuis had little interest in the German research model. Much like Wilkins, Dupuis dismissed the notion that the inductive method could properly be applied by means of laboratory experiments. In his view, scientific theories were far too complex to be grasped from the standpoint of a small number of experiments:

> People who know little or nothing of science will cry out, Experiment, experiment; nothing is to be learned but by experiment. . . .
>
> A science needs theory – a long experience in teaching a science taught me that experiment alone cannot impart a knowledge of science, and that experiment at random unites utility to amusement to about the same extent as a game at base-ball. . . . How absurd then to put a lot of apparatus into the hands of boys, and to ask them to work out the explanation of phenomena. To do so requires the matured mind of men, and even ten of these fail for every one that succeeds. The theories must be given to the boy dogmatically, and then rendered probable by experiment.

87 Wilkins, "Observations," 88–90. Physics, on the other hand, Wilkins exempted from these concerns: "[T]he leading facts of physics may be presented to the mind as both to communicate knowledge and to develop the entire mind-nature, the character. Here by an appeal to common, every-day phenomena, there by simple and cheap experiments, here by Socratic questioning, there by direct dogmatic teaching, all other means being out of our power, or by a combination of these, we have in physics a most valuable educational agent" (88).

> In all that I have said I do not mean that experimenting will do a lad any harm. . . . But I do mean to assert that scientific experiment is not as great an educator of the young mind as many people suppose it to be.[88]

Lone voices of dissent like Dupuis's and Wilkins's were ultimately drowned out by the pervasive enthusiasm for the inductivist approach as it was legitimated not only in the high school science curriculum and textbooks, but also in wider efforts to apply the scientific method to other branches of knowledge.[89] For critics like Goodwin, the problem was not the experimental method itself but that it had not sufficiently permeated classroom teaching.

As the 1890s progressed, concern shifted to yet another question: could science truly engage students' interest? The somewhat naïve confidence in the intrinsic appeal and fruitfulness of the study of science that had coloured the advocacy of the early 1880s was beginning to fade. Coleman, the author of the "Plea for more science," had displayed typical optimism when he had emphasized the "keen pleasure all boys feel in practical work in science." Coleman had expressed confidence that introducing practical science into schools would "to many, make school life as attractive as it is now repulsive."[90] Eleven years later, by contrast, another contributor to the *Educational Monthly* starkly concluded that "The present age has outlived the sanguine hopes once expressed for the regenerating influence of scientific study in the education of boyhood. He must be superior to reason or experience who still believes that natural science will transform learning from a pain to a pleasure in boys' eyes. . . . Natural science possesses no special charm for boys."[91]

7. EDUCATIONAL INTROSPECTION AND THE DECLINE OF MENTAL DISCIPLINE

Although student interest had been an important ingredient in the rationales for science teaching that were advanced in the 1880s, it took on renewed importance in the 1890s. As in other Canadian provinces and the United States, high schools in Ontario faced drastic enrolment increases in the late nineteenth century, thanks in part to population growth and immigration. Between 1883 and 1904, enrolments more than doubled from approximately 11,500 students to well over 27,000. This trend was even more pronounced in urban areas: Toronto's high schools, for instance, saw an eightfold increase in enrolments between 1870 and

88 N.F. Dupuis, "Annual Convocation of Queen's University," *Canada Educational Monthly* 8 (1886), 383–384.

89 For an overview of the adoption of the inductivist approach in American high schools as well as Karl Pearson's advocacy for broader application of the scientific method, see Rudolph, 352–354.

90 Coleman, "A Plea for More Science," 148.

91 J.E. Welldon, "The Educational System in Public Schools," *Canada Educational Monthly* 13 (1891), 224–225.

1900.[92] Even so, Ontario's high schools remained highly selective and academic: in 1900, fewer than ten percent of fifteen- to nineteen-year-olds attended high school.[93]

As the high schools expanded, their curriculum faced increased scrutiny. In particular, educators debated whether high school work should be made more relevant the needs of the expanding student body. Such re-examination of the curriculum was hardly unique to Ontario. Foreign commissions like the inquiry of the British Association for the Advancement of Science into the teaching of chemistry (1889) and the report of the Committee of Ten in the United States (1893), which was circulated to every school principal in Ontario, helped stimulate educational introspection in Ontario. In 1895, Minister Ross addressed these pressures in a carefully worded statement seemingly designed to appease both the traditionalist and the reform-minded camps:

> When the High School system of the province was first inaugurated, its primary object was to prepare pupils for the learned professions and especially for the University. While in that respect our High Schools amply fulfil their original purpose, in later years the course of education which they provide has been considered a desirable qualification for various other pursuits in life. Many young men in preparing for mercantile life or for agriculture take advantage of the High School, perhaps not so much because of the direct training which it gives for their intended calling as for the superior culture which it provides.

As the Minister noted, more than twice as many pupils were leaving high school for careers in business or in agriculture than for university or the professions.[94] Nevertheless, as his statement indicates, the mission of the high schools remained resolutely tied to broader cultural ideals rather than to the provision of occupational skills.

Proponents of school science, however, had long negotiated the balance between the cultural and utilitarian advantages of studying science, playing up one or the other as needed. Within the pronouncements of Department officialdom, as we have seen, any practical benefits of studying science had been firmly subordinated to its value as mental training. But science teachers themselves did not always tow the Department line. In 1890, they founded the Science Teachers' Association of Ontario (STAO), which would convene annually in Toronto at the meeting of the Ontario Educational Association. The STAO provided a forum for teachers to reflect on the shape and purpose of the curriculum and to lobby the Minister and the universities for needed reforms. Teachers frequently emphasized on the need to make science courses more practical. The botany course, for instance, needed to devote more time to agriculture and horticulture, proposed one science teacher.[95] Moreover, it needed to move away from classification and to-

92 Stamp, *Schools of Ontario*, 40.

93 P. Axelrod, *The Promise of Schooling: Education in Canada, 1800–1914* (Toronto: U of Toronto Press, 1997), 62.

94 "Diffusion of High School Education," in *Report of the Minister, 1895*, xx.

95 A. Stevenson, "A Wider Botany," 205.

ward ecology, physiology, and "the economic importance of our ordinary plants," claimed another.[96]

Meanwhile, faculty psychology – and accordingly the tenet of mental discipline that had given science a foothold in the traditional curriculum – was under fire. The first major blow to faculty psychology was its failure to hold up to empirical tests by American psychologists William James and Edward Thorndike. But as Herbert Kliebard has argued, faculty psychology gave way primarily because of the many social changes of the 1890s that made the traditional high school curriculum seem obsolete.[97] Like their American counterparts, Ontario educators turned to the ideas of the new breed of educational psychologists like G. Stanley Hall and John Dewey, who were seen as having put educational theory on a scientific basis while supplanting the speculative notions of faculty psychology. In 1889, Director of Normal Schools James McClellan published a textbook in applied psychology that was based heavily on Dewey's first book, *Psychology* (1887), and that introduced teachers-in-training to the concepts of scientific psychology.[98] In 1894, Hall visited Toronto to deliver a series of lectures on child study, an event that prompted the Ontario Education Association to form a child study section.[99] By 1897, Elora high school science teacher Norman MacMurchy could assert that "the old idea that the mind is made up of separate parts is being discarded, the modern view being that the mind is a unit and should be developed as such." The consequences for the teaching of science, MacMurchy realized, were profound. Educators and policymakers could no longer assume that studying subjects like chemistry or physics trained the mind in a general sense for the broader pursuits of everyday life, as Huxley and others had claimed. Science conferred *scientific* knowledge and skills, and nothing more. MacMurchy reflected:

> If this view of mental science [i.e., the mind as an indivisible unit] is correct, the old doctrine that the work of the mind in any direction develops power that may be used equally well in all directions, is wrong. To put the case broadly, no person will maintain that the study of physics will prepare a person for the practice of law as well as if he had read jurisprudence. . . . [S]tudy in any particular line will limit our faculties to development in that direction. If this is so, surely no subject should be studied merely for the discipline alone it may be supposed to give. . . . The old idea of formal discipline by certain subjects is losing ground, and those subjects which will have a direct value in giving the pupil knowledge that will be of service to him in after life will in the future receive more prominence.[100]

96 J.B. Turner, "Science," *Canada Educational Monthly* 23 (1901), 196–197.

97 H.M. Kliebard, *The Struggle for the American Curriculum, 1893–1958*, 3rd ed. (New York: RoutledgeFalmer, 2004), 6.

98 Although *Applied Psychology* was widely published with Dewey listed as co-author, a wide range of evidence suggests that the McLellan was in fact the sole author. McLellan and Dewey did however later collaborate on a textbook on the teaching of arithmetic, *The Psychology of Number* (1895). See Jo Ann Boydston, "A Note on *Applied Psychology*," in *The Early WWorks of John Dewey: 1882–1898,* ed. Jo Ann Boydston (Carbondale: Southern Illinois UP, 1969), xiii-xix.

99 Stamp, *Schools of Ontario,* 52.

100 N. MacMurchy, "Nature Study in Public Schools," *Canada Educational Monthly* 19 (1897), 168.

In other words, the curriculum more than ever needed to be *practical* – not in the hands-on, empirical sense intended by the Department regulations ten years prior, but in its straightforward relevance to students' interests and ambitions. The value of any subject on the program was necessarily tied to its content, and so this content needed to relate to pupils' everyday lives and future pursuits.

Inspired largely by the ideas by American educationists, Ontario's "New Education movement," as historian Robert Stamp has called it, had its greatest impact within the elementary schools. Another major contributor to the reformist momentum was Ontario public school inspector James Hughes. Drawing on the ideas of Johann Pestalozzi and Friedrich Froebel, Hughes had campaigned successfully throughout the 1880s for the institution of kindergartens across Ontario. For Hughes, kindergarten was not just one year of school, but a pedagogical method that extended to the whole elementary school. At its centre of Hughes's philosophy was Froebel's principle of the self-activity of the child.[101] In MacMurchy's view, this principle applied equally to high school laboratory science:

> If we consider the kindergarten we find that the child is there active and not passive; his activity is a self-activity. . . . The senses are being employed with a definite end in view for him and thus they are being cultivated. The power gained for the child is a power to use power. This is the reason, or at least one of the main reasons, why we in our high schools have our pupils perform their own experiments in chemistry and physics and do not perform them ourselves. That can only become a part of the child's knowledge which he has obtained by a free action of his perceptive faculties and thus made his own."[102]

Ultimately, however, New Education made few inroads beyond the elementary schools, with some conservative critics dismissing its reforms as "Yankee frills."[103] Despite MacMurchy's pronouncement about the demise of faculty psychology, mental discipline kept a firm hold on educational thought in Ontario. Twenty years later, a professor in the faculty of education at Toronto would report with exasperation that "elsewhere, the theory [of formal discipline] is discredited, but Canada clings to it with mid-Victorian tenacity."[104]

8. CONCLUSION

A recurring theme in the history of nineteenth-century Canadian education is the ongoing tension between the bureaucratic authority of the Province and local trustees, parents, and teachers over management of the curriculum. The standardization and reform of science teaching during the 1880s represents an era of tightening central control over Ontario's high schools, largely by virtue of the influ-

101 Stamp, "James Hughes," in *Profiles,* 199–200.

102 MacMurchy, "Nature Study," 168.

103 Stamp, *Schools of Ontario*, 52, 71.

104 P. Sandiford, "Education in Canada," *Comparative Education*, ed. P. Sandiford (London: J.M. Dent, 1918), 367.

ence of the inspectorate, and particularly John Seath. In large part due to Seath's outspoken support for the ideal of "practical" and experimental science teaching, the Department of Education enforced teaching methods, oversaw the purchasing of equipment and the construction of laboratory spaces, and regulated the adoption of new Canadian-authored science textbooks. Although Seath's sometimes severe reports and abrasive manner caused an initial outpouring of complaints, the Department could ensure that science was being taught to its specifications by withholding grants and shaming non-compliant schools with low inspection grades. Its efforts to promote experimental science teaching were no doubt eased by the advocacy work of science teachers in the pages of the Canada Educational Monthly.

The pedagogical changes that were instituted in Ontario schools were remarkably closely aligned with the cultural shift that was underway within university science departments. The changing norms of professional science as practised in the universities entailed a wholesale revision of the perceived role of science in a general education. Seath's remarks in 1889 on the schools' successful implementation of "what is *now known as science*" – in contrast to the science of five years prior – illustrate how quickly laboratory practice redefined the popular understanding of what it meant to learn science. Nonetheless, it is clear that the research model inherited from the universities did not make a seamless transition into the high schools. The firsthand experience of teachers such as Wilkins shows the challenges encountered in converting a simplified experimental method into fruitful pedagogical practice. Wilkins and Dupuis challenged core assumptions of experimental pedagogy – the notion, for instance, that it inculcated rigorous reasoning skills (Wilkins charged that it sanctioned rash conclusions) or that it fostered a "scientific habit of mind" (Dupuis claimed it flouted the true processes of theory formation). But they were aiming at a receding target: the priority of mental discipline was gradually giving way to the perceived need to equip growing numbers of non-university-bound students with useful knowledge and skills. And while the principles of formal discipline were slow to be fully relinquished within the high schools of Ontario, educators at the turn of the century increasingly drew on the charged rhetoric employed by educationists from the United States, where demographic pressures on the school system were even more pronounced.[105] Amid this educational ferment, the hands-on, empirical methods of laboratory science, practical in more ways than one, were there to stay. The research laboratory had become an enduring symbol of scientific progress and expertise, a model that school science would continue to emulate. Harmonized as easily with the learning theories of psychologists and reformers in the early twentieth century as they had been with the intellectual and humanist goals of the traditional curriculum in the 1880s, the methods of laboratory science proved to be amazingly adaptable to the changing mission of the high school.

105 See Rudolph, 354–362.

REFERENCES

Althouse, John G. *The Ontario Teacher.* Toronto: Ontario Teachers' Federation, 1967.

Anonymous. "The Relation of Science to Culture." *Canada Educational Monthly* 7 (1885): 137–140.

Anonymous. "The Teaching of Scientific Method." *Canada Educational Monthly* 13 (1891): 260–265, 300–304.

Archives of Ontario. Series RG 2–42. Department of Education select subject files.

Axelrod, Paul. *The Promise of Schooling: Education in Canada, 1800–1914.* Toronto: U of Toronto P, 1997.

Boydston, Jo Ann. "A Note on *Applied Psychology.*" In *The Early Works of John Dewey: 1882–1898*, edited by Jo Ann Boydston, xiii–xix. Carbondale: Southern Illinois UP, 1969.

Coleman, A.P. "A Plea for More Science." *Canada Educational Monthly* 2 (1880): 146–148.

Croal, A. G. "The History of the Teaching of Science in Ontario, 1800–1900." D.Paed diss., University of Toronto, 1940.

Donnelly, James F. "The 'Humanist' Critique of the Place of Science in the Curriculum in the Nineteenth Century, and its Continuing Legacy." *History of Education* 31.6 (2002): 535–555.

Dupuis, N.F. "Annual Convocation of Queen's University." *Canada Educational Monthly* 8 (1886): 377–385.

Gage, A.P., and C. Fessenden. *The High School Physics*. Toronto: W.J. Gage, 1887.

Gidney, R.D., and W.P.J. Millar. *Inventing Secondary Education: The Rise of the High School in Nineteenth-Century Ontario.* Montreal: McGill-Queen's, 1990.

Gingras, Y. *Physics and the Rise of Scientific Research in Canada.* Montreal: McGill-Queen's, 1991.

Goodson, Ivor. "Subjects for Study: Aspects of a Social History of Curriculum." *Journal of Curriculum Studies* 15.4 (1983): 391–408.

Goodwin, W.L. "A School of Science." *Canada Educational Monthly* 9 (1887): 85–88.

Goodwin, W.L. "The High School Curriculum in Science." *Canada Educational Monthly* 13 (1891), 88–93.

Huxley, T.H. "On the Educational Value of the Natural History Sciences [1854]." In *Science and Education,* 40–63. New York: P.F. Collier & Son, 1902.

Kliebard, Herbert M. *The Struggle for the American Curriculum, 1893–1958*. 3rd ed. New York: RoutledgeFalmer, 2004.

Knight, A.P. *High school Chemistry.* Toronto: Copp Clark, 1887.

Knight, A.P. "The High School Curriculum." *Canada Educational Monthly* 6 (1884): 285–286.

Layton, David. *Science for the People.* London: George Allen & Unwin, 1973.

Loudon, J. "The Evolution of the Physical Laboratory." *University of Toronto Monthly* 8 (1907): 42–47.

Loudon, J. "The Universities in Relation to Research." *Science* 15.391 (1902): 1001–1009.

MacMurchy, N. "Nature Study in Public Schools." *Canada Educational Monthly* 19 (1897): 167–170.

McGill, A. "The Natural Sciences in Relation to the Work of the Higher Schools." *Canada Educational Monthly* 4 (1882): 197–204.

Moyer, Albert E. "Edwin Hall and the Emergence of the Laboratory in Teaching Physics." *Physics Teacher* 14 (1976), 96–103.

Olesko, Kathryn M. "German Models, American Ways: The 'New Movement' Among American Physics Teachers, 1905–1909." In *German Influences on Education in the United States to 1917,* edited by Henry Geitz, Jürgen Heideking, and Jurgen Herbst, 129–153. Washington: German Historical Institute, 1995.

Ontario. "Act Respecting High Schools and Collegiate Institutes." In *Revised Statutes of Ontario, 1887*, vol. 2, chap 226. Toronto, 1887.

Ontario Department of Education. *Manual of Suggestions for Teachers of Science.* Toronto: L.K. Cameron, 1910.
Ontario Department of Education. *Report of the Minister of Education, 1885.* Toronto, 1886.
Ontario Department of Education. *Report of the Minister of Education, 1885/1886.* Toronto, 1887.
Ontario Department of Education. *Report of the Minister of Education, 1887.* Toronto, 1888.
Ontario Department of Education. *Report of the Minister of Education, 1888/1889.* Toronto, 1890.
Phillips, Melba. "Early History of Physics Laboratories for Students at the College Level." *American Journal of Physics* 49 (1981), 522–527.
Robertson, W.J. "The Mathematical Condition of Our High Schools." *Canada Educational Monthly* 15 (1893): 163–171.
Rudolph, John L. "Epistemology for the Masses: The Origins of 'The Scientific Method' in American Schools." *History of Education Quarterly* 45.3 (2005): 341–376.
Sandiford, Peter. "Education in Canada." In *Comparative Education: Studies of the Educational Systems of Six Modern Nations*, edited by Peter Sandiford, 343–437. London: J.M. Dent, 1918.
Sheppard, Keith, and Gail Horowitz. "From Justus von Liebig to Charles W. Eliot: The Establishment of Laboratory Work in U.S. High Schools and Colleges." *Journal of Chemistry Education* 83.4 (2006): 566–570.
Smith, Julian A. "Charles Potter, Optician and Instrument Maker." *Journal of the Royal Astronomical Society of Canada* 87.1 (1993): 14–33.
Spaulding, V.M. "Botany in the High School." *Canada Educational Monthly* 12 (1890): 294–298.
Spencer, Herbert. *On Education: Intellectual, Moral, Physical.* New York: Appleton, 1866. First published in 1861 by Williams and Norgate.
Spotton, H.B. *High School Botany: Elements of Structural Botany with Special Reference to Canadian Plants*. Rev. ed. Toronto: Gage, 1888.
Spotton, H.B. "The Inductive Method." *Canada Educational Monthly* 2 (1884): 175–177.
Stamp, Robert M. "John Seath." In *Profiles of Canadian Educators*, edited by R.S. Patterson, J.W. Chalmers, and J.W. Friesen, 235–247. Toronto: D.C. Heath, 1974.
Stamp, Robert M. *The Schools of Ontario, 1876–1976.* Toronto: U of Toronto P, 1982.
Stevenson, A. "A Wider Botany for High Schools." *Canada Educational Monthly* 17 (1895): 205–209.
Tomkins, George S. *A Common Countenance: Stability and Change in the Canadian Curriculum.* Vancouver: Pacific, 2008.
Turner, J.B. "Science." *Canada Educational Monthly* 23 (1901): 115–117, 196–197.
Welldon, J.E. "The Educational System in Public Schools." *Canada Educational Monthly* 13 (1891): 224–225.
Wilkins, D.F.H. "Observations Regarding the Teaching of Science in Our High Schools." *Canada Educational Monthly* 9 (1887): 88–90.
Wilson, J.M. "On Teaching Natural Science in Schools." In *Essays on a Liberal Education*, edited by F.W. Farrar, 241–291. London: Macmillan, 1867.
Wright, R. Ramsay. *High School Zoology: An Introduction to Zoology for the Use of High Schools*. Toronto: Copp, Clark, 1889.
Wright, R. Ramsay. "Introduction of Zoology into High Schools." *Canada Educational Monthly* 10 (1888): 254–256.

CHANGING IMAGES OF THE INCLINED PLANE, 1880–1920: A CASE STUDY OF A REVOLUTION IN AMERICAN SCIENCE EDUCATION

Steven Turner

> "It was in the period between 1880 and 1920 that the American high school assumed its familiar shape and characteristics." "The high school, like all other institutions, was caught in a vast complex of change, or as the terminology of the times would have expressed it, of reform. It was an age of criticism directed against established orders. The established order of the high school was identified as the academic program."[1]
>
> Edward A. Krug, *The Shaping of the American High School 1880–1920*

INTRODUCTION

The fundamental transformation of American secondary education that Krug refers to is well established in the narrative of education history. So too is the fact that this transformation contained within it a profound revolution in the way that science – and especially physics – was taught. Prior to 1880 high school physics was taught exclusively by the "text" or "lecture/demonstration" method, where the findings of science were presented in classroom lectures to essentially passive students. But by 1920 this method had been completely replaced by a more active system that tried to involve students in the learning process, in part through the use of classroom problems, but particularly through the use of experiments that the students performed themselves. This new "laboratory method" of teaching also included changes in content. For example, the old curriculum had devoted considerable attention to the topic of static electricity, but the new curriculum reduced this topic dramatically and instead devoted much more effort to teaching about batteries and current electricity. In part, this reflected the growing importance of electricity as a technology, but there was also significant emphasis in the new curriculum on teaching students how to reason "scientifically."[2]

1 Edward A. Krug, *The Shaping of the American High School 1880–1920*, Madison: The University of Wisconsin Press, 1969, xi.

2 A note on terminology: During the period of this article, *Natural Philosophy* came to be associated with the old curriculum and new textbooks almost invariably used *Physics* in their titles after about 1890. However accounting for this difference contributes little to the story, so where it occurs in this article, the term *Natural Philosophy* should be considered equivalent to *Physics*. Also, use in this article of phrases like "secondary education" and

The introduction of student experiments into the secondary physics classroom affected nearly every aspect of the way that science was taught. Teachers needed to be trained in the new method and persuaded to use it. New textbooks had to be written, published, marketed and purchased, as did student notebooks and laboratory manuals. Experiments required materials and apparatus, which in turn needed laboratory and storage space. And school boards needed to raise money to pay for it all. It was a large and expensive undertaking, yet our understanding of how this revolution occurred has been almost exclusively based on the speeches, writings and committee reports of a relatively small group of actors – the period's professional educators.[3]

While these sources tell an important part of the overall story, the failure to also consider the actions of authors, publishers, school apparatus makers, school boards or even teachers is methodologically problematic. This article attempts to broaden the existing narrative by integrating the rich and largely ignored material culture of science education – such things as textbooks, lab manuals, student notebooks, science teaching instruments and scientific instrument catalogs. Surprisingly, much of this story can be seen in the changes to the design and depiction of a single, venerable and otherwise unremarkable teaching instrument: the inclined plane.

THE INCLINED PLANE

Literally no more than a flat surface with one end elevated, the inclined plane is physically a simple concept. Yet the utility of applications like the ramp and the wedge have made it an object of study since antiquity. Thomas Young traced interest in it as far back as the 3rd century BC writings of Philo of Byzantium, where it was one of the "simple machines" that formed the basis of the earliest studies of Mechanics.[4]

The inclined plane also appears at the very beginnings of science itself. Lacking the precise clocks necessary to study falling bodies, Galileo used an inclined plane – a flat board, set at an angle – to "dilute" gravity and slow its effects enough to allow him to study the acceleration of the marbles he rolled down it. And not long after Galileo, a different version of the inclined plane was used to

"high school" should be understood to represent not only public high schools but private academies and preparatory schools. The distinctions between these institutions were not well defined and varied over time, but were not consequential to the topic of this article. For more on these institutions see: Krug, *The Shaping of the American High School*, 3–4.

3 Schiro calls this group the *scholar academics*; see Michael Schiro, *Curriculum Theory: conflicting visions and enduring concerns*, Sage Publications: Thousand Oaks CA, 2007.

4 Thomas Young, *A Course of Lectures on Natural Philosophy and the Mechanical Arts*, Vol. I, Taylor and Walton, London, 1845, 185–6.

explore some of the questions that eventually led to the law of the conservation of energy – although a modern understanding of this concept was still far away.[5]

But in the physics classroom the main interest in the inclined plane has always been practical: to show how the principles of physics explain the common experience of *mechanical advantage*. The phrase refers to the common experience that it is much easier to push a heavy object up a ramp than to lift it directly to the same height. Accordingly, the most commonly used textbook illustration of the inclined plane has traditionally been some variation of a barrel being rolled up a ramp (Fig.1).

133. What is an Inclined Plane? — ***The inclined plane is a surface sloping so as to make an oblique angle with the direction of the force to be overcome.***

FIG. 35.

In most cases, it is used to aid in lifting bodies

Fig. 1: A typical illustration of the "mechanical advantage" of the inclined plane. From Avery, First Principles of Natural Philosophy, 1884, 86.

This has the advantage of being both a good example of the explanatory power of physics (for use in the "Mechanics" section of the textbook) and a way to introduce the concepts of *forces*, *work* and *energy*, which were often discussed later in the text. For a variety of reasons, demonstrations of the inclined plane work well for teaching mechanical advantage, but it was rarely used to actually demonstrate forces. This concept, as it applied to the inclined plane, was generally taught with drawings and was analyzed geometrically (Fig.2). To the extent that it was demonstrated at all, the concept of forces was generally presented on some variation of the "forces table."[6]

5 Erwin N. Hiebert, *Historical Roots of the Principle of Conservation of Energy*, Madison, The State Historical Society of Wisconsin, 1962.

6 Elroy M. Avery, *First Principles of Natural Philosophy*, New York: American Book Company, 1884, 86; Elroy M. Avery, *Elements of Natural Philosophy, a text-book for high schools and academies*, New York: Sheldon and Company, 1885, 106.

***LM* represent a plane inclined**

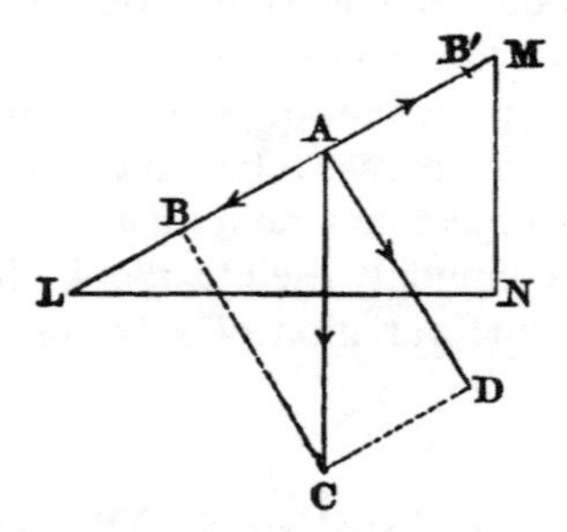

Fig. 2: Typical drawing of the "forces" on the inclined plane. From Avery, Elements of Natural Philosophy, 1885, 106.

This emphasis on demonstrating the "mechanical advantage" of the inclined plane can be seen in early European science teaching instruments, and by 1800 the essential form of the instrument seems to have been set (Fig.3). It consists of a flat surface (the inclined plane), hinged on one end, with a graduated arc to measure its angle. A moveable weight cart on the plane, connected by a pulley to another weight, provided a way to measure the work required to move the cart as the position of the inclined plane was changed, and it could be shown that as the angle of the plane became shallower, less force was required to move the weight up it – although the force had to be applied over a longer distance to get it to the same height.[7]

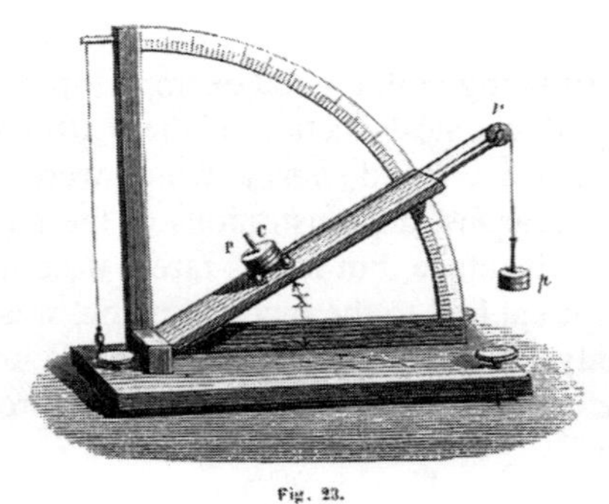

Fig. 3: A typical style of inclined plane apparatus used in 18th and 19th century European science instruction. From Salleron, Notice sur les Instruments de Precision, 1864, 16.

7 J. Salleron, *Notice sur les Instruments de Precision*, Paris, 1864, 16.

THE INCLINED PLANE IN AMERICA: LECTURE/DEMONSTRATION

For most of the 19th century, American colleges tended to use science teaching instruments that had been made in Europe, and this included the inclined plane.[8] But by the 1840s American instrument makers were offering simplified and much more economical inclined planes for use in schools and academies (Fig.4).

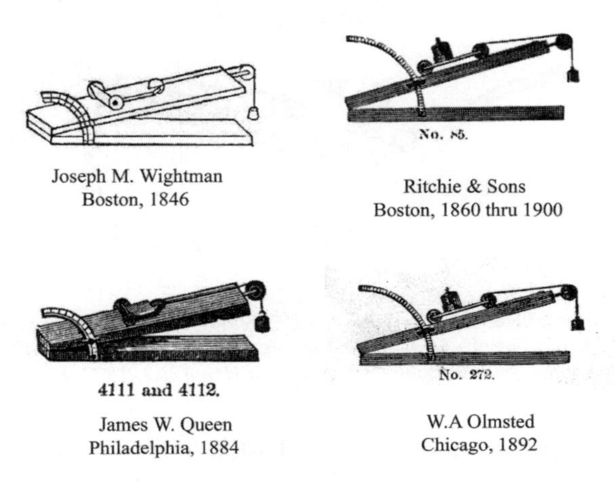

Fig. 4: Standardized images of the inclined plane found in the catalogs of 19th century American scientific instrument makers. These images were not used by companies making the new student apparatus.

There were several designs, but the most popular was made from two boards, hinged together at one end, with a graduated metal arc and a binding screw to hold them in the desired position. A pulley, weight pan and small three-wheeled cart completed the apparatus. While materials and details of the construction seem to have varied, the silhouette image of it quickly became standardized and was widely used in both catalogs and introductory physics textbooks for nearly fifty years[9] (Fig.5).

Although few of these instruments survive, inspection of the ones that do reveals that they were probably not intended for either precise measurements or use by students. The arc scale is only graduated in increments of ten degrees and the construction (and finish) would not have stood up to rough handling. Instead, it seems clear that these instruments were made for occasional use in demonstrations by teachers, and that these demonstrations were not intended to produce a

8 For photographs of historic instruments, see the "Mechanics" page at the site "Instruments for Natural Philosophy", by Thomas B. Greenslade, Jr.: http://physics.kenyon.edu/EarlyApparatus/Mechanics/Inclined_Plane/Inclined_Plane.html.

9 E.S. Ritchie, *Physical Apparatus*, Boston, 1877, 11; Rolfe and Gillete, *Handbook of Natural Philosophy for School and Home Use*, New York, 1872, 85; and Dorman Steele, *The Chautauqua Course in Physics*, New York, 1889.

series of precise measurements, but rather to provide a few select examples to illustrate and confirm the content of their lectures.[10] This "lecture/demonstration" method of instruction was an effective way to teach science if the teacher was well educated and supplied with the necessary instruments, facilities, and demonstration skills that the method required.[11]

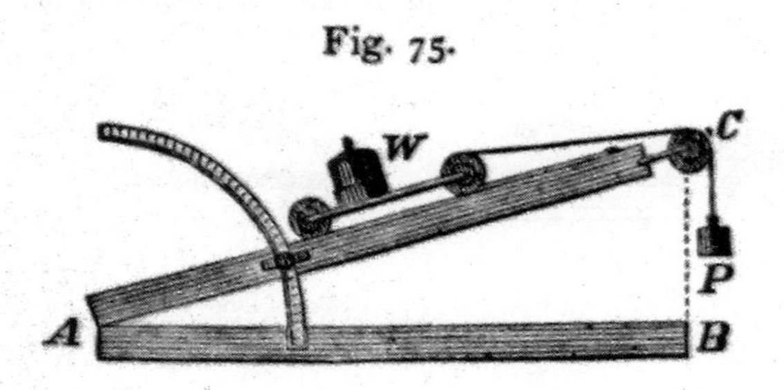

Fig. 5: This particular image was widely used in 19th century American physics texts, but only in those teaching by the old "lecture/demonstration" method. From Rolfe and Gillete, Handbook of Natural Philosophy for School and Home Use, 1872, 85.

Unfortunately, few American schools could afford this level of instruction and by the 1880's there was widespread dissatisfaction with the way that introductory physics was being taught. Too often, schools that had money preferred to spend it for demonstrations that were rather more entertaining than instructive. Boston science teacher Alfred Gage railed against the expensive and "trifling electric playthings" that had come to represent science instruction in many high schools. Recalling his own experience with secondary science, Harvard Professor John Trowbridge wrote that after completing the physics class: "We remembered a

10 These comments are based on the author's inspection of the only two known surviving examples, both in private collections. Both are mahogany and stamped: "E.S. Ritchie and Sons, Boston." The smallish scales are divided and marked in only ten degree increments and the overall construction is notably insubstantial compared to the student instruments designed by Gage and Hall.

11 The "lecture/demonstration" method continued to be used in European and American universities well into the twentieth century, so it would be an over-simplification to say that it was an ineffective way to teach science. This is, after all, essentially the lecture form that Davy and Faraday pioneered at the Royal Institution and in the right hands these sorts of presentations can be extremely effective. But this form of instruction requires a knowledgeable, dynamic presenter and an ample supply of apparatus. University professors also had mechanics and assistants to maintain and assemble their apparatus. American high school teachers had none of this. For more on the support universities provided see: Paul Forman, John L. Heilbron, and Spencer Weart, "Physics circa 1900, Personnel, funding, and Productivity of the Academic Establishments," *Historical Studies in the Physical Sciences, Volume 5*, Russell McCormmach (ed.), 1975, Princeton University Press: Princeton, 40–5, 83–6.

tuning-fork, an electrical machine, and a big electro-magnet which lifted the smallest boy in school, and that was all we remembered of natural philosophy."[12]

Trowbridge may have been lucky. Many schools could afford neither teachers nor apparatus, and for these schools physics instruction had become a short "cram subject" that was included in the class schedule solely to fill the student's heads with the necessary facts and formula to pass college admission tests. After that, they could be – and usually were – quickly forgotten. A favorite target of the critics was Steele's *Fourteen Weeks in Physics,* the most widely used "cram text." Here both lectures and demonstrations were largely dismissed and the student was expected to master physics by sheer power of memorization. The text's numerous illustrations were intended to substitute for the apparatus and demonstrations that money and time did not allow. As one educator noted, "Here [with Steele] we have a mere feat of memory, of just as much value, perhaps, as the committing to memory of so many lines of 'Paradise Lost,' certainly no more. In many, perhaps most, cases the teacher is as ignorant of the subject as the scholar and must have the text-book continually open in order to recognize if the answers are correct."[13]

The situation was formally acknowledged in 1880, when Professor F.W. Clarke submitted "A Report on the Teaching of Chemistry and Physics in the United States" to the Commissioner of the United States Bureau of Education. Based on a widely distributed questionnaire, the report summarized the state of secondary chemistry and physics instruction at that time. It noted that while the laboratory method of instruction was being introduced in many colleges, the use of student labs in high schools and academies was almost non-existent. Despite his strong opinion that "Three months of laboratory work will give more real insight into any science than a whole year's study of the printed page" Clarke found that only 11 of the 607 secondary schools sampled offered a physics course with some laboratory work and that only 4 of those 11 schools offered physics for a full school year.[14]

A NEW WAY OF TEACHING SCIENCE

Clarke's report reflected a widely held belief that the answer to America's science education problems involved the use of student experiments. Yet despite a broad agreement on this principle, there was little agreement about the kinds of experi-

12 Alfred P. Gage, *A Text-Book on the Elements of Physics for High Schools and Academies*, Ginn, Heath & Co., Boston, 1882, v; John Trowbridge, "The Study of Physics in the Secondary Schools," *Popular Science Monthly*, XV (1879), 159.

13 Joel Dorman Steele, *Fourteen Weeks in Physics*, New York: A.S. Barnes and Company, 1878; Jas. Lewis Howe, "The Teaching of Science," *Science*, Vol. 19, No. 481 (Apr. 22, 1892), 233.

14 Frank Wigglesworth Clarke, "A Report on the Teaching of Chemistry and Physics in the United States, by, S.B., Professor of Chemistry and Physics in the University of Cincinnati," Bureau of Education Circular No. 6–1880, Washington: Government Printing Office, 1881, 17.

ments and how (or even where) they should be performed. Student laboratories for chemistry were known to be successful, and in 1869 the first American physics laboratory for college students had been established at MIT – with widely heralded results. But the need – and affordability – of laboratories for high school students was by no means established.[15]

It was a topic of great interest and in the decade between 1875 and 1885, textbook publishers brought out at least a dozen new introductory physics texts, all written by Americans and more than half describing some version of student experiments. Tellingly, none of these "experimental" texts used the now familiar image of the inclined plane with the three-wheeled cart.[16] These texts all shared a common belief in the value of the so-called "inductive" or "object teaching" method of instruction, which emphasized the importance of a direct student experience. However they differed significantly in the details of what constituted an experience and what should be learned from it. Some of the new texts listed "experiments" that were so qualitative as to be almost unrecognizable as such by modern standards. In Baker's *Natural Philosophy*, the complete experiment to illustrate crystallization consisted of the instruction: "Exp. 15. Place a drop of a solution of ammonium chloride on a piece of glass; a beautiful arborescence will soon appear." To be fair, other writers suggested experiments that were more original

15 By the middle of the 19th century student experiments had become an essential part of chemistry instruction in Great Britain, and this example was not lost on American educators. When MIT's Barton Rogers announced plans for the new School of Industrial Science, he described four new laboratories: Physics and Mechanics, General Chemical Analysis and Manipulation, Metallurgy and Mining, and Industrial Chemistry. Of the four, the Rogers Laboratory of Physics opened first, in 1869. It was largely the product of Edward C. Pickering and was widely admired. Pickering later authored an influential laboratory text based on the course. See: David Knight, "Communicating Chemistry: The Frontier between Popular Books and Textbooks during the First Half of the Nineteenth Century," *Communicating chemistry: textbooks and their audiences, 1789–1939*, Anders Lundgren and Bernadette Bensaude-Vincent eds, Canton, MA: Science History Publications/USA., 2000, 191–2; William Barton Rogers, *Scope and Plan of the School of industrial Science*, Boston: John Wilson and Son, 1864; Edward C. Pickering, *Elements of Physical Manipulation, Part I*, Boston: Houghton Mifflin, 1873.

16 The new "experimental" texts were: C.L. Hotze, *First Lessons in Physics: for use in the upper grades of our common schools*, St. Louis: Central Publishing Company, 1876; Thomas Baker, *Elements of Natural Philosophy: based on the experimental method*, Philadelphia: Porter and Coates, 1881; A.P. Gage, *A Text-Book on the Elements of Physics for High Schools and Academies*, Boston: Ginn, Heath and Co,1882; Edwin J. Houston, *The Elements of Natural Philosophy for the use of schools and academies*, Philadelphia: Eldredge and Brother, 1879; John Trowbridge, *The New Physics, a manual of experimental study for high schools and preparatory schools for college*, New York: D. Appleton and Company, 1884; Elroy M. Avery, *First Principles of Natural Philosophy, a text-book for common schools*, New York: American Book Company, 1884; Elroy M. Avery, *Elements of Natural Philosophy, a Text-Book for High Schools and Academies*, New York: Sheldon and Company, 1885.

and notably more challenging, but few required that anything be measured and in many cases it's quite apparent that the students were mere observers.[17]

One concern that all these authors shared was a desire to minimize the cost of the apparatus required for their experiments. Some preferred simple experiments that used common materials. C.L. Hotze declared that "Costly apparatus is unnecessary. A pencil, a marble; a piece of board, of India-rubber, of wire; glass tubes, and other objects of trifling expense are sufficient, for our purposes even preferable." At the other extreme, Harvard professor Trowbridge's textbook was one of the few to advocate that students make actual measurements. He encouraged teachers and students to use advanced apparatus and to make most of it themselves. While his suggestions were often clever, it's doubtful that many teachers were prepared to make their own spherometers or galvanoscopes as he suggested.[18]

The most successful of the new "experimental" textbook authors – and the one who provided the first practical model for the new method of physics instruction – was A.P. Gage, author of *Elements of Physics* (1882). Gage was a Boston high school teacher and based his text on the introductory course that he had developed for his own class. His course continued to evolve and by the time he published his second textbook, *Introduction to Physical Science* (1887), it contained many features that would later become common. It reserved demonstrations involving expensive or complicated apparatus for the teacher to perform in the classroom. There were a large number of student experiments and these were performed in a separate student laboratory furnished with "rude pine tables and crude apparatus bearing the scars, scratches, and other marks of use." The student apparatus that Gage had designed were similarly simple, rugged and inexpensive. Surprisingly, the total cost to furnish the student physics lab in Gage's own school had been just three hundred dollars.[19]

The question of where the apparatus would come from was a critical one. School funds were limited and while it was easy to tell teachers to make their own equipment, the reality was that science teachers varied widely in their ability and willingness to build apparatus that their schools failed to provide. Commercial school instrument makers, like Queen in Philadelphia, had the skill to make instruments, but these were often expensive and the companies had their own ideas about what constituted proper design and quality. Further, they were not always

17 The terms "inductive" and "object teaching," as well as the context in which they are used, come from the preface of A.P. Gage, *A Text-Book on the Elements of Physics for High Schools and Academies*, 1884, iii; Thomas Baker, *Natural Philosophy based on the Experimental Method*, 48.

18 Hotze, First Lessons in Physics, viii; Trowbridge, *The New Physics, A Manual of Experimental Study for High Schools and Preparatory Schools for College*, 44, 100.

19 A.P. Gage, *Introduction to Physical Science*, Boston: Ginn and Co., 1887; see also Smith and Hall, *The teaching of chemistry and physics in the secondary school*, New York: Longmans, Green, 1902. 269–70; A.P. Gage, *A Text-Book on the Elements of Physics*, Boston: Ginn and Co., 1886, v; Steven Turner, "The Reluctant Instrument Maker: A.P. Gage and the introduction of the student laboratory," *Rittenhouse*, Vol. 18, No. 2, 40–61.

responsive in offering the new instruments that textbooks called for, generally preferring to stay with established designs.

Gage soon discovered that if he wanted other schools to adopt his course (and buy his textbooks) they needed to be able to buy the apparatus his course required. But he soon discovered that in simplifying the apparatus to reduce its cost he had eliminated the profits necessary to entice commercial manufacturers to make it. And so shortly after publishing his first textbook he reluctantly started the A.P. Gage Company, a scientific instrument business to provide the "cheap and efficient" apparatus needed to follow the plan of his book. Gage had clearly not planned for this and initially advised customers to simply send their orders to him at his high school. But as sales of his textbook increased, his publishers soon agreed to display and sell the instruments for his course in their showroom. In time he opened his own showroom and his company became an important source of apparatus for his and other textbooks that advocated student experiments. By 1884 demand was so strong that he had branch offices in New York and Chicago and was publishing an illustrated catalog of his apparatus.[20]

Gage was energetic and idealistic, and he benefited from being in Boston, which had long been a center of American education reform. He was also fortunate to be a teacher in the Boston Public School system. Its Superintendent, E.P. Seaver, was an early supporter of "object teaching" and clearly encouraged Gage's innovations. The combination of these factors may well explain why secondary student laboratories became widespread in the United States at least a decade before they became common in Europe.[21]

HARVARD, EDWIN HALL AND THE "DESCRIPTIVE LIST"

There was another center of educational reform in Boston – Harvard College. Harvard had a long history of involvement with curriculum innovation and its energetic president, Charles Eliot, was one of the few "college men" to show an interest in secondary education. In 1885 he was instrumental in forming the New England Association of Colleges and Preparatory Schools, which was largely concerned with revising and standardizing college admission requirements, and easing the passage from secondary to higher education. At this time only a small percentage of secondary school graduates actually went on to attend college, but the topic of admission requirements was a sensitive one. On the one hand, a schools' reputation depended at least partly on preparing its graduates for further

20 Turner, "The Reluctant Instrument Maker," *Rittenhouse*; A.P. Gage, *A Text-Book on the Elements of Physics*, 1882, back advertisement; Alfred P. Gage, *A Text-Book on the Elements of Physics*,1886, 447; Alfred P. Gage, *A Text-Book of the Elements of Physics*, 1884, back advertisement.

21 A.P. Gage, *A Text-Book on the Elements of Physics*, 1882, preface.

study, but conversely, many of the academies had their own goals and traditions and were reluctant to let the colleges dictate their course-work.[22]

It was against this background that in 1886, as part of Eliot's ongoing effort to improve the quality of incoming students, Harvard announced a new entrance requirement in physics. According to an announcement placed in the Harvard Catalog, beginning in 1887 applicants for admission would be expected to have completed a course of at least forty experiments "in the subjects of mechanics, sound, light, heat and electricity" and would be tested on some of these at the Harvard lab as part of the admission process. Students were also expected to have kept a laboratory notebook which would count as part of the examination. While the laboratory test was not mandatory, it was strongly encouraged. Although the traditional written test remained an option, its difficulty had been "considerably increased."[23]

Harvard's admission policy changes were particularly important because they affected not only the curricula of the relatively small number of east-coast schools and academies that provided most of Harvard's students, but also the curriculum of any school that hoped to send a student to Harvard. The actual implementation of the new physics requirement fell to Edwin Hall, a newly appointed physics instructor who would later become famous as the discoverer of the "Hall Effect." Although initially reluctant to enter the realm of secondary school pedagogy, Hall would go on to become a major figure in secondary education and a leading advocate for the laboratory method. However he seems to have initially underestimated the difficulty of the change he was suggesting.

In 1881 Harvard had recommended that schools use Arnott's *Elements of Physics*, a traditional lecture/demonstration text that Hall would later describe as having "873 pages and, though it was without an index, its table of contents covered 13 pages. It dealt with an enormous variety of particulars. But it did not, I believe, give one problem of any kind for solution by the reader." Hall intended the laboratory course to replace Arnott's text, but unfortunately the Harvard announcement failed to designate a replacement text and, as Hall later admitted, expecting schools to suddenly abandon this familiar course for an unspecified laboratory method was "a very wide swing of the educational pendulum – too wide, indeed."[24]

So in 1887, on "one very hot day in Room 41 of the laboratory with only the help of the assistant janitor" Hall personally tested the laboratory skills of 91 prospective students. The results were disappointing. As Hall described it, lacking a

22 Edward A. Krug, *The Shaping of the American High School 1880–1920*, Madison: The University of Wisconsin Press, 1969, 2–3, 6–7, 18–21.

23 Edwin H. Hall and Joseph Y. Bergen, *A Text-Book of Physics, Largely Experimental, on the Basis of the Harvard College "Descriptive List of Elementary Physical Experiments"*, New York: Henry Holt and Company, 1891, introduction.

24 Neil Arnott, *Elements of physics or natural philosophy*, 7th edition, edited by Alexander Bain, and Alfred Swaine Taylor, London: Longmans, Green and Co., 1876; Alexander Smith and Edwin H. Hall, *The Teaching of Chemistry and Physics in the Secondary School*, London: Longmans, Green and Company, 1902, 269–71.

specific program to follow "every teacher was left to follow pretty nearly his own plan and devices. The candidates who offered experimental physics came, like the traditional beggars, 'some in rags and some in shags and some in velvet gowns.'" Presumably, most of the students in velvet gowns were from schools already using Gage's laboratory course.[25]

Undeterred, President Eliot instructed Hall to prepare detailed descriptions of the 40 experiments that would be required and these were published towards the end of 1887 as a pamphlet titled: "Descriptive List of Experiments in Physics Intended for Use in Preparing Students for the Admission Examination in Elementary Experimental Physics." Commonly referred to simply as the "Descriptive List," these "exercises" – as Hall called them – were widely circulated and Hall soon became a leading advocate for the laboratory method. As one writer put it, "through Hall's effort and Harvard's prestige, the 'Descriptive List of Elementary Physical Experiments' increased in popularity and soon a number of other books based on it appeared on the market." The list represented a "standard of work" by which future textbooks and curriculum changes could be judged.[26]

These last statements may go too far. Hall and Harvard would go on to play key roles in the movement to adopt student experiments, but it seems important to point out that their role was less to initiate the transformation of secondary physics teaching than to assume leading positions in a movement that was well underway. As we have seen with the success of A.P. Gage, New England's secondary schools were already energized with the fervor of educational change. Eliot's many reforms at Harvard, particularly those related to admission requirements, were certainly intended to put pressure on area schools to adopt laboratory teaching, but the schools themselves were also initiating reform. Indeed, considering the hasty way that it was implemented, Harvard's decision to install a laboratory requirement may best be seen as an effort to catch up with a reform movement that had started without it.

Influential as it was, the Descriptive List never rose above the status of a pamphlet. It was an evolving document, frequently updated and revised several times. Hall initially recommended that teachers supplement the list with several other textbooks. For the classroom he recommended the texts of A.P. Gage or Elroy Avery, another innovative author. For help in the laboratory, he suggested Trowbridge's *New Physics* and Worthington's *Physical Laboratory Practice*. The list's initial importance was that it called attention to the issue of physics reform and associated it with an important institution. Additionally, Hall was a strong advocate of quantitative student experiments and he structured the Harvard List to reflect this. Over time, this influenced other authors to include more student mea-

25 Albert E. Moyer, "Edwin Hall and the emergence of the laboratory in teaching physics," *The Physics Teacher*, 14, 2 (Feb 1976), 99–100; Hall & Bergen, *A Text-Book of Physics,* 1891, iii.

26 Edwin H. Hall, *Descriptive List of Experiments in Physics*, Harvard University, Cambridge, 1887; Moyer, "Edwin Hall," *The Physics Teacher*, 96–103; note: there were several versions of the Harvard List, with slightly different titles, increasing numbers of experiments and other updates. The title that Moyer refers to first appeared in 1889.

surements in their experiments, although few took this approach as far as Hall did.[27]

In 1891 Hall teamed with Boston school teacher Joseph Bergen and finally brought out his own text, *A Text-Book of Physics, Largely Experimental*. This was intended to be the official textbook of the Harvard course and it even included a complete copy of the Descriptive List. By this time more than half the students applying for admission to Harvard were choosing to take the physics laboratory examination instead of the written test. That so many students now felt competent in the laboratory indicates that more and more of the secondary schools were adopting the laboratory method of teaching. And with demand for apparatus constantly growing, more and more companies were entering the student apparatus market. When the Harvard List first came out, Hall was able to name just six companies that sold apparatus for his course, with only two of them carrying everything the course required. But just four years later, in 1891, there were fourteen companies on the list, with at least five of them providing complete sets of Harvard apparatus.[28]

So by the time that Hall & Bergen's textbook was published, there was already a robust national market for student laboratory apparatus. More and more businesses were making and selling the apparatus, and a growing number of schools were adopting the new method of teaching. The appearance of Hall's *Text-Book* was the first complete presentation of the Harvard course and the timing of its release makes it seem plausible that, as has been suggested, it set the "standard of work" for all future textbooks. But was this really the case? Were the ideas and methods advocated by Harvard really so influential and widely copied? It is difficult to know what was actually taught in the classroom, but as we will see, the design of Hall's inclined plane experiment had a distinctive feature that makes it easy to identify. Images of the inclined plane thus provide clues as to how this teaching instrument was designed and used during this period.

27 Note: originally 54 pages long, by 1897 the Descriptive List had grown to 92 pages. The last printing appears to have been the 4th revised edition, published in 1912. Avery, *Elements of Natural Philosophy,* 1885; note: Avery was more noted for his student problems than for student experiments, but this was still an important innovation and his course was easily adapted to experiments). Hall, *Descriptive List*, 1887, 4; John Trowbridge, *The New Physics, A Manual of Experimental Study for High Schools and Preparatory Schools for College*, New York: Appleton and Co., 1884; A. M. Worthington, *Physical Laboratory Practice*, Rivingstons, London, 1886.]

28 Edwin H. Hall, Joseph Y. Bergen, *A Text-Book of Physics, Largely Experimental*, New York: Henry Holt and Company, 1891; *Harvard Descriptive List*, 1887, appendix; Hall & Bergen, *A Text-Book*, 1891, iv, 380–1; E.S. Ritchie & Sons, *Laboratory and Experimental Apparatus for High Schools and Colleges*, Boston, 1890.

THE STUDENT INCLINED PLANE EXPERIMENT

While the inclined plane had not been prominent in the old curriculum, Gage considered it to be important and from the very first he included it as one of his student experiments (Fig.6).

Fig. 6: This is Gage's drawing of his inclined plane experiment. Note the use of a simple four-wheeled cart. From Gage, Elements of Physics, 1882, 137.

He initially described using a "toy carriage" which, with a pulley, a weight pan and a board could be used to create a simple, but functional instrument. From this the students could discover for themselves the principles of mechanical advantage. Just as images of the three-wheeled cart became the standardized representation of the inclined plane in the old lecture/demonstration texts, images showing the use of *four*-wheeled carts for this experiment quickly became a distinguishing characteristic of the new "experimental" texts (Fig.7). The use of four-wheeled carts for student experiments seems to have begun with A. P. Gage and was probably based solely on the availability of inexpensive toy cars. But the striking similarity of the images in other textbooks to those in Gage's book may also indicate how influential his ideas were.[29]

29 A.P. Gage, *A Text-Book on the Elements of Physics*, 1882, 137; John D. Quackenbos, Alfred M. Mayer, Silas W. Holman, Francis E. Nipher, and Francis B. Crocker, *Appleton's School Physics*, New York: American Book Company, 1891, 153; Edward R. Shaw, *Physics by Experiment, an Elementary Text-Book for the Use of Schools*, New York: Effingham Maynard & Co., 1891, 24; Allen, *Laboratory Exercises in Elementary Physics, edition for teachers*, New York: Henry Holt and Company, 1892, 170.

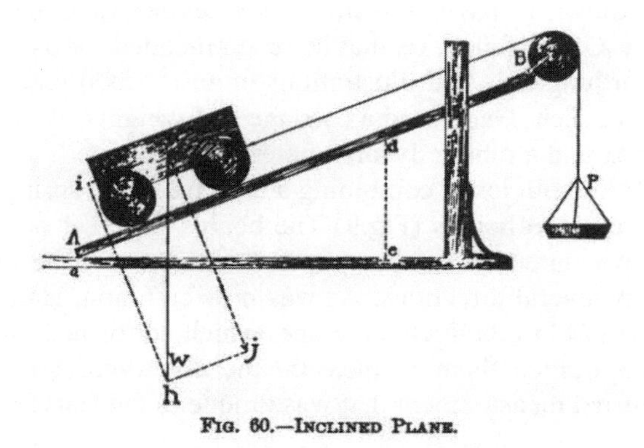

FIG. 60.—INCLINED PLANE.

keeping the cord always parallel to the plane. Note how

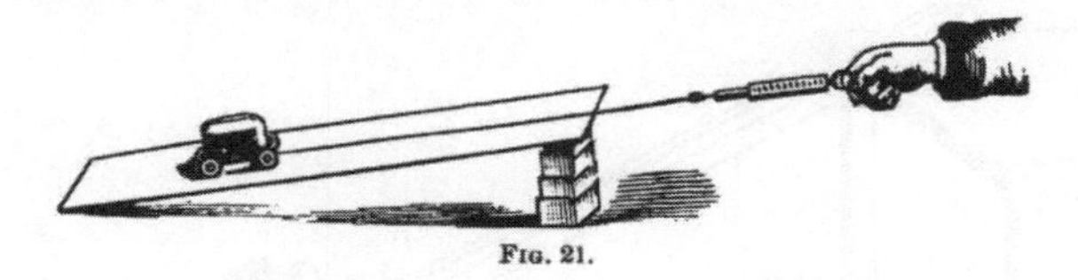

FIG. 21.

many pounds of force it requires to draw up the weight.

Fig. 7: Gage's use of four-wheeled carts was widely copied in other text-books and became one of the symbols of the "laboratory" movement.
The top image is from Appleton's School Physics, 1891, 153. And the lower image is from Allen's Laboratory Exercises in Elementary Physics, 1892, 170.

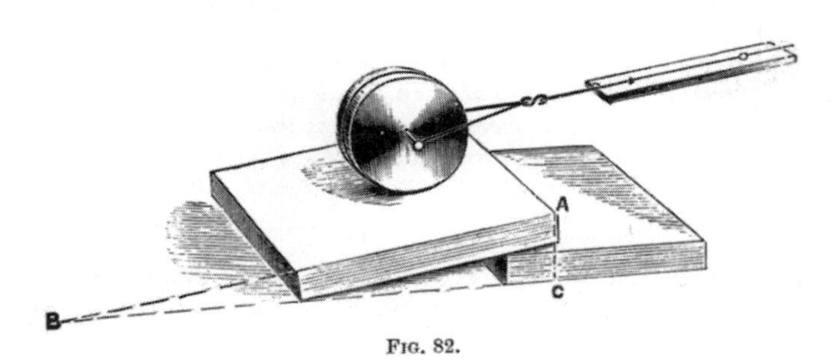

FIG. 82.

Fig. 8: Inclined plane experiment from Worthington's A First Course of Physical Laboratory Practice, 1887, 116. Hall may have gotten the idea of using a spring balance from this experiment.

Edwin Hall is also known to have consulted other textbooks as he designed the Harvard experiments. Of the four texts that he recommended for use with the List, only Gage's and Worthington's had illustrations of an inclined plane experiment. But where, as we have seen, Gage used a carriage and weights, Worthington used an improvised pill-box and a rubber dynamometer (Fig.8).

Hall borrowed from both texts, combining a carriage and a spring balance, but with a ramp made from *two* boards (Fig.9) The boards were set with a space between them, which was intended to allow students to measure the force required to move the car from several directions. As was now common, Hall had students measure the force *parallel* to the inclined plane, which confirmed mechanical advantage, but he also wanted them to measure the *horizontal* force required to move the car, an unusual measurement that was unique to the Harvard course.[30]

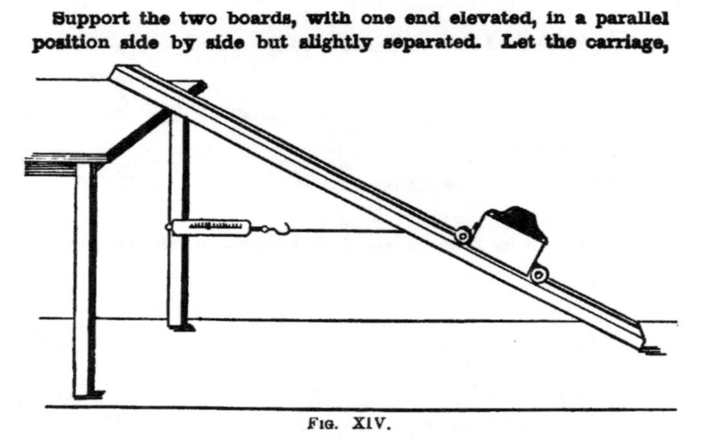
Support the two boards, with one end elevated, in a parallel position side by side but slightly separated. Let the carriage,

FIG. XIV.

when in position to move up and down the incline, bestride the space between the boards, the pull of the balance being applied

Fig. 9: This is Hall's student inclined plane experiment, which was invariably shown making the horizontal measurement. Hall's interest in precision was also noticeable, as students were expected to calibrate their spring balances before starting the experiment and to correct all measurements for friction when it was finished.
From Hall and Bergen, A Text-Book of Physics, 1892, 158.

To keep the cost down Hall initially recommended used roller skates for the "carriage" in his experiment, and he noted that Winslow's Skating Rink in Boston was a source of "Second-hand roller-skates at 75 cents a pair." The same boards and roller skates from this experiment could also be used in his inertia experiment, where two skates with different loads were accelerated by identical rubber strips.

30 Gage, *Elements of Physics*, 137; A. M. Worthington, *A First Course of Physical Laboratory Practice*, Boston: John Allyn, 1887, 116; Hall & Bergen, 1891, 7–8, 157–9.

However the choice of roller skates turned out to be troublesome. Even well-oiled skates tended to run unevenly and the wheels were "frequently so placed as to give a tendency to motion in a circle." Hall recommended removing the roller-skate's wheels and attaching them to a wooden box, but even this solution was not ideal.[31]

In truth, Hall's experiment seems to have had a number of problems. The use of two boards was an awkward arrangement: it took two students to perform the experiment and the boards could still slip or the car could fall into the space between them. None of the other textbooks adopted the experiment, although at least two lab manuals from the period suggested modifications to make the apparatus more stable (Fig.10).

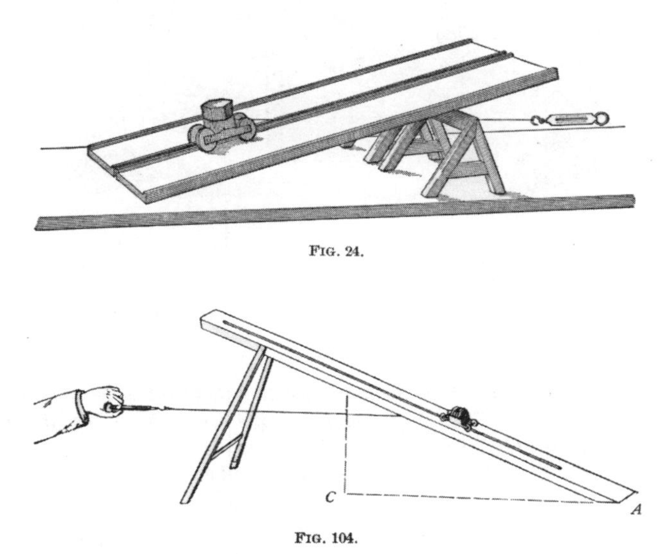

Fig. 10: At least two laboratory manuals suggested ways to improve Hall's inclined plane experiment. The top illustration is from Gage, Physical Experiments, a manual and notebook containing the exercises required for admission to Harvard University, 1897, 33. The lower illustration is from Stone's Experimental Physics, 1897, 284.

31 Harvard University, *Descriptive List of Experiments in Physics*, Cambridge, 1887, 21–2, 28–9, 52; note: the inertia experiment first appeared in Hall's *Elementary Ideas, Definitions and Laws in Dynamics*, Cambridge: The Riverside Press, 1886, 10–11 and may have been inspired by the English text, O.J. Lodge, *Elementary Mechanics*, 1879, 43. Both works are referred to in Edwin Hall, *Descriptive List of Elementary Physical Experiments*, Harvard College, 1889, 36. See also Hall & Bergen, 1891, 132.

However modifications to the apparatus did not make the experiment any easier, and in the 1897 revision of his textbook Hall quietly dropped the student experiment in favor of an inclined plane demonstration by the teacher. He also changed the apparatus, making it much more robust and replacing the carriage with a precisely made brass roller[32] (Fig.11).

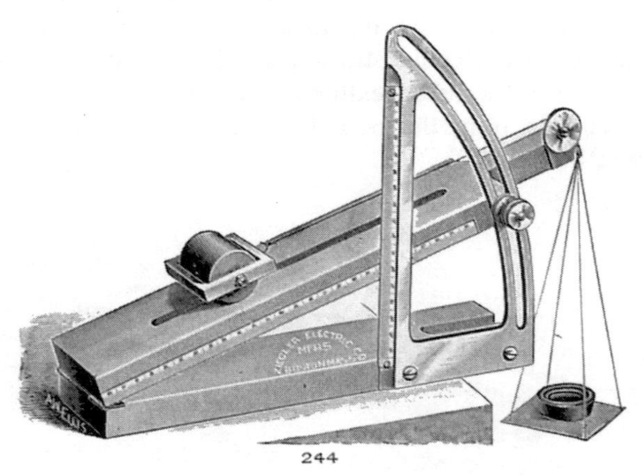

244. **Adjustable Inclined Plane,** shown about one-third actual size in cut. The roller is made of brass and accurately turned. It weighs with its frame just 16 ozs. The scale is graduated into millimeters. Great care has been exercised in making this piece. Without weights 4.25

Fig. 11: This is the inclined plane demonstration apparatus that Hall designed for teachers. This image is from the 1896 Ziegler Electric catalog, page 35. A similar image also appeared in Hall and Bergen, A Text-Book of Physics, 1897, 176. Later illustrations usually show this apparatus being used with a Hall's carriage.

When Hall abandoned the student experiment, he also dropped the use of roller skates. But he had also used them for the dynamics experiment, which he considered to be important. To replace the skates he now designed a special brass "carriage" with conical bearings, a low center of gravity and a swiveling front axle (Fig. 12). Quickly dubbed "Hall's Carriages," they came to be one of the icons of student experiments and were ubiquitous in physics classrooms throughout the twentieth century. Ironically, although Hall designed these cars for his acceleration experiment (and had dropped the entire student inclined plane experiment from his course), these carriages were almost immediately marketed as

32 Gage, *Physical Experiments, a manual and note book containing the exercises required for admission to Harvard University*, Boston: Ginn and Co., 1897, 33; William Abbott Stone, *Experimental Physics*, Boston: Ginn & Co, 1897, 284; Hall & Bergen, *Text-Book* (rev), 1897, 176; Ziegler Electric Company, *Catalogue No. 8, of Physical Apparatus and Supplies*, Boston, 1896, 35.

an "Improved Carriage for the Inclined Plane"[33] (Fig.13). It was an association that stuck, although the sturdy little carts were useful for a variety of other student experiments.

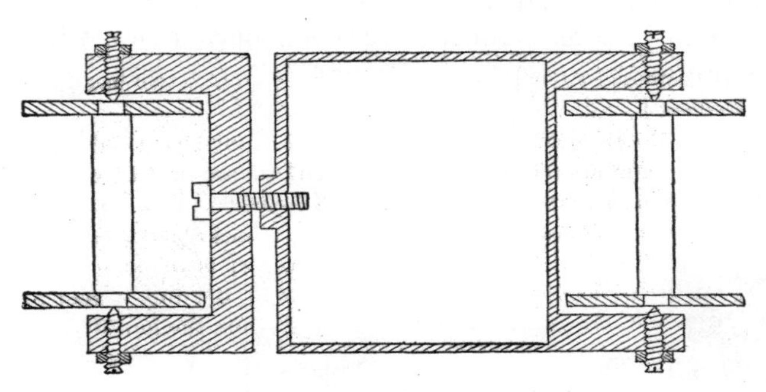

Fig. 12: Hall designed this carriage (here seen in plan view) for his acceleration experiments, but it was widely advertised and used for inclined plane experiments. Image from Hall & Bergen, A Text-Book of Physics, 1897, 575.

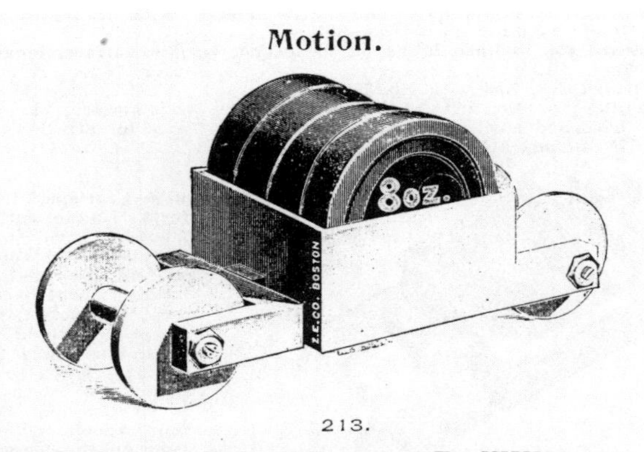

Fig. 13: This advertisement appeared in 1896 and may be the first image of "Hall's Carriage." From the very beginning it was associated with inclined plane experiments – although Hall had not intended this use. From the Ziegler Electric Company, Catalog, 1896, 29.

33 Hall & Bergen, 1897, 176, 294–6 and 575; the earliest reference that I have found to "Hall's Carriage" appears in LeRoy Cooley, *The Student Manual of Physics for the Study Room and Laboratory*, New York: American Book Company, 1897, 54.

Despite Hall's decision not to use it, the student inclined plane experiment continued to appear in most of the other textbooks, lab manuals and instrument catalogs – minus, of course, any mention of Hall's measurement of the horizontal force. Even the inclined plane apparatus that Hall designed for teachers – which he intended to be used with a brass roller – was commonly shown with a "Hall's carriage" instead (Fig.14).

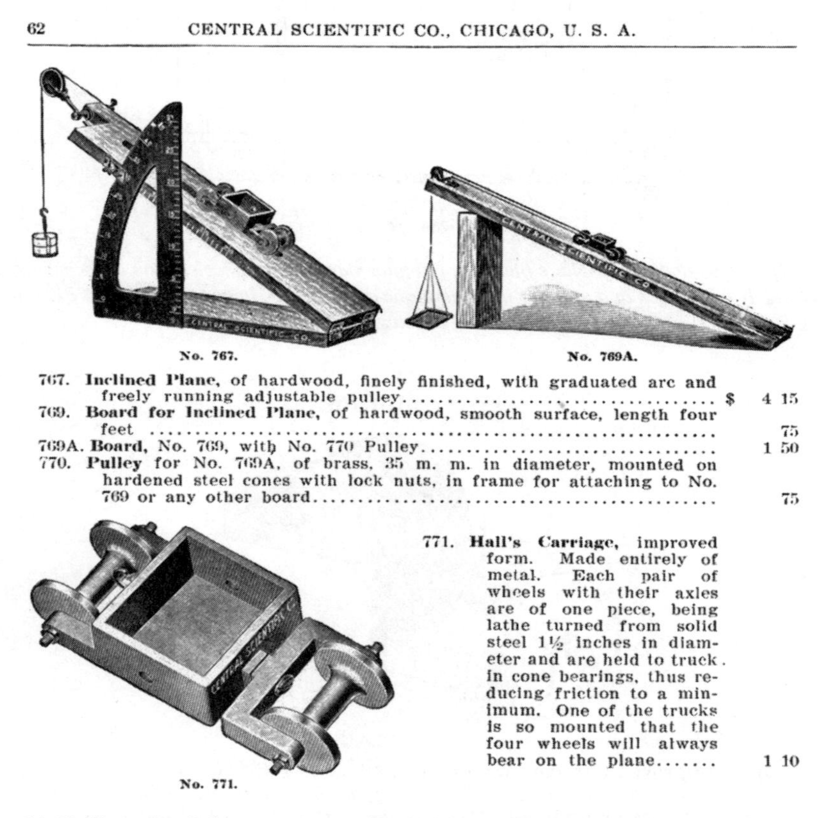

62 CENTRAL SCIENTIFIC CO., CHICAGO, U. S. A.

No. 767. No. 769A.

767. **Inclined Plane**, of hardwood, finely finished, with graduated arc and freely running adjustable pulley.................................. $ 4 15

769. **Board for Inclined Plane**, of hardwood, smooth surface, length four feet 75

769A. **Board**, No. 769, with No. 770 Pulley................................ 1 50

770. **Pulley** for No. 769A, of brass, 35 m. m. in diameter, mounted on hardened steel cones with lock nuts, in frame for attaching to No. 769 or any other board.. 75

771. **Hall's Carriage**, improved form. Made entirely of metal. Each pair of wheels with their axles are of one piece, being lathe turned from solid steel 1½ inches in diameter and are held to truck in cone bearings, thus reducing friction to a minimum. One of the trucks is so mounted that the four wheels will always bear on the plane....... 1 10

No. 771.

Fig. 14: Hall's inclined plane apparatus, shown with a Hall's Carriage. The Hall's Carriage was much more popular for this experiment than the roller mechanism that Hall had designed for it. From the Central Scientific Company, Catalog M, Physical and Chemical apparatus for Science Laboratories, 1909, 62.

While it is difficult to judge how Hall's teacher demonstration was actually used in the classroom, it is interesting that I have seen dozens of Hall's carriages in various instrument collections, but I have seen only one of Hall's rollers.[34]

34 Central Scientific Company, *Catalog M, Physical and Chemical Apparatus for Science Laboratories*, Chicago, 1909, 62; Franklin T. Jones and Robert R. Tatnall, *Laboratory*

On the whole, at least in the case of the inclined plane, the assertion that either the Harvard List or Hall's textbook constituted a "standard" for subsequent textbooks does not appear to be supported by the evidence. At least in the case of the inclined plane, while schools may have felt compelled to try (or at least consider) Hall's suggestions, teachers, textbook writers and even instrument makers seem to have felt free to modify and even ignore them.

THE NEA: THE COMMITTEE OF TEN AND THE COMMITTEE ON COLLEGE ENTRANCE REQUIREMENTS

Besides Edwin Hall and Harvard, the other key actor usually associated with the transformation of secondary physics has been the National Education Association. The NEA is important to this story because of its concern with the problem of "uniformity" – the desire to coordinate the secondary and college curriculums – and because of its strong advocacy for increased science instruction in secondary schools. Lacking explicit authority to mandate change, the NEA used reports from prestigious committees, along with its connections to the U.S. Bureau of Education, to exert its influence – which has traditionally been thought to be considerable. As the noted educator Harold Rugg wrote in 1927, the committees of the 1890s "exerted a tremendous influence in shaping the school curriculum. The prestige of their reports was so great that, once published, their recommendations were copied into entrance requirements of universities and they constituted the outline to which textbooks had to correspond if the authors and publishers expected widespread adoption. Both state and local, town and city systems came to base their syllabi definitely upon the recommendation of the committee."[35]

One of the most important NEA committees from this period was the famous Committee of Ten. It was chaired by Harvard's influential President, Charles Eliot and included the U.S. Commissioner of Education, William T. Harris, as one of its members. Formed in 1892, the committee was charged "to hold a conference of school and college teachers of each principal subject which enters into the programmes of secondary schools in the United States and into the requirements for admission to college." The conference consisted of 9 subcommittees, ten members each, and the one that considered the secondary physics curriculum included A.P. Gage. Not surprisingly, their report came out strongly in favor of both the laboratory method and the full year's instruction that implementing student experiments required. The report went so far as to include a list of 51 recommended student

Problems in Physics to accompany Crew and Jones' "Elements of Physics", New York: The Macmillan Company, 1912.

35 Rugg quotation from Michael Shiro, *Curriculum Theory: conflicting visions and enduring concerns*, Thousand Oaks: Sage Publications, 2008, 31.

experiments and these had a striking resemblance to the experiments in the Harvard List, as well those in Gage's *Elements of Physics*.[36]

The committee's final report was submitted in December 1893 and, although harshly criticized in some quarters, William T. Harris declared it "the most important educational document ever published in this country." As Commissioner of the Bureau of Education, Harris was able to see that the report was quickly and widely distributed. He had 30,000 copies printed and distributed free to schools, school board members and other interested parties. After these copies were gone, which happened in just a few months, the NEA made arrangements for a second edition – which had sold over 10,000 additional copies by 1901. Looking back a decade later, Harris confirmed his original estimate of the significance of the report: 'The scheme of studies recommended by the Committee of Ten as Secondary School studies to the National Education Association has become the model for all secondary or high schools, public and private."[37]

Important as the report may have been, the NEA and the Bureau of Education had limited ability to compel schools and colleges to implement it. Almost as soon as the report was published, the NEA membership began to discuss this and at the 1895 meeting a paper was read entitled "What Action Ought to Be Taken by Universities and Secondary Schools to Promote the Introduction of the Programs Recommended by the Committee of Ten?" Discussions based on that paper led, in 1896, to the formation of yet another committee, the Committee on College Entrance Requirements. This committee, which included Edwin Hall as chair of the physics section, continued to meet and file progress reports until 1899, when it submitted its final report.[38]

The 1899 report of the Committee on College Entrance Requirements contained essentially no new suggestions about teaching physics. It basically repeated what had been in the Committee of Ten report and as Hall himself later said, the list of student experiments in the supplement was "little more than the Table of Contents of the Harvard Descriptive List and two paragraphs taken, almost without change, from the introduction to that list." He would later maintain that the Harvard List (and, by association, his *Textbook of Physics*) had received "some-

36 *Report of the Committee on Secondary School Studies, Appointed at the meeting of the National Educational Association, July 9, 1892, with the reports of the Conferences arranged by this Committee and held December 28–30, 1892,* Washington: Government Printing Office, 1893; C. Riborg Mann, *The Teaching of Physics for Purposes of General Education*, New York: Macmillan Company, 1912, 60; *Report of the Committee of Ten on Secondary School Studies*, New York: American Book Company, 1894, 126; Hall and Bergen, 1891, 157; Gage, *Elements*, 1886, 137.

37 Quotation is from a "Letter of Transmittal" by W. T. Harris, Commissioner, Bureau of Education, to the Secretary of the Interior, published as part of the *Report of the Committee on Secondary School Studies,* 1893; for more on criticisms of the report, see Krug, 66–92; second Harris quotation from Henry J. Perkinson, *Two Hundred Years of American Educational Thought*, Lanham, MD: University Press of America, 1987, 153.]

38 George E. DeBoer, *A History of Ideas in Science Education*, New York: Teachers College Press, Columbia University, 1991, 50.

thing approaching the authority of official utterance in the various publications of the National Educational Association during the past ten or twelve years."[39]

At the time, some educational authorities took exception to Hall's claim of official endorsement, but modern historians have tended to accept it.[40] They point to the creation, in 1900, of the College Entrance Examination Board and, to a lesser extent the school accreditation movement, as the mechanisms by which the national switch to the laboratory method of teaching (which is equated with the Harvard physics course) was effected:

> "The establishment in 1899 of the College Entrance Examination Board (CEEB) ensured even wider adoption of the revised Harvard course of experiments by the secondary schools of the nation. With Charles Eliot (and, to a lesser extent, Edwin Hall) involved in the CEEB, the Harvard position again prevailed. The statement on physics in the 'Definition of Requirements' is almost identical to prior Harvard dictums and was in fact based on the earlier report of Hall's NEA subcommittee. If a student were applying to a college that chose to participate in the CEEB's system of uniform tests, then he was obliged to master the experiments in the 'Descriptive List.'"[41]

This scenario would be convincing if it could be shown that the Harvard course was actually widely used at this time – or if use of the course dramatically increased after the CEEB report was published. However this does not appear to have been the case. In the decade following the establishment of the CEEB, only two new editions of Hall's textbook were printed – one in 1902 and one in 1905. This would seem to indicate a steady, but not rapidly increasing use. Yet during this same period at least fifteen <u>other</u> secondary physics texts were also published – and several of these appeared in multiple editions just like Hall's.[42] The production of so many competing texts represented a significant investment by both the

39 Smith & Hall, 1902, 327, 329.

40 Mann, *The Teaching of Physics*, 63; Krug 146–168; DeBoer, 62–4.

41 Moyer, 101–2.

42 The four printings of Hall & Berger's *A Text-Book of Physics* were: 1891, 1897 (rev), 1902, 1905 (rev); the fifteen competing texts were: Charles B. Thwing, *An Elementary Physics for Secondary Schools,* Boston: B.H. Sanborn & Co., 1900; G.A. Wentworth and G.A. Hill, *A Text-Book of Physics,* Boston: Ginn & Co., 1900; Henry Crew, *The Elements of Physics for Use in High Schools,* New York: Macmillan & Co., 1900; C. Hanford Henderson and John F. Woodhull, *Elements of Physics,* New York: D. Appleton & Co., 1900; Alfred P. Gage, *The Principles of Physics,* Boston: Ginn & Co., 1902; Amos T. Fisher and Melvin J. Patterson, *Elements of Physics, Experimental and Descriptive,* Boston: D. C. Heath & Co., 1902; Alfred P. Gage, *Introduction to Physical Science (rev),* Boston: Ginn & Co., 1902; Frank W. Miller and August F. Foerste, *Elementary Physics*, New York: Charles Scribner's Sons, 1903; Lothrop D. Higgins, *Lessons in Physics*, Boston: Ginn & Co., 1903; Joseph A. Culler, *A Text-Book of Physics for use in Secondary Schools,* Philadelphia: J.B. Lippincott Company, 1906; Robert A. Millikan and Henry G. Gale, *A First Course in Physics,* Boston: Ginn & Co., 1906; Henry S. Carhart and Horatio N. Chute, *Physics for High School Students,* Boston: Allyn and Bacon, 1907; Henry Crew and Franklin T. Jones, *Elements of Physics for Use in High Schools (rev),* New York: Macmillan & Co., 1909; Frederick R. Gorton, *A High School Course in Physics,* New York: D. Appleton & Co., 1910; Charles R. Mann and George R. Twiss, *Physics,* Chicago: Scott, Foresman & Co., 1910; note: this list is cursory and likely not complete.

authors and the publishers. The production of these texts – few of which attempted to link themselves to the Harvard course – calls into even sharper question how well regarded the Harvard course actually was.

One indication about this comes from reading the actual report of the Committee on College Entrance Requirements, where we find the surprising statement that Hall's *Committee on Physics* "did not submit a regular report signed by the members of the committee." Indeed, the Chair of the overall Committee later complained of a general "lack of harmony in the subcommittee as to matter and method." Further investigation reveals that at least three of the other members of the committee had authored their own physics texts and had fundamental differences with Hall's approach. When the committee failed to reach an agreement as to the best method of instruction, Hall simply submitted a version of his *Descriptive List*, along with a statement that "Comments by the members of the committee, in case they dissented from any part of this, were to be sent at once to the chairman of the Committee on College Entrance Requirements. It may be assumed that the list met with the approval of those who did not so indicate dissent." The only member to formally dissent was Professor Carhart of the University of Michigan. He submitted a list of forty of his own experiments, fourteen of which were added – without comment – to the very end of the report.[43]

Hall was a respected scientist and his position at Harvard, combined with his close connections to Eliot, made him a formidable figure. It would be surprising to find any direct criticisms of him or of the *Descriptive List* and none has yet come to my attention. However the Harvard Course had a reputation not only for being difficult, but also for being expensive. In 1883 Gage estimated the cost of a complete student laboratory for his course to be about $300. In 1896 the cost was still was still well under $400. By contrast, in 1897 Hall estimated the cost of apparatus for his complete course to total almost $1,300. With much less expensive alternatives readily available, cash-strapped American schools would have needed a good reason to purchase Hall & Bergen's textbook. Did the college entrance requirements, as has been suggested, force high schools to specifically adopt the Harvard course?[44]

A surprisingly strong answer can be found in one of the preliminary reports of the Committee on College Entrance Requirements. In 1896 the committee sent a survey to all the major American colleges and universities and of the 56 institutions responding, all but 18 had some sort of physics requirement for admission – which generally took the form of recommending a preferred text. Unfortunately, these texts were only identified by author, but the results are revealing nonetheless. Gage's textbooks were by far the most popular, being listed 28 times. Avery's texts came second (12) and Carhart and Chute's were third (9). Even in-

43 The other text authors were: Professor H.S. Carhart, University of Michigan; Prof. E.L. Nichols, Cornell; E.D. Pierce, Hotchkiss School, Lakeville, Conn.; see also National Education Association, *Report of Committee on College Entrance Requirements, July, 1899*, Published by the NEA, 1899, 180, 182; DeBoer, 51.

44 Gage, *Elements*, 1883, v–vi; Ziegler, *Catalog*, 1896, 221–3; Hall & Bergen, 1897, vi.

cluding all references to the "Harvard List," Hall & Bergen's text was only mentioned seven times and finished a distant fourth. So less than three years after the Committee of Ten's report we find strong evidence that colleges were indeed pressuring high schools to switch to the laboratory method of teaching, but that few of them specifically called for the Harvard course. But if – as now seems likely – the Harvard course was not the embodiment of the new laboratory method but merely one of many laboratory textbook options, then the alleged power of the NEA reports to effect specific educational reform seems suspect.[45]

Certainly, as all the NEA committees and reports demonstrate, there was widespread interest in educational reform, and student laboratories were widely seen as modern and progressive. But there was also significant opposition to the changes being proposed. In 1901, G. Stanley Hall, the influential psychologist and president of Clark University argued for a partial return to the old Natural Philosophy curriculum, asserting that the average student "would like to see hundreds of demonstrative experiments made in physics and the liberty to repeat most of them himself, without being bothered about mathematics." Finally, the proposed power of the college entrance requirements is challenged by the fact that very few high school graduates went on to attend college during this period. High schools were being strongly encouraged to adopt laboratory teaching, but at no point were they actually required to do so. How then did the change so quickly become so complete? Why is it that, even a single decade later, we fail to find pockets of educational conservatism where the old natural philosophy course still prevailed?[46]

"THE NATIONAL PHYSICS COURSE"

One possible answer comes from examining the activities of the firms marketing the student laboratory apparatus to schools. As we have seen, the student laboratory movement emerged in Boston and the firms initially offering the student apparatus were all based there. These companies fell into two groups: the established scientific instrument firms like Thomas Hall (founded 1857) and E.S. Ritchie & Sons (founded 1867), and new firms like A.P. Gage (1883), The Educational Supply Company (1886), and Franklin Educational (1888). Unlike the established firms, which had diverse product lines and sold their goods to a wide range of customers, the new companies had limited manufacturing capacity, con-

45 "Report of the Committee on College Entrance Requirements," *The School Review*, Vol. IV, No. 6 (June 1896), 395–403, note: there were also a handful of references to some of the old non-laboratory texts, notably: J. Dorman Steele, *Fourteen Weeks in Physics*, New York: A.S. Barnes & Co., 1878; William G. Peck, *Introductory Course of Natural Philosophy: for the use of schools and academies, edited from Ganot's Popular physics*, New York: A.S. Barnes & Co., 1881.

46 G. Stanley Hall, "How Far is the Present High-School and Early College Training Adapted to the Nature and Needs of Adolescents?" *School Review* (December, 1901), 649–651.

centrated almost exclusively on student laboratory apparatus and dealt almost exclusively with teachers and school boards.

To give some sense of the size of the market for these goods, in 1896 Carhart and Chutes' *Elements of Physics* was being advertised as being "now in successful use in more than one thousand schools." Assuming an average class size of just ten students, this means that at least ten thousand texts had been sold – in addition to apparatus for a thousand labs. As we have seen, this was not the most popular text. Even allowing for a certain amount of exaggeration, the total number of American schools already using student labs by this time likely ran to several thousand.[47]

Student instruments were an ideal product for these new companies. They were simple and easy to manufacture, and the market for them was not well established. Traditional American scientific instrument manufacturers had invested large sums in specialized machinery and skilled workers, and traditionally gained prestige (and presumably clients) by producing elaborate precision instruments for colleges and universities. Like the European instrument firms (who they modeled themselves after) they produced instruments with considerable ornamentation and with highly polished brass and wood surfaces. By comparison, the new student instruments were crude and inaccurate. Instead of precision and finish these instruments placed a premium on durability and ruggedness. They essentially represented a new class of scientific instruments and the traditional makers had trouble working down to their standards. Indeed, while Hall recommended Ritchie & Sons in his Descriptive List, he repeatedly warned teachers to specify "that no unnecessary finish is to be given to such apparatus. Otherwise the prices will be unduly high." He even suggested that "Teachers who have large classes may do well to order first one set only of such articles as are new to them and to use these articles as models in having others made under their own supervision." The country's largest scientific instrument firm, Queen & Company, initially showed little interest in making student apparatus, and Hall never recommended them as anything other than a source of "optical apparatus." They appear to have considered student instruments beneath their standards and their 1893 *Catalogue of Philosophical and Electrical Apparatus*, which neither mentioned nor carried any of the new laboratory apparatus, sniffed that "New instruments are being designed from time to time to meet new requirements, and it has always been our policy to make such apparatus if it really proves better than instruments already in use for the same purpose."[48]

This attitude contrasted markedly with that of the new companies, which were much more aware of changes in the market and were prepared to act quickly. The Franklin Educational Company was particularly aggressive. It had been founded by George A. Smith just a year after the publication of the first *Descriptive List*. Smith was the former purchasing agent for the city of Boston and for years he had

47 Ziegler, *Catalogue*, 1896, back cover advertisement.

48 *Descriptive List*, 1887, 51–2, 54; Queen and Company, *Catalogue of Philosophical and Electrical Apparatus*, Philadelphia, 1893, preface.

regularly traveled to Europe to purchase scientific instruments for the Boston schools. He obviously saw the business opportunities presented by the student laboratory movement and when the Committee of Ten issued its report late in 1893 he moved to take advantage of it. In early 1894 he published a special new "Catalogue of the National Course in Physics: Elementary Physical Experiments recommended by 'The Committee of Ten.'" Inside the catalog, in a letter to teachers titled "National Recognition of the Revolution in Teaching Physics" he went on to state that "The work suggested [by the Committee of Ten] is directly in line with what the Franklin Educational Company have been advocating, and the apparatus required is what we have been supplying for some time. We are therefore prepared to co-operate with instructors to our mutual advantage. To this end we issue this catalogue, embodying the experiments recommended in the Report of the 'Committee of Ten.' This invaluable document is issued by the U.S. Bureau of Education, and as to 'Educational Values,' is, as Dr. Harris says, 'The most important educational document ever published in this country.' Every teacher should read it."[49] (Fig.15)

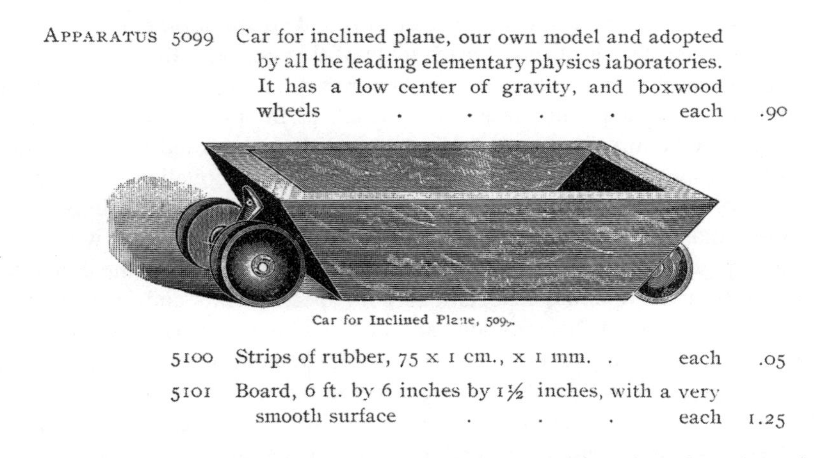

APPARATUS 5099 Car for inclined plane, our own model and adopted by all the leading elementary physics laboratories. It has a low center of gravity, and boxwood wheels . . . . each .90

Car for Inclined Plane, 5099.

5100 Strips of rubber, 75 x 1 cm., x 1 mm. . each .05

5101 Board, 6 ft. by 6 inches by 1½ inches, with a very smooth surface . . . each 1.25

Fig. 15: Car for the Harvard experiments, based on Hall's suggestion of attaching skate wheels to wooden boxes. From the Franklin Educational Company, Catalog of the National Course in Physics, 1894, 20.

The phrase "National Course in Physics" appears to have been Smith's invention and some educators would later take great exception to its implied claim of official sanction. But it was an effective marketing tool and other manufacturers soon adopted it. In 1896, the Ziegler Electric Company (successors to the A.P. Gage Company) issued a catalog of apparatus "especially adapted to such works as

49 Earnest Child, *The Tools of the Chemist, Their Ancestry and American Evolution*, New York: Reinhold Publishing Corporation, 1940, 197; Franklin Educational Company, *Catalogue of the National Course in Physics*, Boston, 1894. 2.

Gage's series of text-books on Physics, and 'Physical Laboratory Manual and Note Book,' Hall & Bergen's 'Physics,' as well as to the National Course of Physics, as recommended by the famous 'Committee of Ten.'" In 1901, the newly formed L.E. Knott Company brought out a dedicated catalog of apparatus for "The National Physics Course" that ran to 88 pages. This time the course was credited as conforming to the report of the College Entrance Requirement Committee and was described on the cover as being "A course acceptable to any college having a Physics requirement." But where the committee's report had only listed 61 experiments, this catalog listed 128 separate pieces of apparatus, and the student inclined plane experiment – which Hall had deleted from the Descriptive List in 1897 and was not one of the report's experiments – was clearly shown. By 1905 the Columbia School Supply Company, another new company, was advertising an elaborate set of instrument parts that would "perform all the experiments in the National Physics Course and over 300 additional ones."[50]

Although it has often been thought to be so, the National Physics Course was *not* the Descriptive List. This is clear from the numerous instrument catalogs that carried apparatus for *both* the Harvard course and the National Physics Course. While the National Course did roughly correspond to the *topics* of the Harvard course – as did nearly all of the physics texts of this period – the exact composition of the "course" varied between manufacturers, and they felt free to offer their own versions of the apparatus required to teach it. In some cases, like Franklin Educational, the "course" was a set of apparatus that supported several different texts. In this respect it followed the example of another innovation of this period, the laboratory manual. Almost as numerous and diverse as the textbooks themselves, these manuals were designed to provide teachers and students with helpful advice and practical suggestions for performing experiments in the laboratory. They generally tried to avoid associations with specific texts and instead tried to concentrate on details of the apparatus and best methods of performing the actual experiments. Allen's *Laboratory Physics* (1892), for example, could be used with five different texts. The Franklin catalog adopted the same format, organizing its catalog by topics, listing the experiments generally used to teach that topic, and finally describing the apparatus needed to perform each one. Each experiment also had a list of textbooks to which it applied. For example, the inclined plane was one of 12 experiments listed under the "Mechanics of Solids" section and eight different textbooks were listed as calling for experiments for which the apparatus could be used. It is worth noting that the catalog described the inclined plane ex-

50 Charles Riborg Mann, *The Teaching of Physics for Purposes of General Education*, New York: Macmillan Company, 1912, 65; Ziegler Electric Company, *Catalogue No. 8, Physical Apparatus and Supplies*, 1896; Child, *Tools of the Chemist*, 197; L.E. Knott Apparatus Company, *The National Physics Course*, Boston, 1901 (reprinted 1904); advertisement for Columbia School Supply Corporation, *School Science and Mathematics*, Vol. V, No. 1 (Jan 1905).

periment as "Work done in moving bodies up an inclined plane" and omitted the horizontal measurement that the Harvard course still specified at this time.[51]

While some later National Physics Course catalogs chose not to refer to individual textbooks, they all felt free to include items that would support general laboratory work, such as lab furniture, specially printed student notebooks, selected advanced apparatus and books – both reference books and textbooks. It was common to find lists of physics textbooks at the end of these catalogs and without exception they included only those that advocated the use of student experiments. More often than not, these texts were available at discounted prices and would invariably be shipped for free.[52]

In general, schools buying apparatus from the companies advertising the National Physics Course could expect to receive instruments that were inexpensive, rugged, and would efficiently meet the requirements of most of the newer physics texts. Importantly, and unlike most of the established scientific instrument firms, the new companies did not carry apparatus that would (almost certainly) not be needed with the new teaching method – such as the elaborate static electricity and polarization apparatus that had been so prominent in the old Natural Philosophy texts. These demonstrations had no place in the new laboratory curriculum. In this context it may be useful to consider the National Physics Course as an early example of the modern concept of "branding," in that it associated intangible qualities such as "modern," "approved" and even "trustworthy" with the companies that offered it. More than just a source of teaching apparatus, these companies came to be associated with the reform movement itself.

The National Physics Course may also be seen as a reaction against what many must have seen as the continuing dithering of professional educators. Gage's first text had been published in 1882 and, despite the promise of consensus contained in the NEA reports already discussed, the debate over the best way to teach physics continued into the early 20th century with yet another round of meetings and reports, this time under the guise of the National Commission on the Teaching of Physics (1906) and the associated "New Movement Among Physics Teachers." A process of curriculum reform taking over twenty years might have been tolerable if the number of schools was relatively stable, but this was far from the case.[53]

A factor curiously absent from the standard narrative of this period has been the unprecedented growth of student enrollment. In 1870 the total number of students in all the public high schools in America came to about 80,000, with a somewhat larger number estimated to have been in private high schools and academies. But the country was entering a period of constantly increasing immi-

51 Charles R. Allen, *Laboratory Exercises in Elementary Physics*, New York: Henry Holt and Company, 1892, vi; Franklin, *Catalog*, 22.

52 see for example the catalogs of Franklin, 1894, 42; Ziegler, 1896, 230.

53 This topic is outside the scope of this article, but an introduction can be found by referencing: Paul Monroe, *A Cyclopedia of Education*, New York: The Macmillan Company, 1913, 713; and "The New Movement Among Physics Teachers, *School Science and Mathematics*, Vol. VI (December, 1906), circular IV.

gration and this, combined with ever more inclusive public education, caused high school enrollment to spike. By 1889 the number of public high school students had swelled to over 200,000 (in 2,526 schools) with nearly 95,000 more students in 1,632 private secondary schools and academies. After 1890 the total number of high school students basically doubled every ten years and by 1920 it had ballooned to nearly two and a half million. During this same period America went on a school building boom. Between 1900 and 1910 alone, over ten thousand new high schools were constructed. Between 1889 and 1921 total annual expenditures across the country for elementary and secondary public schools went from $143 million to $1.4 billion – a ten-fold increase![54]

Federal funding for schools was almost nonexistent during this period, so the rapid expansion of enrollment often seriously limited the funds available for student laboratory apparatus. Despite widespread support for laboratory teaching, many schools had to build up their apparatus collections gradually. Journal articles advised teachers that "if you have only a small sum to spend, confine your purchases at first to apparatus illustrating elementary physics. Buy a few pieces of a kind and get as many kinds as your appropriation will permit. Next year you may buy more to supplement what you already have." In response, instrument makers worked hard to bring down the cost their wares and by 1901 the cost of apparatus for the first twenty-five experiments of the National Physics Course was less than seven dollars per student. Teachers also had wide flexibility in the way they taught different topics and were able to save money by eliminating some experiments altogether. Instrument makers accommodated this by designing student notebooks with loose-leaf pages. Teachers were then able to buy only the pages that described experiments they actually used.[55]

Instrument companies aggressively marketed these sets of apparatus, the cheapest of which were priced to encourage even poor schools to take up laboratory teaching. Ziegler Electric offered a set that would furnish an entire lab for just $100, but would "answer very well for illustrating a large proportion of the leading experiments." Even critics of the National Physics Course were forced to admit the role that these sets played in the adoption of laboratory teaching: "These outfits were sold extensively throughout the country, and exerted a powerful influence toward fixing the nature of the course and that of each experiment, and toward encouraging schools to introduce laboratory work of the kind specified, because they furnished an easy means of doing so, and one which did not require the teacher to be too much of a specialist in physics."[56]

54 Krug, 5; Sidney Rosen, "A History of the Physics Laboratory in the American Public High School (to 1910), *Physics History from AAPT Journals* (edited by Melba Newell Phillips), College Park: American Association of Physics Teachers, 1985, 189; Thomas D. Snyder, ed., *120 Years of American Education: A Statistical Portrait*, Washington DC: U.S. Department of Education, 1993, 36–7.

55 Irving P. Bishop, "Individual Laboratory Work in Physics," *American Education*, Vol. V, 1902, 334; Knott, *National Physics Course*, 1901; L.E. Knott Apparatus Company, *Catalog 17, A Catalogue of Physical Instruments*, Boston, 1912, 30.

56 Ziegler, 1896, 224; Mann, *Teaching of Physics*, 1912, 63–4.

By 1910, few of the companies that sold school science apparatus had been in existence when the student laboratory movement started. As demand for science apparatus grew, the companies producing them went through a period of rapid growth and consolidation. In Boston, the firms E.S. Ritchie, Thomas Hall, A.P. Gage, Ziegler Electric, Franklin Educational and Hall Scientific had all been absorbed by the newly incorporated L.E. Knott Company. In Chicago a handful of companies, including Central Scientific, were similarly dominant. These large, aggressive, and very successful firms used their profits to expand into other instrument markets and put enormous pressure on the established science instrument companies. Prime among these was the venerable James W. Queen Company of Philadelphia. In 1888 *Scientific American* had called it "the largest and most comprehensive [company] of its kind in the United States or in the world," but as we have seen Queen was extremely slow to enter the student laboratory market. In 1903 they finally began to market student laboratory apparatus, but by then it was too late and they were never able to recover their former dominance. By 1910, the company was just a shadow of its former size and its output of instruments was largely limited to electrical apparatus. There were multiple causes for the company's rapid decline, including bad management, an economic downturn and the departure of key personnel, but certainly their early failure to enter the rapidly growing market for student apparatus was a major contributing factor.[57]

The distinction between the old and new companies even extended to their use of instrument images. The traditional companies continued to use the old lecture-demonstration images and the new firms only showed apparatus for the experimental courses. In the case of the inclined plane, the old image was the apparatus with 3-wheeled cart and the experimental apparatus was usually the Hall's carriage. In an extensive search of instrument maker's catalogs, I have found only one instance where both of these images appear in the same catalog – the Henry Heil Company of St. Louis (Fig.16). In terms of education apparatus, this was a medium-sized company in a secondary market and their relative isolation may have given them the time necessary to adapt; this was the only traditional scientific instrument company able to survive the transition to laboratory teaching. They continued to use the traditional inclined plane apparatus until 1909, but dropped it from subsequent catalogs.[58]

As much as any other, this act marked the effective conclusion of the switch to laboratory teaching. By 1910 no scientific instrument company in America carried either the textbooks or the apparatus needed to teach the old non-laboratory course. After 1910, even if schools had *wanted* to use the old teaching method, they would have been unable to buy the equipment necessary to do so.

57 "The Manufacture of Scientific Apparatus, James W. Queen & Company," *Scientific American*, 58 (April 28, 1888), 258–9 and cover; Queen & Company published a small advertising brochure in 1903 with an image of Hall's inclined plane apparatus, but this image did not appear in any of their regular catalogs.

58 Henry Heil Chemical Company, *Illustrated Catalog and Price List of Chemical and Physical Apparatus*, St. Louis, 1909.

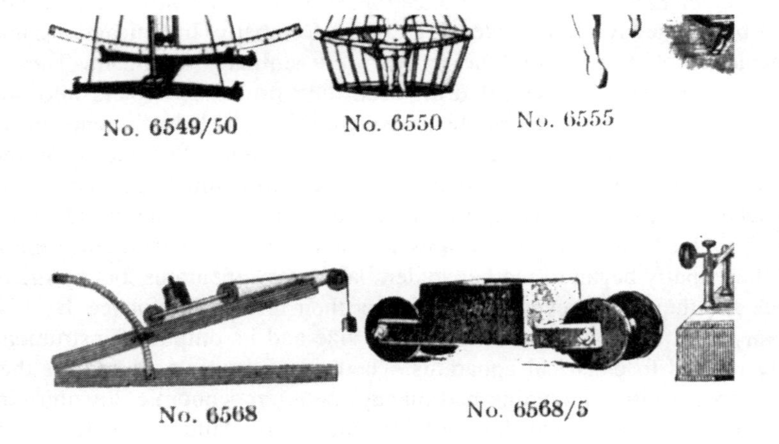

Fig. 16: Page from the 1909 catalog of the Henry Heil Company of St Louis. This is the only known example of images associated with the old teaching method being presented in the same catalog as images of the new "experimental" method. The old image did not appear in later catalogues. From the Henry Heil Chemical Company, Illustrated Catalogue and Price List of Physical Apparatus, 1909, 81.

CONCLUSION

This article began with a criticism of the narrow range of sources traditionally used to study science education history. This is not a new complaint and few education researchers would argue against the desirability of broadening the scope of these studies to include more information about the activities of teachers and actual classroom practices. But such studies have been slow in materializing and science education research continues to rely too heavily on the writings and speeches of a relatively small group of elite educators and administrators.

This study takes a different approach and also considers the material artifacts of education. These are mostly commonplace items and include such things as textbooks, lab manuals, student notebooks, teaching instruments and school supply catalogs. By looking closely at the ways that these items were made, marketed, used and understood it is possible to infer much about actual classroom practices and the actions of teachers and schools.

One important benefit of this kind of detailed study is that it lends itself to comparative analysis. For instance, during the period of this story Canada experienced much of the same immigration and expansion as the United States but, as Hoffman's revealing study of Ontario high schools shows [this volume] a centralized administration produced dramatically different results in the schools.

Drawing on a wider range of historical sources can also influence our understanding of the basic history of science education itself. As we include Gage and some of the other text writers in the narrative, the assumed dominance of Hall and the Harvard course come under question. This article presents evidence for a reconsideration of Hall's importance and if Hall is now seen as a regional figure with limited national influence, this fundamentally changes our understanding of American education during this period. It also challenges any work based on Hall's presumed influence.[59]

This is a topic that deserves more work but, if the analysis in this article is confirmed it calls into question three other aspects of our basic understanding of this period. The first concerns the origin of the "laboratory movement": was it really a "top-down" initiative, as previously assumed? The second question relates to the role of teachers and apparatus manufacturers in establishing the new curriculum: were teachers and apparatus manufacturers really the passive recipients of guidance from committees? Finally, what was the real influence of the NEA and similar groups during this period? Was there, as evidence now suggests, a dissociation between educational leadership and classroom practice during this period? These are fundamental questions and they need to be seriously considered before we can feel confident in our understanding of this important time.

REFERENCES

Allen, C.R.: 1892, *Laboratory Exercises in Elementary Physics, edition for teachers*, Henry Holt and Company, New York.

Arnott, N.: 1876, *Elements of physics or natural philosophy*, 7th edition, ed. by Bain, A. & Taylor, A.S., Longmans, Green and Co., London.

Avery, E.M.: 1884, *First Principles of Natural Philosophy, a text-book for common schools*, American Book Company, New York.

Avery, E.M.: 1885, *First Principles of Natural Philosophy*, American Book Company, New York.

Avery, E.M.: 1885, *Elements of Natural Philosophy, a text-book for high schools and academies*, Sheldon and Company, New York.

Baker, T.: 1881, *Elements of Natural Philosophy: based on the experimental method*, Porter and Coates, Philadelphia.

Bishop, I.P.: 1902, 'Individual Laboratory Work in Physics', *American Education*, 5.

Carhart, H.S. & Chute, H.N.: 1907, *Physics for High School Students*, Allyn and Bacon, Boston.

Central Scientific Company: 1909, *Catalog M, Physical and Chemical Apparatus for Science Laboratories*, Chicago.

Child, E.: 1940, *The Tools of the Chemist, Their Ancestry and American Evolution*, Reinhold Publishing Corporation, New York.

59 See, for example: John L. Rudolph, "Turning Science to Account," *Isis*, 2005, 96, 359–60; Olesko, "German Models, American Ways," 135–6; Albert E. Moyer, *American Physics in Transition*, Los Angeles: Tomash Publishers, 1982, 66–7.

Clarke, F.W.: 1881, 'A Report on the Teaching of Chemistry and Physics in the United States, by, S.B., Professor of Chemistry and Physics in the University of Cincinnati', *Bureau of Education Circular No. 6–1880*, Government Printing Office, Washington.

Cooley, L.: 1897, *The Student Manual of Physics for the Study Room and Laboratory*, American Book Company, New York.

Crew, H.: 1900, *The Elements of Physics for Use in High Schools*, Macmillan & Co., New York.

Crew, H. & Jones, F.T.: 1909, *Elements of Physics for Use in High Schools* (rev), Macmillan & Co., New York.

Culler, J.A.: 1906, *A Text-Book of Physics for use in Secondary Schools*, J.B. Lippincott Company, Philadelphia.

DeBoer, G.E.: 1991, *A History of Ideas in Science Education*, Teachers College Press, New York.

Fisher, A.T. & Patterson, M.J.: 1902, *Elements of Physics, Experimental and Descriptive*, D. C. Heath & Co., Boston.

Franklin Educational Company: 1894, *Catalogue of the National Course in Physics*, Boston.

Gage, A.P.: 1882/1883/1884/1886, *A Text-Book on the Elements of Physics for High Schools and Academies*, Ginn, Heath & Co., Boston.

Gage, A.P.: 1887/1902 (rev), *Introduction to Physical Science*, Ginn and Co., Boston.

Gage, A.P.: 1897, *Physical Experiments, a manual and note book containing the exercises required for admission to Harvard University*, Ginn and Co., Boston.

Gage, A.P.: 1902, *The Principles of Physics*, Ginn & Co., Boston.

Gorton, F.R.: 1910, *A High School Course in Physics*, D. Appleton & Co., New York.

Hall, E.H.: 1886, *Elementary Ideas, Definitions and Laws in Dynamics*, The Riverside Press, Cambridge.

Hall, E.H.: 1887, *Descriptive List of Experiments in Physics*, Harvard University, Cambridge.

Hall, E.H. & Bergen, J.Y.: 1891/1897 (rev)/1902/1905, *A Text-Book of Physics, Largely Experimental, on the Basis of the Harvard College "Descriptive List of Elementary Physical Experiments"*, Henry Holt and Company, New York.

Hall, G.S.: 1901, 'How Far is the Present High-School and Early College Training Adapted to the Nature and Needs of Adolescents?' *School Review* (December, 1901).

Henderson, C.H. & Woodhull, J.F.: 1900, *Elements of Physics*, D. Appleton & Co., New York.

Henry Heil Chemical Company: 1909, *Illustrated Catalog and Price List of Chemical and Physical Apparatus*, St. Louis.

Hiebert, E.N.: 1962, *Historical Roots of the Principle of Conservation of Energy*, The State Historical Society of Wisconsin, Madison.

Higgins, L.D.: 1903, *Lessons in Physics*, Ginn & Co., Boston.

Hotze, C.L.: 1876, *First Lessons in Physics: for use in the upper grades of our common schools*, Central Publishing Company, St. Louis.

Houston, E.J.: 1879, *The Elements of Natural Philosophy for the use of schools and academies*, Eldredge and Brother, Philadelphia.

Howe, J.L.: 1892, 'The Teaching of Science', *Science*, Vol. 19, No. 481.

Jones F.T. & Tatnall, R.R.: 1912, *Laboratory Problems in Physics to accompany Crew and Jones' "Elements of Physics"*, The Macmillan Company, New York.

Knott Apparatus Company: 1901/1904, *The National Physics Course*, Boston.

Knott Apparatus Company: 1912, *Catalog 17, A Catalogue of Physical Instruments*, Boston.

Krug, E.A.: 1969, *The Shaping of the American High School 1880–1920*, The University of Wisconsin Press, Madison.

Lodge, O.J.: 1879, *Elementary Mechanics including hydrostatics and pneumatics*, W. & R. Chambers, London.

Mann, C.R. & Twiss, G.R.: 1910, *Physics*, Scott, Foresman & Co., Chicago.

Mann, C.R.: 1912, *The Teaching of Physics for Purposes of General Education*, Macmillan Company, New York.

Miller, F.W. & Foerste, A.F.: 1903, *Elementary Physics*, Charles Scribner's Sons, New York.

Millikan, R.A. & Gale, H.G.: 1906, *A First Course in Physics*, Ginn & Co., Boston.
Monroe, P.: 1913, *A Cyclopedia of Education*, The Macmillan Company, New York.
Moyer, A.E.: 1976, 'Edwin Hall and the emergence of the laboratory in teaching physics', *The Physics Teacher*, **14**, 2.
NEA: 1893, *Report of the Committee on Secondary School Studies, Appointed at the meeting of the National Educational Association, July 9, 1892, with the reports of the Conferences arranged by this Committee and held December 28–30, 1892*, Government Printing Office, Washington.
NEA: 1894, *Report of the Committee of Ten on Secondary School Studi*es, American Book Company, New York.
NEA: 1899, *Report of Committee on College Entrance Requirements, July, 1899*, Published by the NEA.
Olesko, K.: 2006, "German Models, American Ways: The 'New Movement' among American Physics Teachers, 1905–1909". In H. Geitz, J. Heideking, & J. Herbst (eds.), *German Influences on Education in the United States to 1917*, Cambridge: Cambridge University Press, 2006, 129–153.
Peck, W.G.: 1881, *Introductory Course of Natural Philosophy: for the use of schools and academies, edited from Ganot's Popular physics*, A.S. Barnes & Co., New York.
Perkinson, H.J.: 1987, *Two Hundred Years of American Educational Thought*, University Press of America, Lanham, MD.
Quackenbos, J.D., Mayer, A.M., Holman, S.W., Nipher, F.E., & Crocker, F.B.: 1891, *Appleton's School Physics*, American Book Company, New York.
Queen and Company: 1893, *Catalogue of Philosophical and Electrical Apparatus*, Philadelphia.
Ritchie, E.S.: 1877, *Physical Apparatus*, Boston.
Ritchie & Sons: 1890, *Laboratory and Experimental Apparatus for High Schools and Colleges*, Boston.
Rolfe, W.J. & Gillete, J.A.: 1872, *Handbook of Natural Philosophy for School and Home Use*, Woolworth, Ainsworth and Co., New York.
Rosen, S.: 1985, 'A History of the Physics Laboratory in the American Public High School (to 1910)', *Physics History from AAPT Journals* (ed. by Melba Newell Phillips), American Association of Physics Teachers, College Park.
Salleron, J.: 1864, *Notice sur les Instruments de Precision*, Paris.
Schiro, M.: 2007, *Curriculum Theory: conflicting visions and enduring concerns*, Sage Publications, Thousand Oaks, CA.
Shaw, E.R.: 1891, *Physics by Experiment, an Elementary Text-Book for the Use of Schools*, Effingham Maynard & Co., New York.
Smith, A. & Hall, E.: 1902, *The teaching of Chemistry and Physics in the Secondary School*, Longmans, Green and Co., New York.
Snyder, T.D. (ed): 1993, *120 Years of American Education: A Statistical Portrait*, U.S. Department of Education, Washington DC.
Steele, J.D., 1878, *Fourteen Weeks in Physics*, A.S. Barnes and Co., New York.
Steele, D.: 1889, *The Chautauqua Course in Physics*, A.S. Barnes and Co., New York.
Stone, W.A.: 1897, *Experimental Physics*, Ginn & Co, Boston.
Thwing, C.B.: *An Elementary Physics for Secondary Schools*, B.H. Sanborn & Co., Boston.
Trowbridge, J.: 1879, 'The Study of Physics in the Secondary Schools', *Popular Science Monthly*, **XV**.
Trowbridge, J.: 1884, *The New Physics, a manual of experimental study for high schools and preparatory schools for college*, D. Appleton and Company, New York.
Turner, S.: 2004, 'The Reluctant Instrument Maker: A.P. Gage and the introduction of the student laboratory', *Rittenhouse*, Vol. 18, No. 2.
Wentworth, G.A. & Hill, G.A.: 1900, *A Text-Book of Physics*, Ginn & Co., Boston.
Worthington, A.M.: 1886, *Physical Laboratory Practice*, Rivingstons, London.

Worthington, A.M.: 1887, *A First Course of Physical Laboratory Practice*, John Allyn, Boston.
Young, T.: 1845, *A Course of Lectures on Natural Philosophy and the Mechanical Arts*, Vol. I, Taylor and Walton, London.
Ziegler Electric Company: 1896, *Catalogue No. 8, of Physical Apparatus and Supplies*, Boston.

REFORMING AMERICAN PHYSICS PEDAGOGY IN THE 1880S: INTRODUCING 'LEARNING BY DOING' VIA STUDENT LABORATORY EXERCISES

Richard L. Kremer

1. INTRODUCTION

Since the 1970s, historians of American physics have been debating the significance of Europe for the launching of the "American physics community." Much of the debate, sometimes inflected with nationalist overtones, focused on theoretical physics and on whether American physics had "matured" on its own during the 1920s, via American students doing postdocs in Europe, or whether it reached "maturity" only in the 1930s with the influx of European (often Jewish) refugees fleeing fascism. All sides agreed, however, that by the 1930s, the new American generation "had virtually closed the historic gap in quality between European and American research," and that with the rise of legendary teachers like Oppenheimer, physics training available in American graduate schools had reached European levels.[1]

This debate, however, generally ignored the earlier (pre-1914) period and did not consider undergraduate physics pedagogy at American universities and colleges. In this paper, I want to examine the most significant reform of physics teaching in nineteenth-century America, the introduction of student laboratory exercises in the 1880s. To what extent did this reform develop in response to

1 Daniel J. Kevles, *The physicists: The history of a scientific community in modern America* (New York: Knopf, 1978), 219. Cf. Stanley Coben, "The scientific establishment and the transmission of quantum mechanics to the United States, 1919–1932," *American historical review* 76 (1971): 442–60; Nathan Reingold, "Refugee mathematicians in the United States of America, 1933–41: Reception and reaction," *Annals of science* 38 (1981): 313–38; Paul Hoch, "The reception of central European refugee physicists in the 1930s: USSR, UK, USA," *Annals of science* 40 (1983): 217–46; Albert Moyer, "History of physics," *Osiris* 1 (1985): 163–82; Alexi Assmus, "The creation of postdoctoral fellowships and the siting of American scientific research," *Minerva* 31 (1993): 151–83. For studies of American high school physics teachers seeking, circa 1900, new approaches to student laboratory instruction that would be independent from both German patterns and the recommendations of American university physicists, see Sidney Rosen, "A history of the physics laboratory in the American public high school (to 1910)," *American journal of physics* 22 (1954): 194–204; Kathryn M. Olesko, "German models, American ways: The 'new movement' among American physics teachers, 1905–1909," *German influences on education in the United States to 1917*, eds. Henry Geitz, Jürgen Heideking and Jürgen Herbst (Cambridge: Cambridge University Press, 1995), 129–53.

American perceptions of European pedagogical innovation? Did the many Americans who traveled abroad in the 1870–80s to study physics bring back any particular pedagogical styles that they then introduced into American universities and colleges? In particular, did American visitors to the new German physics institutes, created after 1870 in what historian David Cahan has called the "institutional revolution," bring back German teaching methods along with their stories of German beer and hikes in the Alps? In the mid 1880s, John Tyndall told the American traveler (a recent immigrant from the Austro-Hungarian empire), Michael Pupin: "You will find in the Berlin laboratory the very things which my American and British friends and I should like to see in operation in all college and university laboratories in America and the British Empire."[2] But what exactly did the American travelers see in Berlin? Did America's student laboratory movement have a German origin?

I shall explore these questions by considering, as a case study, the teaching of physics at Dartmouth College, one of North America's more venerable institutions of higher education. Founded in 1769 as the ninth and final colonial college, Dartmouth by the mid-nineteenth century had about 250 students and in many ways typified the American college at that time. Since few records have been preserved that could document Dartmouth's nineteenth-century classrooms from the inside (student or faculty notes), I shall survey the textbooks in natural philosophy and physics required by Dartmouth's faculty. These changing texts, as well as the changing physics curriculum being taught, will enable us to analyze Dartmouth's introduction of student laboratories in the early 1880s. Central to this story will be a two-volume laboratory manual published in 1873–76 by the MIT physics professor, Edward C. Pickering. This manual, more than German models, shaped the "learning by doing" of Dartmouth's physics students at the end of the nineteenth century.

2. TEACHING PHYSICS BY "DEMONSTRATION AND ILLUSTRATION" IN MID-19TH-CENTURY AMERICA[3]

In his masterful survey of nineteenth-century physics, historian Iwan Rhys Morus devoted a chapter to the "science of showmanship," describing the spectacular electrical demonstrations performed by itinerant lecturers at urban sites for entertainment, by professionalizing staff at new educational ventures such as London's Royal Institution, and after mid century at the national exhibitions and fairs.[4] Yet Morus ignored perhaps the most common site for scientific spectacle after 1850,

2 David Cahan, "The institutional revolution in German physics, 1865–1914," *Historical studies in the physical sciences* 15.2 (1985): 1–65; Michael Pupin, *From immigrant to inventor* (New York: Scribners, 1926), 208.

3 Quoting from Dionysius Lardner, *Hand-books of natural philosophy and astronomy*, 3 vols. (Philadelphia: Blanchard and Lea, 1851–54), i:7.

4 Iwan Rhys Morus, *When physics became king* (Chicago: University of Chicago Press, 2005), Chapt. 4.

viz., the hundreds of physics classrooms at universities, colleges, and even secondary schools. In these pedagogical settings across Europe and North America, growing numbers of teachers and professors offered instruction in what came to be called "physics," using textbooks and commercially-made apparatus designed to support spectacular classroom demonstrations of natural phenomena and laws. Reading those physics textbooks and perusing instrument makers' catalogs do not necessarily allow us to deduce what happened in those classrooms; but we can follow the national and international circulation of pedagogical ideals by examining such sources. Spectacular physics, we shall see, was the mid-century ideal for American classrooms.

To the best of my knowledge, no one has attempted to survey the history of nineteenth-century physics textbooks used either in European or American institutions. However, historians have emphasized common patterns of science teaching at American colleges through at least the 1860s (before the rise of universities started to differentiate higher educational institutions there).[5] To capture the flavor of this widely practiced pedagogy, I will consider the teaching of "physics" at Dartmouth College.[6] European patterns, we shall see, provided some important exemplars for classroom instruction at Dartmouth. But we must also remember the fundamental differences between European (especially German) and American higher education in the nineteenth century. Unlike the German universities where students after 1820 increasingly could earn PhDs and freely select courses in which to enroll, American institutions rarely awarded PhDs before the 1880s and generally prescribed a single curriculum for all the undergraduate students.[7]

In 1850, the term "physics" replaced "natural philosophy" in Dartmouth's printed catalog, the annual volume that described the required curriculum for students' four years at the college. Despite the terminological change, what counted as "physics" remained broad. Thus in 1850–51, Dartmouth's third-year students (juniors) all studied "physics" for the entire year, which involved three separate classes in natural philosophy, geology and zoology, and astronomy and mineralogy. The prescribed natural philosophy textbook for 1850–51 did not change. Denison Olmsted's two-volume *Introduction to natural philosophy, designed as a textbook for the use of students in Yale College* had been used at

5 Stanley M. Guralnick, *Science and the ante-bellum American college* (Philadelphia: American Philosophical Society, 1975).

6 For information on Dartmouth's curriculum, including the required textbooks for each class, see the annually published *Catalogue of the officers and students of Dartmouth College* (Hanover: Dartmouth Press). For general overviews, cf. Sanborn C. Brown and Leonard M. Rieser, *Natural philosophy at Dartmouth: From surveyor's chains to the pressure of light* (Hanover, New Hampshire: University Press of New England, 1974); David Pantalony, Richard L. Kremer and Francis J. Manasek, *Study, measure, experiment: Stories of scientific instruments at Dartmouth College* (Norwich, Vermont: Terra Nova Press, 2005).

7 In 1861, Yale awarded the first three PhDs in the United States, including one in physics to Arthur Williams Wright for a dissertation on celestial mechanics. See Ralph P. Rosenberg, "The first American doctor of philosophy degree," *Journal of higher education* 32 (1961): 387–94.

Dartmouth since 1833 and would continue in use through 1855.[8] Olmsted taught natural philosophy and astronomy at Yale from 1825 until his death in 1859. Most of his first volume deals with mechanics; the second volume treats hydrostatics, pneumatics, electricity, magnetism and optics. In his preface, Olmsted admitted that he had borrowed most of his mechanics from Bewick Bridge's *Treatise on mechanics* (London, 1814) and Henry Kater's and Dionysius Lardner's identically titled text (London, 1830). Indeed, large sections of Olmsted's text, diagrams and quantitative examples are copied verbatim from the British authors.[9] Bridge and Olmsted offered mathematical treatments of motion in one plane, employing algebra and trigonometric functions, and frequent numerical problems.[10] From Kater and Lardner, Olmsted borrowed his extensive discussions of practical machines, balances, pulleys, clockworks, gears, fly wheels, etc.

Mechanics, Olmsted urged, was to be recited like mathematics "and to be demonstrated on the black board." Numerical problems "compel [the student] to think for himself; they lead him to a just understanding of the principles demonstrated; and they teach him how to reduce his knowledge to practice." However, mathematics disappears from Olmsted's second volume on the other branches of physics. For those topics, lectures and "numerous experiments" were to "amplify" and "illustrate" the text, which was written, Olmsted allowed, in a "freer and more popular style, adapted to be read with much greater rapidity."[11] Rather than the abstract, geometric diagrams that illustrate the principles of motion in volume one, Olmsted's second volume is filled with engravings showing sectional and axonometric views of classroom demonstration apparatus—an air pump, siphon, Coulomb's torsion balance, electric machine, Leyden jar, magnetic dip circle, Gregorian telescope, and a handsome fold-out sheet depicting "Watt's double acting steam engine."[12] But Olmsted did not provide instructions on how to per-

8 First published in 1831–32, Olmsted's two-volume textbook appeared in five editions and several more printings, by both New Haven and New York publishers, before the death of its author in 1859. A modified title in the New York edition of 1847 captures the gradual emergence of the term "physics" at mid century: *Introduction to natural philosophy, designed as a textbook in physics for the use of students in Yale College*.

9 After a Cambridge education, Bridge had taught mathematics and natural philosophy from 1806–16 at the East–India Company's college at Haileybury. Kater, after a military career in India, became vice president of the Royal Society and conducted extensive practical work on precision metrology for that body. Lardner, who had earned his M.A. at Trinity College, Dublin, was an exceedingly popular lecturer and writer in London during the 1820–30s who also had served as the first professor of natural philosophy and astronomy at the newly founded University College in that city.

10 From the section on projectile motion: "A body is projected at an angle of 15° with the horizon, with a velocity of 40 feet in a second; find its range, greatest altitude, and time of flight." Denison Olmsted, *An introduction to natural philosophy, designed as a textbook for the use of students in Yale College*, 2 vols. (New Haven: Hezekiah Howe & Co., 1831–32), i:107; Bewick Bridge, *A treatise on mechanics, intended as an introduction to the study of natural philosophy*, 2 vols. (London: T. Cadell and W. Davies, 1814), i:207.

11 Olmsted, *Natural philosophy*, i:iii–iv.

12 Only in the final section on optics did Olmsted use abstract, geometrical diagrams, in the manner of Newton's *Opticks* and many eighteenth-century texts on optics.

form classroom demonstrations with such apparatus. I have found only one instance in which he explicitly alluded to his own classroom experience. When noting that a Leyden jar of thinner glass can store more electric charge, Olmsted added: "The writer of this treatise had a large jar constructed of very thin glass: it took an extraordinary charge, and when discharged gave a report like that of an ordinary Battery; but it was fractured by the first experiment."[13]

A second textbook, authored by Olmsted for use in secondary (pre-collegiate) schools and reprinted much more frequently than his college text, would by 1844 include a supplement "containing instructions to young experimenters, together with a copious list of experiments in natural philosophy, accompanied by minute directions for performing them, and by numerous engravings of apparatus."[14] By "young experimenters" Olmsted meant young schoolteachers who often lacked the "natural aptitude for performing philosophical experiments." Since public experiments must be neat, elegant and invariably successful to capture the "confidence" of the pupils, Olmsted provided 60 pages of practical advice on how to stock a workshop and build simple apparatus, and on how to cut glass, solder, cement, fit corks, etc. For more complex items like air pumps or microscopes, he urged the purchase of commercially made instruments.[15] After advising teachers how to store and clean their apparatus, Olmsted then described how to perform more than 60 demonstration experiments, following the order of the branches of physics as presented in both his college and school texts, illustrating the apparatus with many woodcuts. Only several of these experiments yield quantitative results; Olmsted sought to demonstrate natural phenomena, not to measure them. With only $100, Olmsted suggested, a teacher could acquire the basic apparatus for many of the experiments in his list. $500 would support the purchase of more costly items like an Atwood's machine. With $1000, a teacher could demonstrate

13 Olmsted, *Natural philosophy*, ii:129.

14 Denison Olmsted, *A compendium of natural philosophy ... To which is now added a supplement containing instructions to young experimenters with a copious list of experiments accompanied by minute directions for performing them*, 20th ed. (New Haven: S. Babcock, 1848). First published in 1833, this text would appeaer in more than 20 editions, revised editions and enlarged editions published in New Haven, New York, and Charleston, South Carolina through the 1850s. I have seen identical versions of Olmsted's "supplement" in editions printed in 1848, 1851 and 1853; I have not seen the earliest edition of 1844, but it appears to have the same number of pages as the editions I have seen and presumably is essentially identical to those editions.

15 "Philosophical instrument makers can be found in most of our large cities, who are able to supply to order the various articles of apparatus essential to perform these experiments. A large portion of our cuts are loaned expressly for this work, by Mr. Joseph M. Wightman, 33 Cornhill, Boston, who has the corresponding articles always on hand and whose skill and fidelity may be fully relied on." Wrightman's first printed catalog had appeared in 1838. See Denison Olmsted, *Rudiments of natural philosophy and astronomy, designed for the younger classes in academies and for common schools, with numerous engravings illustrative of philosophical experiments* (New Haven: S. Babcock, 1844), 367; Joseph M. Wightman, *Illustrated catalogue of philosophical, chemical, astronomical and electrical apparatus* (Boston: G.W. Light, 1838).

physics with "completeness and elegance." Surely, Dartmouth faculty must have consulted Olmsted's secondary school text as they taught the college version for more than two decades and regularly conducted classroom demonstrations with a growing instrument collection.[16]

Interestingly, Olmsted's 1844 instructions for classroom spectacle predate the appearance of the earliest well-known European guides to physics classroom demonstrations. In 1850, a lyceum (secondary school) physics teacher from Freiburg i. Br., Joseph Frick, authored a 426-page manual, *Physikalische Technik, oder Anleitung zur Anstellung von physikalischen Versuchen und zur Herstellung von physikalischen Apparaten mit möglichst einfachen Mitteln*, published by Vieweg in Braunschweig. Seven ever-larger German editions of Frick's *Anleitung* would appear before 1909; two English translations were published in Philadelphia in 1861 and 1878. Like Olmsted, Frick intended his book to accompany a standard physics textbook, the third German edition of Pouillet-Müller, *Lehrbuch der Physik und Meteorologie* (also published by Vieweg, in 1847).[17] Identical woodcuts illustrate both volumes. In his preface, Frick emphasized that his experiments were intended to demonstrate "known laws" and not to enable "original research." Like Olmsted, Frick listed three criteria for demonstration experiments; they should be easy to conduct, inexpensive with home-made apparatus, and *anschaulich*, a multi-valenced term with Kantian roots referring to intuitive, demonstrative, visual modes of knowing. In more detail than Olmsted, Frick instructed his readers in the craft skills of soldering, glass blowing, glass cutting, and metal and wood turning. He indicated how to set up a *Laboratorium*, or workshop outfitted with 50 tools he listed, and urged Gymnasium teachers to procure a dedicated lecture hall with a large "experimental table" and a neighboring storeroom for the apparatus.[18] Most of the 269 apparatus described by Frick were to be constructed by the teachers; but like Olmsted, Frick urged readers to purchase the

16 Several heavily used editions of Olmsted's *Compendium* remain in Dartmouth's libraries.

17 Johann Müller's translation and reworking of Claude Pouillet's *Élémens de physique expériementale et de météorologie* would appear in eleven ever-expanding German editions from 1842–43 through 1925–34. English translations appeared in London (1847), Philadelphia (1848), a Dutch translation in Amsterdam (1852); the 2d German edition was jointly published by Vieweg in Braunschweig and Rudolph Garrigue in New York. Pouillet's French original first appeared in Paris as two volumes bound as four in 1827–30; six additional French editions appeared by 1856. Pouillet-Müller, as it was called, was one of the most used elementary physics textbooks during the second half of the nineteenth century.

18 Joseph Frick, *Die physikalische Technik oder Anleitung zur Anstellung von physikalischen Versuchen und zur Herstellung von physikalischen Apparaten mit möglichst einfachen Mitteln* (Braunschweig: Vieweg, 1850), vii–xi, 1–7. The second edition added detailed, richly-illustrated chapters on working with glass and metal. Joseph Frick, *Die physikalische Technik oder Anleitung zur Anstellung von physikalischen Versuchen und zur Herstellung von physikalischen Apparaten mit möglichst einfachen Mitteln*, 2d enlarged and improved ed. (Braunschweig: Vieweg, 1856). For an intellectual history of *Anschaulichkeit* and its extended resonance in nineteenth-century German discussions of physics, mathematics and pedagogy, see Salvo d'Agostino, "Il difficile ricupero dell'anschaulichkeit di Goethe nell-opera di Helmholtz," *Nuncius* 20 (2005): 401–15.

precision apparatus (barometers, balances, thermometers, microscopes, telescopes, magnets, lenses and prisms) from specialized mechanics. To acquire a complete set of this apparatus would require, Frick estimated, about 700–800 Gulden plus an ongoing expense of about 100 Gulden/year. Like Olmsted's, Frick's demonstrations were to be spectacular, not quantitative.[19]

The earliest French manual for classroom physics demonstrations appeared at the same time, Julien Fau's and Charles Chevalier's two-volume *Nouveau manuel complet du physicien-préparateur, ou, Description d'un cabinet de physique* (Paris: Roret, 1852).[20] Like Frick, these authors (a popular textbook writer and one of Paris's leading instrument-makers) followed Pouillet's widely-used text, *Eléments de physique expérimentale* (Paris, 1827). However, they assumed all the apparatus would be purchased from makers like Chevalier. The *Nouveau manuel* offers no advice about building apparatus. Rather for each experiment, it states the physical principle to be demonstrated, lists the required apparatus (illustrated in a separately bound "atlas" of 88 plates), and then describes how to conduct the demonstration, complete with many practical tips. For example, when showing electromagnetic induction with a bar magnet and a coil "it is necessary to move the magnet quickly, because the deviation [of the galvanometer] is nearly null and almost insensible when the operation is done slowly."[21]

Judging from these German, French and American manuals, physics teaching at both the secondary and university (or college) levels was extensively demonstrative by the middle of the nineteenth century. Extant inventories indicate that most American colleges by that time had spent at least $2000 on physics apparatus for classroom demonstrations; by 1862, Dartmouth had invested nearly $3000 in what it still called "philosophical apparatus."[22] In 1876, Dartmouth represented itself at the Centennial Exposition in Philadelphia with a series of photographs of its physics demonstration apparatus (see Fig. 1). Before the construction of dedicated physics laboratory buildings at the end of the century, an institution's physics demonstration apparatus signified its commitment to science.

19 For a list of the 269 apparatus described, see Frick, *Anleitung* 1850, 417–22. At 1850s exchange rates, Frick's 800 Gulden would have equaled about $1700; i.e., Frick's estimated cost for a set of physical demonstration apparatus was nearly double Olmsted's.

20 Apparently no further editions were published.

21 Julien Fau and Charles Chevalier, *Nouveau manuel complet du physicien-préparateur, ou, description d'un cabinet de physique*, 2 vols. (Paris: Roret, 1853), i:3, 293.

22 Guralnick, *Science and the ante-bellum American college*, 71–73; Pantalony, Kremer and Manasek, *Study, measure, experiment*, 226. Already by the 1830s, Boston's School Committee had prescribed a canonical set of philosophical apparatus and demonstration experiments for all its secondary schools. See Richard Green Parker, *The Boston school compendium of natural and experimental philosophy, embracing the elementary principles of mechanics, hydrostatics, hydraulics, pneumatics, acoustics, pyronomics, optics, electricity, galvanism, magnetism, electro-magnetism, & astronomy, with a description of steam and locomotive engines*, 2d stereotyped ed. (Boston: Marsh, Capen, Lyon and Webb, 1839).

Fig. 1: Dartmouth's optical apparatus, as photographed by H.O. Bly for the Philadelphia Centennial Exposition, 1876. Bly made additional photographs for pneumatics, heat, acoustics, mechanics, surveying, and electricity (three photos). DCLSC, Photo File, Scientific Apparatus.

Following Olmsted's death in 1859, Dartmouth adopted another American-authored textbook, Benjamin Silliman Jr.'s *First principles of physics, or natural philosophy designed for the use of schools and colleges* (1859).[23] Praised as "the best known of physical textbooks in the country" at Silliman's death in 1885, *First principles* continued the non-mathematical, discursive, and demonstrative pedagogical style of Olmsted's textbook. Very few equations appear (e.g., for the lateral strength of a beam, yet Newton's laws of motion and gravity are presented only in prose[24]); quantitative examples are far and few between, especially after the opening section on mechanics; no trigonometric functions appear, even in the extensive treatment of waves. With 677 mostly axonometric woodcuts on 706 pages of text (see Fig. 2), Silliman depicted hundreds of demonstration apparatus for his readers and provided chatty first-person descriptions of how to perform the demonstrations.[25] To enrich recitations conducted with the textbook, Silliman

23 Silliman's second edition (1860) was reprinted at least eight times by American publishers through 1886.

24 Benjamin Silliman, Jr., *First principles of physics, or natural philosophy designed for the use of schools and colleges* (Philadelphia: Peck & Bliss, 1859), 67, 86–87.

25 E.g., after noting that Torricelli, a discipline of Galileo, first measured atmospheric pressure in 1643, Silliman wrote: "To repeat his experiment, a tube of glass closed at one end of selected, A B, fig. 197, about 32 inches in length. Holding the tube mouth upward, it is completely filled with mercury, and inverted, after closing the orifice with the thumb, with its

placed a total of 1095 numbered "discussion questions" at the bottom margins of the pages, queries that tested readers' basic understanding of the material.[26] However, Silliman, professor of chemistry at Yale and long-time editor of the *American Journal of Science*, offered no explicit instructions about how to teach with his textbook. The only pedagogical reference in his preface justifies the historical asides in his textbook:

> ... every teacher must have observed, in his own experience, that an abstract principle is often fixed in the memory by the power of associated ideas, when it is connected with a date, or item of personal interest, as the attention is awakened by the dramatic, far more than by the didactic. Hence it has been thought judicious to introduce numerous important dates in the history of science.[27]

Did Silliman think that a reference to the date of Torricelli's experiment (1643) was more "dramatic" than a classroom demonstration of that experiment? Or were physics demonstrations so common in American classrooms by 1861 that Silliman felt no need to comment on them? In any case, Dartmouth's junior class from 1861–72 learned its physics via Silliman's text, recitations, lectures and classroom demonstrations.

For the next decade, Dartmouth's students, like thousands of others in North America and Europe, read physics from what was undoubtedly the most widely reprinted elementary physics textbook of the nineteenth and early twentieth century. In 1851, Adolph Ganot, a private teacher of mathematics and physics in Paris, published *Traité élémentaire de physique experimentale et appliquée*, a text that before its author died in 1881 would appear in 18 French editions, be translated into eleven languages and would permanently define the introductory physics classroom as a theater for spectacular demonstrations (interestingly, only one German edition appeared, in 1858).[28] An English version, translated and somewhat reconfigured by Edmund Atkinson, a lecturer in chemistry and physics at the Royal Military College, Sandhurst, first appeared in 1863. The first American edition, reprinting the fourth London edition, appeared in 1870; identical

lower end placed in a vessel containing mercury. The liquid column will, on removing the thumb, immediately fall some distance ..." Silliman, *First principles of physics*, 202.

26 E.g., concerning the measurement of atmospheric pressure: "319. How may it be shown that the atmosphere exercises a pressure upon the earth? Describe the Magdeburgh hemispheres? What may be proved with them? 320. What was Torricelli's experiment? What is the Torricellian vacuum?" Silliman, *First principles of physics*, 202.

27 Silliman, *First principles of physics*, vii–viii.

28 For a rich analysis of Ganot's contexts, see Josep Simon, "Adolphe Ganot (1804–1887) and his textbook of physics," M.Sc. thesis, University of Oxford, 2004; Josep Simon, "The franco-british communication and appropriation of Ganot's *Physique* (1851–1881)," *Beyond borders: Fresh perspectives in history of science*, eds. Josep Simon and Néstor Herran (Cambridge: Cambridge Scholars Publishing, 2008), 141–68.

London and American imprints would continue through the eighteenth edition, printed in 1910 and 1917, respectively.[29]

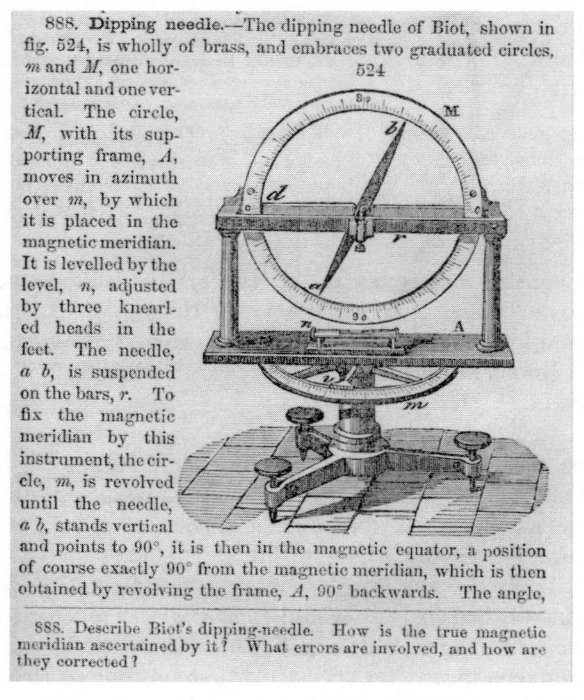

888. **Dipping needle.**—The dipping needle of Biot, shown in fig. 524, is wholly of brass, and embraces two graduated circles, *m* and *M*, one horizontal and one vertical. The circle, *M*, with its supporting frame, *A*, moves in azimuth over *m*, by which it is placed in the magnetic meridian. It is levelled by the level, *n*, adjusted by three knearled heads in the feet. The needle, *a b*, is suspended on the bars, *r*. To fix the magnetic meridian by this instrument, the circle, *m*, is revolved until the needle, *a b*, stands vertical and points to 90°, it is then in the magnetic equator, a position of course exactly 90° from the magnetic meridian, which is then obtained by revolving the frame, *A*, 90° backwards. The angle,

888. Describe Biot's dipping-needle. How is the true magnetic meridian ascertained by it? What errors are involved, and how are they corrected?

Fig. 2: Axonometric illustration of instruments in Silliman, 1859, 555. Note the "discussion" questions at the bottom of the page.

As historian Josep Simon has shown, Ganot maintained close ties to Parisian instrument makers and used woodcuts like Silliman from their printed catalogs for many of the illustrations in his text. Like Silliman, Ganot placed a woodcut on nearly every page of his book (see Fig. 3). Yet as far as I can tell, neither Ganot nor Atkinson offered any general advice to instructors about how to perform classroom demonstrations; and neither mentioned any particular instrument makers or commercial retailers of the demonstration apparatus depicted so effusively throughout the textbook. Yet as did Silliman (as can be seen from Figs. 2 and 3, Silliman had Ganot's early French editions at hand as he wrote and illustrated his own textbook!), Ganot and Atkinson provided hundreds of experimental narratives, in passive voice. One can almost envision the students

29 Although Ganot-Atkinson's *Elementary treatise* would remain in a single volume, the page count grew steadily from 888 pages in 1870 to more than 1225 in the 18th English edition (1917).

following along in their textbooks as the instructor performed the demonstration, guided by Ganot's narrative, at the front of the classroom.[30]

Fig. 3: Axonometric illustration of instruments in Ganot, 1856, 516. Nearly identical versions of this cut appear in all French and English editions of Ganot through the 1880s.

But students using Ganot-Atkinson were not merely entertained by the demonstrations. Already in his early French editions Ganot added at the end of the volume a list of numerical problems with answers, "Sujets de composition de physique donnés a la Faculté des Sciences de Paris." Only simple algebra and logarithms (presented as if readers already understand them) were required to solve the problems; trigonometric functions were not needed. For the early English editions, Atkinson did not include such problems. But his seventh edition (New York, 1875) presents 202 numerical problems in an appendix of 19 pages. In the preface of that edition, Atkinson noted that "such an appendix has from time to time been

30 To give only one example, from Ganot's section on magnetism: "The opposite actions of the north and south poles may be shown by the following experiment: -- A piece of iron, a key for example, is supported by a magnetised bar. A second magnetised bar of the same dimensions is then moved along the first, so that their poles are contrary (fig. 551). The key remains suspended so long as the two poles are at some distance, but when they are sufficiently near, the key drops, just as if the bar which supported it had lost its magnetism ...." Adolphe Ganot, *Elementary treatise on physics, experimental and applied for the use of colleges and schools*, trans. E. Atkinson, 9th ed. (New York: W. Wood and Co., 1879), 581–82. The same text and illustration appear in the earliest French edition I have at hand, Adolphe Ganot, *Traité élémentaire de physique expérimentale et appliquée et de météorologie*, 6th ed. (Paris: L'auteur-éditeur, 1856), 503–4.

urged upon me by teachers and others who use the work. It will, I conceive, be more useful to those students who have not the advantage of regular instruction, affording to them a means of personally testing their knowledge.... He should habituate himself to write out at length the several steps by which the result is obtained, so that he may bring clearly before himself the physical principles involved in each stage."[31] Clearly, for Atkinson conveying the "physical principles," either by demonstration experiments or numerical problems, remained the core of teaching physics.

Some of Atkinson's problems feature physical principles that could not have been easily presented in actual experiments.[32] Others require knowledge of such principles but in experimental contexts that could have been widely available in late nineteenth-century physics classrooms.[33] However neither Ganot's nor Atkinson's numerical questions ever required the students themselves to manipulate apparatus or make physical measurements. In the era of teaching physics by "demonstration and illustration," only instructors wielded the experimental apparatus. The students watched the spectacle and followed along in illustrated textbooks such as Ganot/Atkinson.

3. TEACHING PHYSICS IN AMERICA BY "PRACTICAL EXERCISES" AFTER 1880

Around 1880 the teaching of physics at Dartmouth College changed fundamentally, as it was changing at many colleges, universities and secondary schools across North America. For a variety of reasons that cannot here be discussed, American institutions of higher education embraced two reforms in the 1870–80s.[34] Increasingly, elective courses entered American curricula; individual stu-

31 Adolphe Ganot, *Elementary treatise on physics, experimental and applied for the use of colleges and schools*, trans. E. Atkinson, 6th ed. (New York: William Wood and Co., 1875), vii.

32 For example, question 16: "How far will a heavy body fall in vacuo during the time in which its velocity increases from 40.25 feet per second to 88.55 feet per second?" Ganot, *Physics* 1875, 909.

33 Question 173: "On a table where the earth's magnetism is counteracted, the north pole of a compass needle makes 20 oscillations in a minute under the attraction of a south pole 4 inches distant; how many will it make when the south pole is 3 inches distant?" Ganot, *Physics* 1875, 926.

34 For a useful introduction, see John R. Thelin, *A history of American higher education* (Baltimore: Johns Hopkins University Press, 2004). For the rise of the student laboratory movement, which began in chemistry and by 1880 had entered physics, see Edward C. Pickering, "Physical laboratories," *Nature* 3 (1871): 241; Frank Wigglesworth Clarke, *Report on the teaching of chemistry and physics in the United States* (Washington, D.C.: GPO, 1881); Albert Moyer, "Edwin Hall and the emergence of the laboratory in teaching physics," *Physics teacher* 14 (1976): 96–103; Melba Phillips, "Early history of physics laboratories for students at the college level," *American journal of physics* 49 (1981): 522–27; Keith Sheppard and Gail Horowitz, "From Justus von Liebig to Charles W. Eliot: The establishment

dents began to have some choice in deciding what to study. And some of those new elective courses involved hands-on laboratory work by the students. At Dartmouth starting in 1874–75, seniors could choose to enroll in "optional" courses not required of everyone. First an option for "practical chemistry" appeared in the curriculum, soon followed by "practical astronomy" and in 1879–80 by "practical physics." After a century of watching experimental demonstrations performed by their professors, Dartmouth's undergraduates by 1880 began handling the apparatus themselves. Working late into the night completing the laboratory exercises, they attracted the attention and comment of their peers.[35] By 1886, when Harvard began requiring its incoming students to have completed a course of laboratory exercises in high school, the "student laboratory movement" began to enter American secondary schools. New laboratory spaces, new apparatus, new textbooks, and considerable sums of money would be required to effect this transformation of physics pedagogy. Not that the spectacle of classroom demonstrations had been abandoned; rather, the sites for physics instruction at America's school and colleges had expanded.[36]

Table 1: Dartmouth's Physics Classes, 1882–83[37]

Level	Audience	Title	Length	Textbooks
Physics 1	Junior, required	Elementary physics	77 recitations & 20 lectures	Gage
Physics 2	Junior, elective	Advanced theoretical work & experiments in laboratory	45 exercises, 2 hours each	Pickering, Maxwell, Stone
Physics 3	Junior, elective	Advanced textbook and laboratory work (heat and light)	40 exercises, 2 hours each	Pickering, Stewart, Lommel
Physics 4	Senior, optional	Practical work in laboratory (magnetism and electricity)	44 exercises, 2 hours each	Pickering, Jenkins

of laboratory work in U.S. high schools and colleges," *Journal of chemical education* 83 (2006): 566–70. For an important study of Dartmouth's shift, in the 1880s, to an increasingly professional, discipline-based faculty (curricular changes are not discussed, however), see Marilyn Tobias, *Old Dartmouth on trial: The transformation of the academic community in nineteenth-century America* (New York: New York University Press, 1982).

35 *The Dartmouth* 5 (1883), 10.

36 A systematic survey, in 1879–81, of how physics and chemistry were being taught at American universities and colleges found that 40 of the 135 reporting institutions were offering practical laboratory exercises for students in physics. See Clarke, *Teaching of chemistry and physics*. Dartmouth's curricular changes reflected this national trend.

37 *Catalogue of Dartmouth College and the associated institutions* (Hanover, New Hampshire: Dartmouth College, 1882), 22–23.

At Dartmouth, a new two-track physics curriculum appeared in 1882–83 (see Table 1). In addition to the one course required of all juniors, the curriculum now included two elective courses of laboratory instruction (meaning that students were required to select one of two or three courses for a given time slot) and one optional course in laboratory work for seniors. Each class met five or six times weekly over terms of 16 or 11 weeks' duration.

All juniors were still required to take Physics 1, taught in the earlier demonstrative style with lectures, extensive classroom demonstrations by the professor, recitations or "blackboard work" by individual students in class, and the reading of a simpler version of the Ganot-Atkinson textbook, newly published by Alfred P. Gage (Dartmouth Class of 1859).[38] Like Ganot-Atkinson, Gage provided numerical problems and hundreds of woodcuts of apparatus for demonstration experiments. But unlike the European text, Gage also included instructions for simple experiments to be performed by the students, invariably to reveal phenomena; seldom were quantitative measurements required.[39] In his preface, Gage vigorously called for a new pedagogy for natural science, centered on student laboratory exercises, that he called "object teaching."

> Shall the teacher manipulate the apparatus, and the pupil act the part of an admiring spectator? Or, shall the pupil be supplied with such apparatus as he cannot conveniently construct, always of the simplest and least expensive kind, with which he shall be required, under the guidance of his teacher, to interrogate nature with his own hands? By which method will he acquire the most vigorous growth, and be most likely to catch something of the spirit which animates and encourages the faithful investigator? Can elegantly illustrated works and lucid lectures on anatomy and operative surgery take the place of the dissecting room? Have lecture-room displays proved very effectual in awakening thought and in kindling fires of enthusiasm in the young? Or would a majority of our practical scientists date their first inspiration from more humble beginnings, with such rude utensils, for instance, as the kitchen affords? Is the efficiency of instruction in the natural sciences to be estimated by the amount of costly apparatus kept on show in glass cases, labelled 'hands off', or by its rude pine tables and crude apparatus bearing the scars, scratches, and other marks of use?

Only after students had performed their own experiments should the teacher introduce "statements of definitions and laws," Gage exhorted. "Do not teach pupils to

38 Alfred P. Gage, *A textbook on the elements of physics for high schools and academics* (Boston: Ginn, Heath & Co., 1882). In the preface [vii], Gage thanked a dozen university and high school physics teachers, including Dartmouth's professor of physics, for reading and criticizing proofs of the texbook. A separately paginated four-page notice, bound at the back of the copy I have used, presents testimonials from these faculty, including professors from the University of Michigan, Johns Hopkins, Yale, Wellesley College, and Dartmouth. Gage's text was reprinted nearly every year through at least 1901.

39 For example, from the chapter on current electricity: "Take a strip of sheet of copper and a strip of sheet zinc, each about 10 cm long and 4 cm wide. Take also a tumbler two-thirds full of water, and to it add about two tablespoonfuls of sulphuric acid. Place the zinc strip in the liquid; instantly bubbles of gas collect on the surface of the zinc, break away from it, rise to the surface of the liquid, and are rapidly replaced by others. These are bubbles of hydrogen gas, and may be collected and burned. It is soon found that the zinc wastes away, or is dissolved in the liquid." Gage, *Elements of physics* 1882, 179.

swim before entering the water." [40] An advertisement bound at the end of Gage's book indicated that the author could "furnish teachers, in short notice and at a cost as low as possible, any apparatus described in this text-book." By 1883, Gage and his son had launched their own instrument company, selling simple, rugged, and cheap apparatus for the student laboratory. His textbook, with the motto "Read nature in the language of experiment" embossed on the cover, has been called "the first American text to emphasize student experiments."[41] Within a few years, American textbook publishers would market even more complete infrastructure for student labs; along with textbooks, schools could buy blackboards, laboratory furniture, student apparatus, wall charts, etc. Gage's company, as historian Steven Turner has noted, would become the Zeigler Electric Company, which became the Hall Scientific Company, which became L.E. Knott Apparatus Company in 1907. These firms supplied the equipment for America's student laboratory movement.[42]

For its new laboratory courses (Physics 2–4), Dartmouth used other textbooks, written at a more advanced level than Gage's. As can be seen in Table 1, each of the practical courses used the first widely distributed American manual for student laboratory work in physics, the two-volume *Elements of Physical Manipulation* (1873–76) by Edward C. Pickering, then professor of physics at the newly founded Massachusetts Institute of Technology.[43] Pickering emphasized measuring but even more strongly he emphasized "physical manipulation" which

40 Gage, *Elements of physics* 1882, iii–iv. Gage was silently quoting Clarke, *Teaching of chemistry and physics*, 10: "... much teaching of science preliminary to laboratory practice is like lectures upon swimming before the pupil enters the water." For an earlier, similar exhortation, see "Object teaching and science," *Scientific American* 24 (1871): 39–40.

41 Moyer, "Edwin Hall and the emergence of the laboratory in teaching physics," 98.

42 Steven Turner, "The reluctant instrument maker: A.P. Gage and the introduction of the student laboratory," *Rittenhouse* 18 (2004): 56.

43 A 72-page version had been published in 1872. Pickering's *Elements* would be published in at least six American and London editions through 1886. As far as I can tell, neither French nor German editions were published. Indeed no copies of Pickering's manual, in any language, can be found today in any German library, according to the comprehensive on-line Karlsruher Virtueller Katalog. Interesting, the first widely reprinted German manual for student laboratory exercises, Friedrich Kohlrausch's *Leitfaden der praktischen Physik* (Leipzig: Teubner, 1870), appeared just prior to Pickering's *Elements* and one might wonder whether the American followed the German example. Like Pickering, Kohlrausch began with the mathematics of observational error, offered numbered measuring exercises, and concluded with a set of mathematical and physical tables. However, neither the tables nor the exercises in the two manuals overlap; Kohlrausch's mathematics are more demanding (he introduces least squares, for example); Kohlrausch wrote in passive voice, never providing procedures for students to follow directly; and perhaps most importantly, Kohlrausch never encouraged students to construct their own apparatus. In general, Kohlrausch provided neither images nor detailed discussions of particular instruments. The two manuals did not copy each other. English translations of Kohlrausch appeared in London and New York, in several editions from 1874 through 1900. A visitor to German university physics institutes in 1903 found Kohlrausch's manual everywhere in use. See G.W. Küchler, "Physical laboratories in Germany," *Occasional Reports, India, Department of Education* 4 (1904): 190; Cahan, "Institutional revolution," 48–50.

teaches the student "to think for himself. This should be encouraged by allowing him to carry out any ideas that may occur to him, and so far as possible devise and construct, with his own hands, the apparatus needed ...." Originality was the goal. "The highest aim of every physicist should be to direct, not only his own utmost efforts, but those of his students, toward original investigations of the determination of new facts and laws. Without this, he is liable to become a mere machine, disseminating knowledge, but never advancing it. The whole aim of this book has been in this direction, and without it, we may educate followers, but never leaders in science."[44] Pickering wanted to train future creative scientists; the rhetoric of his textbook did not extol the moral character or civic virtues that experiments convey to the general populace or the practical skills that precision measurement give to future school teachers, language that had been used a generation earlier as German physicists sought to justify student experimentation.[45]

Rather, Pickering's *Elements* instantiate a vision for learning by doing that animate the founding documents of MIT, prepared by William B. Rogers, the geologist, president of the American Association for the Advancement of Science (in 1870), and educational reformer who in 1861 had pushed the Massachusetts State Assembly to charter the new Institute and would become its first president. As outlined in Roger's 1864 "Scope and Plan of the School of Industrial Science," MIT's mission was to offer "instruction in the leading principles of science, as applied to the [useful] arts." In addition to lectures and examinations, the methods of instruction were to include laboratory training in chemical analysis; mechanical and architectural drawing; practical exercises in surveying, levelling, and nautical astronomy; excursions to inspect machines, processes of manufacture, buildings, works of engineering, and mines; and "practice in physical and chemical manipulations."[46] The later Rogers carefully described:

> It will be the object of these exercises to make the student practically familiar with the adjustments and use of the apparatus and agents employed in the more important experiments and processes in natural philosophy and chemistry. With this view, the students, under the direction of their teacher, will be called, by small classes at a time, to execute with their own hands various experiments in mechanics, pneumatics, sound, optics, electricity, and other branches of experimental physics, and to exhibit chemical re-actions, to fit up chemical apparatus, to prepare gases and other products, and demonstrate their properties by suitable experiments, accompanying these manipulations, when required, with an explanation of the apparatus used or of the process or experiment performed.[47]

In the "laboratory for physics and mechanics," continued this document:

44 Edward C. Pickering, *Elements of physical manipulation*, 2 vols. (New York: Hurd & Houghton, 1873–76), i:vi; ii:299.

45 For a classic account of some of the earliest German practical exercises in physics, dating back to the 1840s, see Kathryn M. Olesko, *Physics as a calling: Discipline and practice in the Königsberg seminar for physics* (Ithaca: Cornell University Press, 1991).

46 William B. Rogers, *Scope and plan of the School of Industrial Science of the Massachusetts Institute of Technology* (Boston: John Wilson and Son, 1864), 3, 19.

47 Rogers, *Scope and plan*, 23.

> ... it is proposed to provide implements and apparatus with which the student may be exercised in a variety of mechanical and physical processes and experiments. Thus he may learn practically the methods of estimating motors and machines by the dynamometer, of experimenting on the flow of water and air or other gases, and of testing the strength of the materials used in construction. He may become familiar with the adjustments and applications of the microscope; be practised in observing with the barometer, thermometer, and hygrometer; and, in a room fitted up for photometry, may learn the mode of measuring the light produced by gas and other sources of illumination, and the value of different kinds of burners, lamps, and their appendages.[48]

Rogers' vision of physics was exceedingly practical; experiments in electricity, magnetism, optics, or the mechanics of motion, topics regularly treated in the mid-century textbooks described above, do not appear in his list. And his vision of "manipulations" is student-focused; the "scope and plan" document does not mention lecture demonstrations by faculty. MIT's laboratories and apparatus would be for students' use.

Pickering, who in 1867 (at only 20 years of age!) had replaced Rogers as MIT's physics professor, presented his predecessor's vision in the two volumes of *Physical Manipulation*.[49] For the 200 sequentially numbered experiments, Pickering described the apparatus needed, and the protocol to follow in making the measurements and analyzing the data. For the latter he encouraged both "analytical" (probable error) and "graphical" methods (drawing curves or surfaces for two or three variables, respectively). He offered an appendix of another 100 numbered "problems," presented tersely so that students would be encouraged to devise their own experimental approaches and thereby start to work more

48 Rogers, *Scope and plan*, 24.

49 Due to financial constraints, it was not before 1869–70 that Pickering was able to offer 60 students a full course of laboratory work in a dedicated room with the necessary apparatus. Describing the new venture in April, 1869, another MIT professor remarked that "in time we shall revolutionize the instruction in Physics just as has been done in Chemistry ... Now you know that this idea belongs to our dear President [Rogers], and was one of the prominent features of his plan of the School of the Institute." Several months later, Rogers claimed both personal and institutional priority for the innovation: "It gives me pleasure to refer to the ability with which Prof. Pickering has carried out my views of a Laboratory of Physical Instruction. This is, I believe, the first laboratory of the kind ever established, and, as it furnishes practical training not hitherto attempted in any systematic way, will give the students of the Institute peculiar advantages in their studies, and in future researches in this very important branch of Science." In 1873 Pickering dedicated his *Elements* to "William Barton Robers, the first to propose a physical laboratory." The claim that MIT established the first hands-on student physics laboratory in America quickly became anchored in the literature. Cf. Julius A. Stratton and Loretta H. Mannix, *Mind and hand: The birth of MIT* (Cambridge: MIT Press, 2005), 580–88 (quotes from 583 and 586); T.C. Mendenhall, "Address," *Proceedings of the American Association for the Advancement of Science* 31 (1882): 127–38; Frank P. Whitman, "The beginnings of laboratory teaching in America," *Science* 8 (1898): 201–6; H.M. Goodwin, "Physics at MIT: A history of the department from 1865–1933," *Technology review* 35 (1933): 287–91, 312–13. For a claim of earlier priority for student laboratories, at Rensselaer Polytechnic Institute starting in 1824, see Palmer C. Ricketts, "The first chemical and physical laboratories for the use of individual students," *Review of scientific instruments* 4 (1933), 571–74.

independently.[50] He divided the experiments into sections--general experiments (weights and measures, verniers, placing cross hairs on telescopes, etc.), mechanics of solids, mechanics of liquids and gases, sound, light, heat, mechanical engineering (boilers, dynamometers, fly-wheels, etc.), electricity, meteorology, astronomy, and lantern projections. But the focus of the volumes remains physical manipulation *per se*; they do not attempt to provide a comprehensive picture of physics with all its quantitative, physical laws.

Pickering wrote directly for the student, often in the imperative voice ("Turn the galvanometer around horizontally, reading the two ends of the needle ...."[51]). But he also provided guidance for the laboratory instructor, describing how to organize the logistics of laboratory exercises. The apparatus for each experiment was to be placed on a separate table in the laboratory room.

> A board called an indicator is hung on the wall of the room, and carries two sets of cards opposite each other, one bearing the names of the experiments, the other those of the students. When the class enters the laboratory, each member goes to the indicator, sees what experiment is assigned to him, then to the proper table where he finds the instruments required, and by the aid of the book performs the experiment .... By following this plan an instructor can readily superintend classes of about twenty at a time, and is free to pass continually from one to another, answering questions and seeing that no mistakes are made.[52]

50 For example, no. 282: "Determine the dynamical equivalent of heat from the thermal effects of electric currents by the method of Joule." Pickering, *Physical manipulation*, ii:305.

51 Pickering, *Physical manipulation*, ii:25.

52 Pickering, *Physical manipulation*, i:vi–vii. For an expanded description of his "laboratory system of teaching physics," see Edward C. Pickering, "Report on the physical laboratory of the Mass. Institute of Technology [1870]," *Compilation of the papers on physics written by Professor Edward C. Pickering* (Cambridge: n.publ., 1877), unpag. For a slightly earlier American student laboratory manual, more elementary than Pickering's and designed for use in secondary schools as well as colleges, see Gustavus Hinrichs, *The elements of physics, demonstrated by the student's own experiments* (Davenport, Iowa: Griggs, Watson & Day, 1870). Although Hinrichs wrote his manual to accompany lectures, recitations, and student laboratory exercises, begun in 1865–66 at Iowa State University (so that it contains more exposition of physical principles than does Pickering's manual), Hinrich's description of the arrangement of lab work for groups of 20 students is nearly identical to Pickering's. By 1871–72, Hinrichs was conducting lab exercises for 240 students per semester, in a new laboratory opened in 1867 and explicitly modeled on Liebig's chemical laboratory in Giessen. More research would be required to determine whether Pickering borrowed from Hinrichs. Although Hinrichs did edit a short-lived journal, *The school laboratory of physical science*, 1871–72, to advance his campaign for student laboratories in schools and colleges, his manual was never reprinted and today is very rare. See "Experimental science in schools," *Nature* 4 (1871): 421–22; Gustavus Hinrichs, "On the method of instruction in the elements of physical science," *School laboratory of physical science* 1 (1871): 64–70; Gustavus Hinrichs, "Laboratory news," *School laboratory of physical science* 2 (1872): 38–47. For more on the rather quirky character of Hinrichs, see Carl A. Zapffe, "Gustavus Hinrichs, precursor of Mendeleev," *Isis* 60 (1969): 461–76; James P. Wells, *Annals of a University of Iowa department: From natural philosophy to physics and astronomy* (Iowa City: Department of Physics and Astronomy, University of Iowa, 1980), 27–62.

When he began such laboratory exercises at MIT, Pickering's students had required 1.8 hours on average per experiment; by the time his second volume appeared in 1876, that average had dropped to 1 hour. Pickering urged that the apparatus be cheap and home-made, constructed whenever possible from wood with calibrated scales drawn on paper rather than inscribed in more expensive metal. Pickering was less interested in precision than in the manual processes of manipulation and data generation.

Fig. 4: Illustrating the demonstrator in Stone, 1879, 58.

For the practical physics exercises at Dartmouth in 1882–83, five additional textbooks were used. Physics 2, "advanced theoretical work with experiments in laboratory," followed two British texts. James C. Maxwell's *Matter and motion* (London, 1876; New York, 1878), a popular book first published in London by the Society for Promoting Christian Knowledge, is rather abstract. Using diagrams (in the manner of Newton's *Principia*) rather than equations to represent physical principles, providing no quantitative problems or experiments of any kind, Maxwell's book could not differ further from the style of Pickering. Maxwell offered no pedagogical hints about how to teach physics or how to use his book in that enterprise. But his discussions of vectors and his representation of material systems of points by diagrams would have provided a foundation for a mathematical treatment of analytical mechanics. Physics 2 also used W.H. Stone's *Elementary lessons on sound* (London, 1879; New York, 1891), a Ganot-type book filled with woodcuts illustrating acoustical demonstration apparatus (but unlike Ganot, often including the demonstrator in the image; see Fig. 4).[53] Maxwell's and Stone's texts provided descriptive, non-quantitative prose about the phenomena being manipulated via Pickering's instructions.

Physics 3 at Dartmouth also supplemented Pickering with two European texts. Balfour Stewart's *Elementary treatise on heat* (Oxford, 1866) presents Ganot-type woodcuts of apparatus but unlike the French author offers numerous quantitative examples and graphic representations of principles like the second law of thermodynamics. Stewart's overall structure is conceptual, not experimental. His book begins with the effects of heat on bodies, moves to the laws which regulate the distribution of heat in space, and concludes with the nature of heat (the kinetic theory and the first two laws of thermodynamics). Then director of the Kew Observatory outside London, Stewart in the 1860s undertook extensive researches on radiant heat. Eugen Lommel, the only German author of Dartmouth's prescribed physics textbooks, was a professor of physics at the university in Erlangen. His *The nature of light, with a general account of physics optics* (Leipzig, 1874; London, 1875; New York, 1876) had appeared in the International Scientific Series, a global project of translation and science popularization launched in the early 1870s by Edward L. Youmans, the American enthusiast of Herbert Spencer and evolution.[54] "Mathematical reasonings," wrote Lommel in his preface, "are wholly omitted in the text; where these are required to appear to be desirable for the more thorough and complete knowledge of the phenomena described, they are given in the most elementary form .... I trust that this attempt to render a branch of physics, which at first sight seems from its delicate nature to lie somewhat beyond the grasp of the general public, intelligible, will meet with a kindly reception and consideration at their hands." Not intended for classroom use, Lommel's text offers no problems or experimental protocols. Most of the chapters present the phenomena of reflexion, refraction, dispersion, and absorption "without reference to any theory of the nature of light"; the wave theory is used in later chapters to explain interference, double refraction and polarization.[55]

Physics 4, presumably the most advanced course in Dartmouth's new curriculum, supplemented Pickering with another European text, Fleeming Jenkin's *Electricity and magnetism* (London, 1873; New York, 1873), a book that went through many English and foreign-language editions. Professor of engineering at the University of Edinburgh and well known for his work with William Thomson on submarine cables, Jenkin published his text in the London publisher Long-

53 Stone was a well-known amateur musician and physician in London. From 1875, he lectured on natural philosophy at St. Thomas Hospital, from 1881 on musical acoustics at Trinity College London. He also served as vice president of London's Physical Society from 1878 until his death in 1891. See James D. Brown and Stephen S. Stratton, *British musical biography* (London: William Reeves, 1897), s.v. "Stone."

54 Roy MacLeod, "Evolutionism, internationalism, and commercial enterprise in science: The International Scientific Series, 1871–1910," *Development of science publishing in Europe*, ed. A. J. Meadows (Amsterdam: Elsevier, 1980), 63–93.

55 Eugene Lommel, *The nature of light, with a general account of physical optics* (New York: D. Appleton and Co., 1876), v–vi.

mans' series "Textbook of science adapted for the use of artisans and students in public and other schools." Some Ganot-type woodcuts illustrate apparatus, but the book is also filled with schematic diagrams of electrical circuits, diagrams of magnetic field lines and other phenomena in three-dimensional space. Its mathematics is limited to algebra and the occasional trigonometric function. In his introduction, Jenkin complained that his subject had been too often presented in textbooks as an "incoherent series of facts," as a "long roll of disjointed experiments." Jenkin sought to reverse this presentation. "Not a single electrical fact can be correctly understood or even explained until a general view of the science has been taken and the terms employed defined. The terms which are employed imply no hypothesis, and yet the very explanation of them builds up what may be called a theory." Thus, Jenkin first presented "a general synthetical view of the science" in chapters on electric quantity, potential, current, resistance, and magnetism, and then presented chapters on various apparatus and types of measurement, concluding with practical sections on telegraphs, telegraphic lines and their problems.[56] Jenkin's book offers no problems to solve or experimental protocols to follow. Of all the books used in Dartmouth's physics curriculum for 1882–83, Jenkin's was the most demanding, both conceptually and mathematically; but it was not intended as a guide for practical exercises in laboratory physics.

Dartmouth's new curriculum of practical exercises thus represented an amalgam of conceptual content from European texts and practical content from Pickering's American manual of physical manipulation. The fact that New York editions appeared nearly immediately upon the European publication of the former texts suggests the existence of a ready American market for these works (only Stewart's book on heat never found an American publisher). Clearly Dartmouth was not the only American institution requiring these textbooks for an expanded physics curriculum.

Over their first decade, Dartmouth's new laboratory-based physics courses did not immediately attract large numbers of students. As can be seen in Table 2, the junior electives (Physics 2–3) generally enrolled from one to two dozen students, fewer than half of the junior class. The senior optionals (Physics 4–6), emphasizing independent study, only enrolled a handful of students. However, these enrollment figures are probably quite comparable to those of the medium-sized German universities in this decade.[57] Pickering's expectation, that properly taught laboratory exercises would create physicists, did not immediately bear fruit at Dartmouth. As far as I know, only three seniors wrote "honors theses" in physics during this decade; and only one of those, Edwin B. Frost, Dartmouth 1886, would enter a scientific career, first as an assistant professor of physics at Dart-

56 Fleeming Jenkin, *Electricity and magnetism* (London: Longmans, Green, and Co., 1873), vi–vii.

57 Cf. Cahan, "Institutional revolution," 44–48.

mouth and then as long-time director of the Yerkes Observatory at the University of Chicago.[58]

Table 2: Dartmouth's physics enrollments by year, 1882–1892[59]

	Physics 1	Physics 2	Physics 3	Physics 4	Physics 5	Physics 6
1882–83	56	15	12	4	4	--
1883–84	46	0	5	0	2	1
1884–85	56	13	7	2	0	0
1885–86	64	23	9	4	0	0
1886–87	49	18	2	2	0	0
1887–88	54	--	8	2	2	4
1888–89	58	--	10	2	2	0
1889–90	48	19	13	3	--	--
1890–91	54	24	4	0	2	0
1891–92	63	33	21	1	--	--

What happened in Dartmouth's new laboratory courses? Fortunately, Frost preserved his "Practical Physics Notes" from 1884–85 in a leather-backed, lined notebook neatly written in ink, with frequent graphs, carefully drawn on separate sheets of graph paper and pasted onto the pages (presumably Frost drafted these clean entries from rougher notes he had written in the lab). Frost's notebook records 56 experiments, all but four of which directly follow the procedures and order of those in Pickering's manual. At times Frost borrowed his prose verbatim from Pickering.[60] Frost's drawings of the apparatus also closely mimic the cuts in Pickering's text. But Frost also "learned by doing," often noting practical prob-

58 Frank Olds Loveland, "Magnetism at Hanover, N.H.," Honors thesis, Dartmouth College, 1886; Edwin B. Frost, "On a possible relation between kinetic electricity and the polarization of light by metallic reflection," Thesis for Final Honors in Physics, Dartmouth College, 1886; Benjamin Franklin Ellis, "Magnetism at Hanover," Honors thesis in physics, 1889. Dartmouth College Library Special Collections (henceforth DCLSC), codices 002192, 002194, 002207. Physics 5, "Original Experimentation and Research in Magnetism and Electricity, Theses," and Physics 6, "Elective Practical Work, Theses," were introduced in 1883–84 after four seniors in 1882–83 had taken what was then called Physics 4 twice. See *Catalogue of Dartmouth College and associated institutions*, 1882–83 through 1891–92 for the announced courses.

59 See Merit Roll Lists, DCLSC DA-80, 3297–3305, for the actual enrollments. Courses actually held do not always correspond to those announced in the printed catalogues. A dash indicates no data available in the roll lists.

60 For Frost's third experiment on the "Insertion of cross hairs," cf. Edwin B. Frost, *Practical physics notes*, ms notebook, [9], DCLSC MS–1034, box 4; Pickering, *Physical manipulation*, i:29–30.

lems not mentioned by Pickering. For example, in an experiment with pulleys and weights designed by Pickering to illustrate the "law of the parallelogram of forces," Frost wrote: "Owing to the friction on the pulleys & the stiffness of the thread the measurements are somewhat variable and great accuracy cannot easily be attained" (see Figs. 5 and 6).[61]

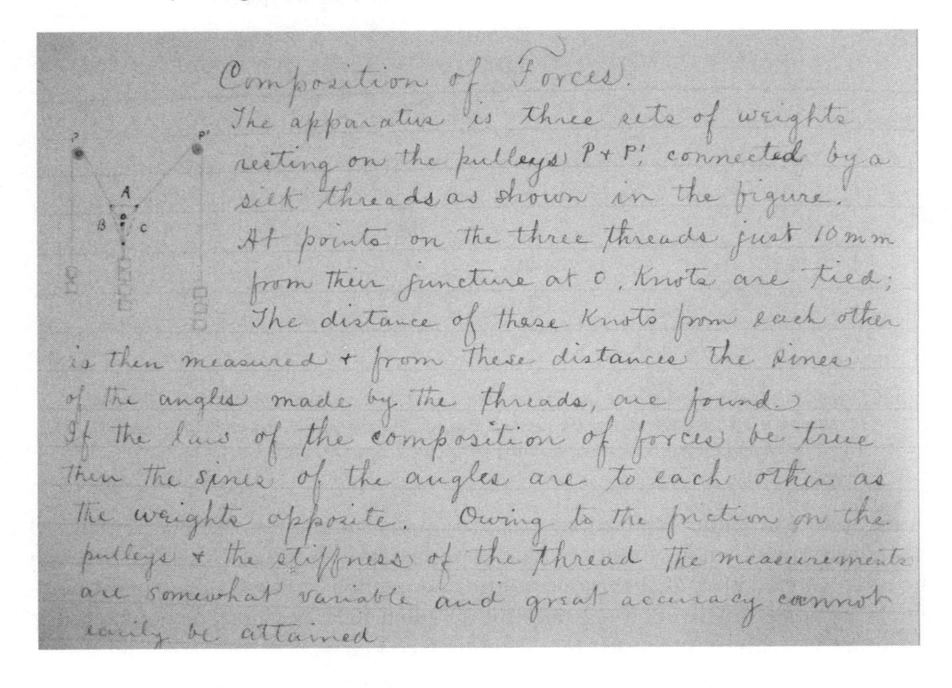
Composition of Forces.
The apparatus is three sets of weights
resting on the pulleys P + P'; connected by a
silk threads as shown in the figure.
At points on the three threads just 10 mm
from their juncture at O, Knots are tied;
The distance of these Knots from each other
is then measured + from these distances the sines
of the angles made by the threads, are found.
If the laws of the composition of forces be true
then the sines of the angles are to each other as
the weights opposite. Owing to the friction on the
pulleys + the stiffness of the thread the measurements
are somewhat variable and great accuracy cannot
easily be attained.

Fig. 5: Frost's notes for an experiment on the composition of forces with weights and pulleys, closely following Pickering's Experiment 23. Frost, Practical physics, [30].

61 Pickering, *Physical manipulation*, i:62; Frost, *Practical physics*, [30].

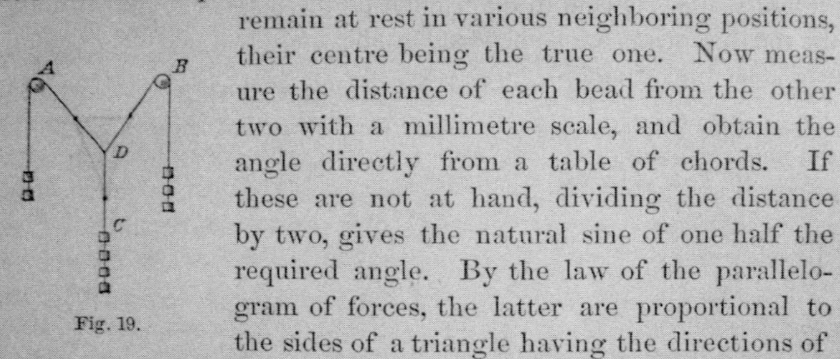
Experiment. Attach weights 2, 3 and 4 to the three cords, and let *D* assume its position of equilibrium. Owing to friction it will remain at rest in various neighboring positions, their centre being the true one. Now measure the distance of each bead from the other two with a millimetre scale, and obtain the angle directly from a table of chords. If these are not at hand, dividing the distance by two, gives the natural sine of one half the required angle. By the law of the parallelogram of forces, the latter are proportional to the sides of a triangle having the directions of the forces. But these sides are proportional to the sines of the opposite angles, hence the sines of the angles included between the threads should be proportional to the forces or weights applied.

Fig. 6: Pickering's instructions for Experiment 23 on the composition of forces. Note that a Dartmouth user has penciled in the central triangle required for a quantitative analysis. Pickering, Physical manipulation, i: 62. DCL Storage QC37 .P59 1873.

Although Frost had completed an exercise on probable error, usually he reduced his measurements more simply by computing mean values (see Fig. 7). Although he often entered both the measured values and a graph of those values in his notebook, at times Frost preserved only the graphed data and did not complete the quantitative analysis suggested by Pickering (see Fig. 8).[62] Hence, it would appear as if Dartmouth's physics students did not scrupulously follow Pickering's instructions; they retained some freedom in their "learning by doing" in the laboratory.

62 Pickering, *Physical manipulation*, ii:83, instructed students to graph the difference between the ambient temperature and the temperature along the bar versus the distance along the bar, and to notice that the relationship was exponential. "See if this is the case, by using as ordinates the logarithms of the excesses of temperature, and abscissas as before, when the result should be a straight line." In his notebook, Frost did not seek this exponential relationship, and even cut his graph paper so that the temperature scale on the vertical axis is missing.

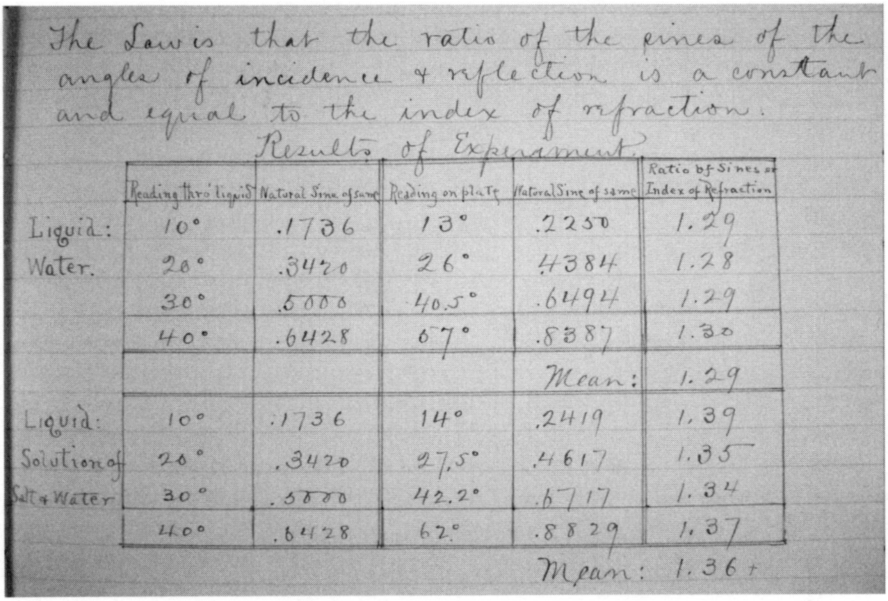

The Law is that the ratio of the sines of the angles of incidence & reflection is a constant and equal to the index of refraction.

Results of Experiment.

	Reading thro' liquid	Natural Sine of same	Reading on plate	Natural Sine of same	Ratio of Sines or Index of Refraction
Liquid: Water.	10°	.1736	13°	.2250	1.29
	20°	.3420	26°	.4384	1.28
	30°	.5000	40.5°	.6494	1.29
	40°	.6428	67°	.8387	1.30
				Mean:	1.29
Liquid: Solution of Salt & Water	10°	.1736	14°	.2419	1.39
	20°	.3420	27.5°	.4617	1.35
	30°	.5000	42.2°	.6717	1.34
	40°	.6428	62°	.8829	1.37
				Mean:	1.36+

Fig. 7: Frost's treatment of measurements in an experiment on indices of refraction. Frost, Practical physics, [88].

Only once, after an experiment not in Pickering on adjusting pendula to beat together, did Frost offer a critical comment: "The reasons for the existence of this experiment are yet to be seen."[63] And only once did Frost's recorded measurements (of the latent heat of fusion by melting snow in a calorimeter) yield such a large error that he had to explain: "This error [73.7° rather than 79°] is readily accounted for by the fact that the snow was not dry."[64] Generally however, Pickering's instructions appear to have yielded reproducible "physical manipulation" and reliable numerical results for the nineteen-year old Frost. As noted above, Frost would go on to pursue an illustrious career as one of America's leading observational astronomers in the first third of the twentieth century. Alas, we lack documentation from the less talented students who sat beside Frost in Dartmouth's new student laboratories; presumably they discarded their notebooks immediately after completing the class.

63 Frost, *Practical physics*, [28].
64 Frost, *Practical physics*, [62].

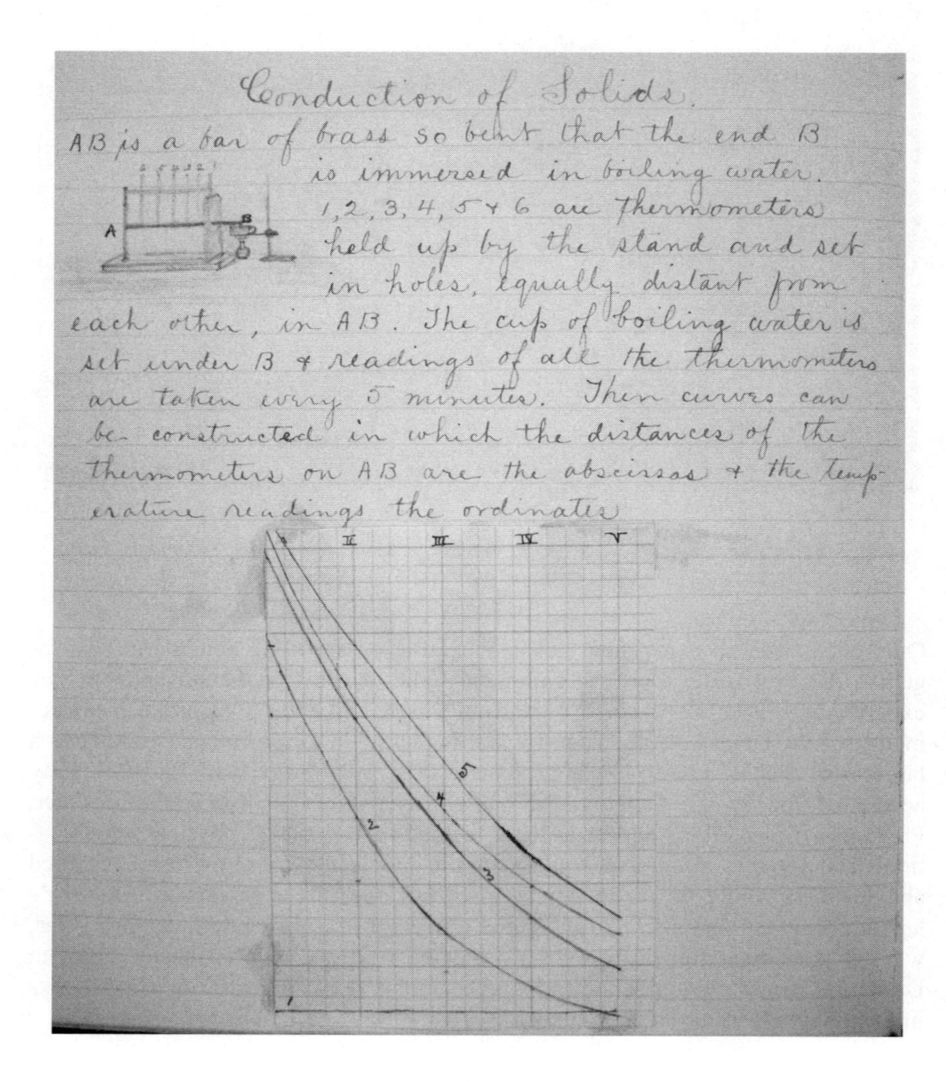

Conduction of Solids.

AB is a bar of brass so bent that the end B is immersed in boiling water. 1, 2, 3, 4, 5 & 6 are thermometers held up by the stand and set in holes, equally distant from each other, in AB. The cup of boiling water is set under B & readings of all the thermometers are taken every 5 minutes. Then curves can be constructed in which the distances of the thermometers on AB are the abscissas & the temperature readings the ordinates.

Fig. 8: Frost's graphical representation of measurements of heat conduction in a brass bar. Note that Frost sliced the units off the vertical axis of his graph. Frost, Practical physics, [73].

4. EUROPEAN INFLUENCES IN THE AMERICAN PEDAGOGICAL SHIFT TO PRACTICAL PHYSICS?

In the 1870–80s just as American universities and schools were introducing student laboratory work into their curricula, another educational reform was creeping into American physics. Americans were starting to obtain PhDs in physics, often from European universities; likewise, increasing numbers of Americans were traveling to Europe to visit university physics institutes. In the 1970s as he was preparing to write his now well-known study on the rise of the American physics community, historian Daniel Kevles assembled prosopographic data on this generation of young men who would become America's first physicists. Defining a "physicist" as anyone who published at least one article in leading American physics journals, Kevles identified 217 subjects who had published between 1870 and 1893. Of these, 55 earned PhDs in physics, 20 of whom received that degree outside the United States. A total of 42 of the 217 spent some time abroad studying physics (i.e., not everyone who traveled completed a foreign PhD).[65] For those travelers who returned to the States to take up teaching careers (about 75 percent of the travelers), did the European experiences shape their approaches to pedagogy? And since they overwhelmingly visited German universities (see Table 3), did the travelers bring back any peculiarly German practices to their subsequent teaching of physics in America? Kevles did not pose such pedagogical questions of his data.

65 Daniel J. Kevles and Carolyn Harding, *The physics, mathematics, and chemical communities in the United States, 1870 to 1915: A preliminary statistical report*, Social Science Working Paper, Nr. 94, typescript (Pasadena: California Institute of Technology, 1975). Nearly half of these 42 early travelers, upon their return to America, published primarily in the fields of geology, chemistry or astronomy. Defining "physicists" in nineteenth-century America remains problematic. Cf. David Cahan, "Helmholtz and the shaping of the American physics elite in the gilded age," *Historical studies in the physical and biological sciences* 35 (2005): 1–34. For a prosopographic study of the roughly 1000 Americans who studied at German universities from 1810 to 1870, men who focused mostly on the humanities (especially philology) rather than the natural sciences, see Carl Diehl, "Innocents abroad: American students in German universities, 1810–1870," *History of education quarterly* 16 (1976): 321–39. For a similar study of about 1100 Americans who studied at Göttingen from the 1840s through 1914 that hesitates to draw hard conclusions about what the travelers brought back, see Konrad Jarausch, "American students in Germany, 1815–1914: The structure of German and U.S. matriculants at Göttingen University," *German influences on education in the United States to 1917*, eds. Henry Geitz, Jürgen Heideking and Jürgen Herbst (Cambridge: Cambridge University Press, 1995), 195–211.

Table 3: American physics travelers' destinations, 1846–95[66]

	PhD earned	Overall visitors	New building
Berlin	5	19	1878
Heidelberg	3	9	1913
Paris	0	5	
Würzburg	3	4	1879
Göttingen	3	3	1905
Leipzig	2	3	1873
Strassburg	1	2	1882
Vienna	0	2	1913
Zürich	0	2	1890
Hannover	1	1	
Tübingen	1	1	1888
Bonn	1	1	1913
Other German cities	0	4	
Other British cities	0	2	
Unknown locations	0	5	

That these early American travelers flocked to German universities is not surprising since physics there was experiencing what historian David Cahan has felicitously called its "institutional revolution." In those years, driven by a complex dynamic of economic and cultural competition among states and a new "research ethos" in the university, German states invested millions of Marks in the construction of new buildings specially designed and outfitted for physics teaching and research. Between 1870 and 1920, every German university acquired a new physics institute, equipped with staff, sizeable budgets, teaching and research laboratories, and large amphitheaters equipped with elaborate technologies for lecture demonstrations to beginning students. Although by 1900, French, British, Italian and American universities would essentially match the German investment in the infrastructure of physics as well as in research productivity, American travelers in the early period nonetheless favored German destinations for their *Wanderjahre*.[67]

Unfortunately, for many of these traveling American physics students, scant documentation remains to reveal what they did in Europe or what they thought

66 Kevles and Harding, *Statistical report*; *American men of science* (New York: The Science Press, 1906); *Dictionary of scientific biography*, 18 vols. (New York: Scribner's, 1970–90); Paul Forman, John L. Heilbron and Spencer Weart, "Physics circa 1900: Personnel, funding, and productivity of the academic establishments," *Historical studies in the physical sciences* 5 (1975): 1–185. I have not gone beyond these standard sources in trying to determine more complete itineraries for the travelers. Some Americans visited more than one university; hence the number of visitors listed in column 3 totals more 42.

67 Forman, Heilbron and Weart, "Physics circa 1900."

about the physics they experienced there. Contemporaneous diaries and letters as well as later autobiographical memoirs often concentrate more on the travelers' impressions of European cultural life (museums, theaters, concerts, architecture, women's fashion) or exigencies of student life (struggles with languages, lodging, food, train travel, mountain climbing in the Alps) than on physics.[68] Nonetheless, we can sort the early travelers into two groups: those who settled, completing a PhD or spending one or more semesters at a given German university; and those who toured, visiting several universities and collecting physics experiences as tourists rather than as enrolled students. I cannot here examine all the scattered reports left by travelers in Kevles' sample; rather I will consider only several examples, documented in various levels of detail.

In 1864, MIT's Board of Governors sent its new president and professor of physics and geology, William Barton Rogers, to Europe to examine "the recent and best arrangements for working laboratories and lecture-rooms" and to purchase architectural and mechanical models for the new Institute's museum.[69] MIT was constructing its first building in Boston's Back Bay, and wanted to benefit from the latest European innovations as it outfitted the new structure. Rogers toured for five months, visiting his brother Henry near Glasgow and, it would appear, mostly museums rather than universities or technical schools. His somewhat sporadic correspondence from these months indicates that he saw the Edinburgh Museum of Science and Art (under construction, opened in 1866, now the National Museum of Scotland), the recently created South Kensington Museum in London (renamed the Victoria and Albert Museum in 1899), the Government School of Design also in South Kensington, the Conservatoire des arts et métiers and the Ecolé central des arts et manufactures in Paris, the Polytechnical School in Karlsruhe found in 1825, and other "schools and museums" not specified in the letters. He apparently spent a month in German areas, but no extant letters document that portion of his tour.

Rogers soon decided that most European institutions had little to teach him.[70] From Paris, he reported: "As far as laboratories and lecture-rooms are concerned I

68 For a richly textured autobiography, see James Morgan Hart, *German universities: A narrative of personal experience* (New York: G.P. Putnam's Sons, 1874). Hart, who earned a PhD in law at Göttingen in 1864 and returned to Germany in the early 1870s, reported nothing about physics or the other natural sciences.

69 Rogers to Henry Darwin Rogers, 13 April 1864, in Emma Savage Rogers, ed., *Life and letters of William Barton Rogers*, 2 vols. (Boston: Houghton, Mifflin and Company, 1896) ii:191–92. As his publications were exclusively geological, Rogers does not appear in Kevles's list of American physicists.

70 The Boston architect, William G. Preston, who designed MIT's first building for Rogers, had traveled in Europe from November 1863 through April 1864, systematically surveying the architecture and outfitting (heating, ventilation, lighting) of museum and school buildings. He praised the arrangements at the Ecoles des arts et métiers in Châlons-en-Campagne and Paris, but was very critical of what he saw in Germany. "The lecture rooms in most of the German schools that I have visited are extremely primitive affairs, there being absolutely no arrangement for ventilation, the seats being uncomfortable, & the heating apparatus frequently consisting merely of a stove placed in the room .... Our proposed hall is certainly

believe we have little to learn either in England or Paris." In a public lecture presented after his return to Boston, Rogers was a bit more generous:

> ... we in this country have a great deal to learn in the arrangement of museums of practical art and science, and much also to learn in regard to the auxiliarities of practical art education. Yet our educational system is in many particulars abreast of the Old World schools, and in the elementary principles decidedly in advance of them. Looking to scientific education and methods of instruction, there is such vitality, quickness of observation and ready, flexible application belonging to our countrymen, that we have already embraced some of the most important ideas introduced in Europe. What is wanted is for American students to give time enough to secure thoroughness in the study of applied sciences.

The South Kensington Museum, Rogers continued, better grouped its natural specimens and artificial productions "for utilitarian and not scenic effect" than did the Conservatoire des arts et métiers. The Ecolé central drew his praise for "laying a broad foundation of scientific study" for "all branches of industry." And he lauded the only German institution he mentioned:

> The Polytechnic Institute at Carlsruhe, which is regarded as the model school of Germany and perhaps of Europe, is nearer what it is intended the Massachusetts Institute of Technology shall be than any other foreign institution. It has an extensive museum of models of all conceivable mechanical combination ... which are the objects of constant study by the pupils.... There are also series of laboratories adapted to the different branches of chemistry. Every part of the establishment is designed for use, and not for show. On the basement floor is a series of workshops, where the students are given pratical instruction.[71]

Nowhere in the sparce reports from his tour did Rogers remark on European physics laboratories; and despite his praise for the Karlsruhe polytechnical school, nowhere did he imply that MIT might borrow either architectural or programmatic ideas from Germany. Touring in 1864 before Germany's institutional revolution would create waves of lavishly outfitted physics laboratories, Rogers revealed himself to be a confident American educational reformer, not a novice eager to emulate European models.

My next American traveler and institution-builder displayed a similar self confidence. In 1875, H.A. Rowland had been hired to establish physics at the newly created Johns Hopkins University in Baltimore.[72] In 1875–76, the 27-year

one of sufficient importance to justify the adoption of all known improvements for the comfort of the audience & it is incumbent upon us I think to make it a model one—at least for the Germans." Preston to Rogers, 26 March 1864, MIT Archives, Rogers Papers MC 1, Box 4, Folder 47.

71 Rogers to Robert Rogers, 26 August 1864, and the *Boston Daily Journal*, 2 December 1864, in Rogers, ed., *Life and letters,* ii:205, 216–18. For an argument that Rogers "sought to import to the United States" a model of the useful arts from the French polytechnical schools, see A.J. Angulo, *William Barton Rogers and the idea of MIT* (Baltimore: Johns Hopkins University Press, 2009), 86–100 (quote at 89). Note that none of Kevles's traveling American physicists are known to have visited the polytechnical institute in Karlsruhe (see Table 3).

72 Hopkins's trustees had instructed Rowland to "mature a plan for a physical laboratory & for the purchase & construction of instruments." Gilman to Rowland, 21 June 1875, Johns Hopkins Library Rare Books and Manuscripts (henceforth JHLRBM), Rowland Papers MS 6. See Samuel Rezneck, "The education of an American scientist: H.A. Rowland, 1848–1901,"

old Rowland traveled to Europe with Hopkins' founding president, D.C. Gilman. For three months, the two men visited British and French universities, scouting for talent to hire, architecture to emulate, scientific instruments to purchase, and practical advice about how to build and run a university physics program. Rowland then spent three months in Berlin, conducting independent research in electromagnetism in Helmholtz's laboratory. For his final two months abroad, Rowland toured German, Austrian and Swiss universities and polytechnics. Since he was explicitly collecting information for his new job, Rowland carefully recorded his impressions, often eager to compare what he saw as national patterns.

In Britain, he complained about the £10,000 spent for "useless ornamentation" on Oxford's new physics building ("as usual the architect had got the best of the physicist") and the lack of apparatus. After a month of visiting British universities and observatories, Rowland summed up:

> I came over here with the notion that I should find everything and everybody better than in America but one month has wrought an entire change. In many things I see we can improve, but in how few! Taking up my special line of inquiry, for instance the workshops for instruments have to me the appearance of museums of antiquity and few of them use steam. Yet the workmen are so much better than ours that their work turned out is not often inferior to ours ... The time is coming when England will find herself behind, though in many things we cannot say that she is at present.... With regard to the departments of physics, and I may add of chemistry, which I have yet seen, I may say there is nothing to compare with what we have in America, not only in apparatus but in ingenious expedients for illustrating points. All the money and brains goes to making a fine building and none is left for other purposes. As to the professors I have yet met, they are all men like the rest of us![73]

In France, Rowland visited only the Collegé de France, noting that "their collection of physical apparatus [is] the best I have yet seen." Their laboratory, however, was small and used only by the faculty.[74]

It was in the German regions that Rowland did his most extended touring, visiting at least eighteen physics institutes from Bonn to Prague to Basel. Judging by his diary, Rowland directed his attention primarily to three features of these institutes: the technologies of demonstration in the large lecture theaters; student laboratory spaces and apparatus; and original research, conducted either by the professors or doctoral students. Upon starting his own research in Helmholtz's Berlin laboratory (i.e., before he had visited most other German universities), Rowland wrote Gilman about these features:

American journal of physics 28 (1960): 155–62; Samuel Rezneck, "An American physicist's year in Europe: Henry Rowland, 1875–1876," *American journal of physics* 30 (1962): 877–86; Cahan, "Helmholtz," 19–22.

73 Rowland to Gilman, 14 August 1875, JHLRBM, Gilman Papers, MS 1. One British physicist, however, did stand out for Rowland. "After seeing Maxwell I felt somewhat discouraged for here I met with a mind whose superiority was almost oppressive but I have since recovered and believe I see more clearly the path in which I hope to excel in the future. It lies midway between the purely mathematical physicist and the purely experimental, and in a place where few are working" (Ibid.).

74 Rowland to Gilman, 27 September 1875, JHLRBM, Gilman Papers, MS 1.

> You were right when you said I would find no lack of scientific spirit here [in Berlin] and the apparatus shows it. In America we have apparatus for illustration, in England & France they have apparatus for illustration and [student] experiment but in Germany they have *only* apparatus for experimental investigation. Our country is hardly ripe for the latter course though I should like to see it pursued to the best of our ability.[75]

After leaving Berlin, Rowland became ever less enthusiastic about the student laboratory exercises he observed (note, however, that most of the newly built physics institutes at German universities would open after Rowland's 1876 tour). In Göttingen, "the physical apparatus ... is not very good ... They have over 40 students in practical physics but where they put them I do not know." In Leipzig, "the collection of apparatus is not very large or valuable and there are only 25 stu[dents] in prac[tical] work. All [physics] Prof. Hankel's students are merely learning and do not make original researches."[76] The Dresden Polytechnic possessed a magnificent new building but "the collection of phys[ical] apparatus [is] not very good. There will probably not be much original work done here." In Prague, "the space devoted to physics is by no means large. The largest room contains the apparatus which is, for the most part, only for demonstration." The Munich polytechnic had a strong collection of apparatus and "about 40 students in lab but only 1 or 2 a year do any original work" (see Fig. 9). At Würzburg, Rowland "[s]aw Kohlrausch who was very pleasant. The University is old and poor but K. expects to have a new laboratory soon. He has no students making original experiments and only room for 10 or 12 in his laboratory. He had a very poor assortment of instruments and none that I had not seen before. There was a large collection of old instruments one or two hundred years old which are about to be thrown away." In March of 1876, Rowland summarized his appraisal of German physics teaching in a letter to his new friend, Maxwell. Yes, many new institutes are being built, but their size was "quite deceptive seeing that the professors and often their assistants live in the buildings." Rowland would have preferred to see "more instruments for use and fewer for illustration. I am surprised to find the instruments of research used here quite poor." Nonetheless, "I am doing my best to study the cause of the great amount of work [research] done in Germany, and the low state of the higher education in America ...."[77]

75 Rowland to Gilman, 5 November 1875, JHLRBM, Gilman Papers, MS 1. For a description of the Berlin physics institute a decade later, see Albert Guttstadt, *Die naturwissenschaftlichen und medicinischen Staatsanstalten Berlins* (Berlin: August Hirschwald, 1886), 135–48.

76 Another American, Edward L. Nichols, spent the 1875–76 year in Leipzig, attending Hankel's physics lectures and Gustav Wiedemann's "physikalische Uebung im Laboratorium." I have not found any sources documenting Nichols' impressions of the German universities he visited. But see Nichols's Collegien-Buch, Cornell University Library, Rare and Manuscript Collections, 14–22–m729.

77 H.A. Rowland, European tour, 1875, ms diary, 59, 63, 69, 80, 87–88, JHLRBM, Rowland Papers, MS 6; Rowland to Maxwell, March 1876, quoted in Rezneck, "Year in Europe," 885.

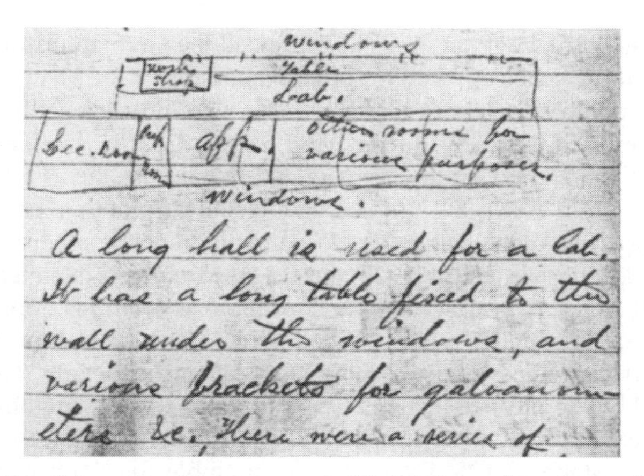
windows
tables
Lab.
app.
other rooms for various purposes.
windows.
A long hall is used for a lab. It has a long table fixed to the wall under the windows, and various brackets for galvanometers &c. There were a series of

Fig. 9: Floor plan of the student laboratory at the Munich Polytechnic, Rowland's Diary, 1876, 80. JHLSC, Rowland Papers, MS 6.

What most impressed Rowland at the German physics institutes was not their theatrical demonstrations in large lectures, their practical laboratory exercises for students, or the sophistication of their apparatus for independent student or professorial research.[78] Rather he was impressed by the idea of independent research, by the small but growing numbers of students who were completing original work for the PhD.[79] Rowland, himself, never earned this degree; but at Hopkins, he would train dozens of American PhDs in physics, men who would then establish graduate programs in physics at universities across the American continent. Rowland's travel reports suggest that he had long considered student laboratory exercises to be part of any university's physics curriculum. Rowland wrote as if he thought that British, German and American universities already had established that com-

78 For a slightly earlier traveler's more enthusiastic description of a newly opened German laboratory, see Gustavus Hinrichs, "A walk through the building of the Polytechnic School at Aachen," *School laboratory of physical science* 1 (1871): 71–77. Born in Schleswig-Holstein, Hinrichs had studied in Copenhagen from 1853–61, and would teach physical science at Iowa State University from 1862–86. Hinrichs concluded his article by exhorting Iowa to follow Prussia: "The schools have made Germany what it is--and still the schools of Germany are developing into higher perfection and greater thoroughness, thus promising still greater results to the Nation. May our State of Iowa be as liberal to its school at Iowa City as the Rhenish provinces of Prussia have been to their school at Aachen!" Yet after visiting German universities in 1873, Hinrichs reported: "In regard to facilities for advanced students, those laboratories are far ahead of any on this side of the Atlantic; but I am sincere in the conviction that many of our American laboratories already now do more good to a greater number of students than is done by similarly endowed laboratories in Europe." Quoted in Wells, *University of Iowa*, 48.

79 A visitor who toured the German physics institutes in 1903 found, on average, only five to six students engaged in original research at each university; Berlin with 30 was the outlier. Küchler, "Physical laboratories," 192. For a contemporaneous British view of expectations for physics laboratories, see Oliver Lodge, "An ideal physical laboratory for a college," *The Electrician* 26 (1890): 32–33, 66–68.

ponent of physics pedagogy; at times he criticized the German institutes he visited for not having more space devoted to student laboratories.

Later Dartmouth travelers to Germany also were not especially impressed by the student exercises they saw. In the early 1880s, Charles Sumner Cook, Dartmouth Class of 1879, traveled to Germany and Britain. As the assistant in physics and astronomy from 1881–87, Cook conducted Dartmouth's practical physics courses in those years (see Table 2). No contemporaneous records are extant to document Cook's travels. In his 1923 autobiography, Cook depicted himself as a tourist rather than a settler, describing in considerable detail his impressions of the leading European physicists whom he had met and heard lecture. But Cook remembered personal idiosyncrasies more than scientific details. Gustav Wiedemann, whose Leipzig physics lectures Cook attended for two months, drew praise only for his German diction. "It seemed that one scarcely needed to study German in order to understand him."[80] Even more memorable for Cook were the general physics lectures by the other Leipzig physics professor, Wilhelm Hankel. Apparently very nearsighted, Hankel had to feel the hands of a clock on the wall of his lecture theater to determine when to conclude his lectures, a practice which resulted, as one might expect, in student pranks that Cook hilariously related in his autobiography. Cook wrote nothing, however, about student laboratory exercises in Germany. Edwin Frost, Dartmouth Class of 1886, succeeded Cook as assistant in physics, and then spent the years from 1990–92 studying physics and astronomy at Strassburg and Potsdam. In his 1933 autobiography, Frost praised Kohlrausch's lectures and "valuable course in practical physics" at Strassburg, but gave no indication of ever seeking to bring German pedagogical practices back to the United States.[81]

An analysis of many more of the early American travelers to German universities would be required before we could conclude that the American pedagogical reforms of the 1870–80s — the introduction of practical laboratory exercises in physics — developed independently of European patterns. However, the few travelers we have considered, Rogers, Rowland, Cook and Frost, were not especially impressed by the German laboratory exercises they saw. And the nearly simultaneous appearance of the student lab manuals by Pickering and Kohlrausch in the early 1870s implies that physicists in both nations had turned to hands-on student exercises at roughly the same time. The next major reform in American physics pedagogy, the establishment of graduate programs to train PhD students in original research, would not emerge until the 1890s; German influences on this reform were considerable. That, however, is a story for another paper.[82]

80 Charles S. Cook, *Autobiography*, 1923, typescript, 27, DCLSC MS–529, Box 1, Folder 12. For a somewhat later description of the Leipzig physics institute, see Otto Wiener, "Das physikalische Institut," *Festschrift zur Feier des 500 jährigen Bestehens der Universität Leipzig*, eds. Otto Kirn and Emil Friedberg, vol. 4/2 (Leipzig: Hirzel, 1909), 24–60.

81 Edwin B. Frost, *An astronomer's life* (Boston: Houghton Mifflin Company, 1933), 68.

82 Cf. Melba Phillips, "Laboratories and the rise of the physics profession in the 19th century," *American journal of physics* 51 (1983): 497–503.

5. ACKNOWLEDGMENTS

I thank David Kaiser, David Pantalony, Josep Simon and participants in the Regensburg Symposium for stimulating conversations about pedagogy and textbooks; my student assistant, Nick Andrews, who “learned by doing” in helping with this research; and staff of the Dartmouth College Library Special Collections, the MIT Archives, and the Johns Hopkins University Library Special Collections. Figures 1–8 are reproduced courtesy of the Dartmouth College Library, Figure 9 is reproduced courtesy of the Johns Hopkins Library Special Collections.

REFERENCES

Angulo, A.J. (2009). *William Barton Rogers and the idea of MIT*. Baltimore: Johns Hopkins University Press.

Assmus, Alexi (1993). “The creation of postdoctoral fellowships and the siting of American scientific research.” *Minerva* 31: 151–183.

Bridge, Bewick (1814). *A treatise on mechanics, intended as an introduction to the study of natural philosophy*. London: T. Cadell and W. Davies.

Brown, James D., & Stratton, Stephen S. (1897). *British musical biography*. London: William Reeves.

Brown, Sanborn C., & Rieser, Leonard M. (1974). *Natural philosophy at Dartmouth: From surveyor's chains to the pressure of light*. Hanover, New Hampshire: University Press of New England.

Cahan, David (1985). “The institutional revolution in German physics, 1865–1914.” *Historical studies in the physical sciences* 15(2): 1–65.

Cahan, David (2005). “Helmholtz and the shaping of the American physics elite in the Gilded Age.” *Historical studies in the physical and biological sciences* 35: 1–34.

Catalogue of Dartmouth College and the associated institutions (1882). Hanover, New Hampshire: Dartmouth College.

Clarke, Frank Wigglesworth (1881). *Report on the teaching of chemistry and physics in the United States*. Washington, D.C.: Government Printing Office.

Coben, Stanley (1971). “The scientific establishment and the transmission of quantum mechanics to the United States, 1919–1932.” *American historical review* 76: 442–460.

d'Agostino, Salvo (2005). “Il difficile ricupero dell'anschaulichkeit di Goethe nell-opera di Helmholtz.“ *Nuncius* 20: 401–415.

Diehl, Carl (1976). “Innocents abroad: American students in German universities, 1810–1870.” *History of education quarterly* 16: 321–339.

“Experimental science in schools” (1871). *Nature* 4: 421–422.

Fau, Julien, & Chevalier, Charles (1853). *Nouveau manuel complet du physicien-préparateur, ou, Description d'un cabinet de physique*. Paris: Roret.

Forman, Paul, Heilbron, John L., & Weart, Spencer (1975). “Physics circa 1900: Personnel, funding, and productivity of the academic establishments.” *Historical studies in the physical sciences* 5: 1–185.

Frick, Joseph (1850). *Die physikalische Technik oder Anleitung zur Anstellung von physikalischen Versuchen und zur Herstellung von physikalischen Apparaten mit möglichst einfachen Mitteln*. Braunschweig: Vieweg.

Frick, Joseph (1856). *Die physikalische Technik oder Anleitung zur Anstellung von physikalischen Versuchen und zur Herstellung von physikalischen Apparaten mit möglichst einfachen Mitteln* (2d enlarged and improved ed.). Braunschweig: Vieweg.

Frost, Edwin B. (1933). *An astronomer's life*. Boston: Houghton Mifflin Company.
Gage, Alfred P. (1882). *A textbook on the elements of physics for high schools and academics*. Boston: Ginn, Heath & Co.
Ganot, Adolphe (1856). *Traité élémentaire de physique expérimentale et appliquée et de météorologie* (6th ed.). Paris: L'auteur-éditeur.
Ganot, Adolphe (1875). *Elementary treatise on physics, experimental and applied for the use of colleges and schools* (E. Atkinson, Trans. 6th ed.). New York: William Wood and Co.
Ganot, Adolphe (1879). *Elementary treatise on physics, experimental and applied for the use of colleges and schools* (E. Atkinson, Trans. 9th ed.). New York: W. Wood and Co.
Goodwin, H.M. (1933). "Physics at MIT: A history of the department from 1865–1933." *Technology review* 35: 287–291, 312–213.
Guralnick, Stanley M. (1975). *Science and the ante-bellum American college*. Philadelphia: American Philosophical Society.
Guttstadt, Albert (1886). *Die naturwissenschaftlichen und medicinischen Staatsanstalten Berlins*. Berlin: August Hirschwald.
Hart, James Morgan (1874). *German universities: A narrative of personal experience*. New York: G.P. Putnam's Sons.
Hinrichs, Gustavus (1870). *The elements of physics, demonstrated by the student's own experiments*. Davenport, Iowa: Griggs, Watson & Day.
Hinrichs, Gustavus (1871). "A walk through the building of the Polytechnic School at Aachen." *School laboratory of physical science* 1: 71–77.
Hinrichs, Gustavus (1871). "On the method of instruction in the elements of physical science." *School laboratory of physical science* 1: 64–70.
Hinrichs, Gustavus (1872). "Laboratory news." *School laboratory of physical science* 2: 38–47.
Hoch, Paul (1983). "The reception of Central European refugee physicists in the 1930s: USSR, UK, USA." *Annals of science* 40: 217–246.
Jarausch, Konrad (1995). "American students in Germany, 1815–1914: The structure of German and U.S. matriculants at Göttingen University," in H. Geitz, J. Heideking & J. Herbst, eds., *German influences on education in the United States to 1917*. Cambridge: Cambridge University Press: 195–211.
Jenkin, Fleeming (1873). *Electricity and magnetism*. London: Longmans, Green, and Co.
Kevles, Daniel J. (1978). *The physicists: The history of a scientific community in modern America*. New York: Knopf.
Kevles, Daniel J., & Harding, Carolyn. (1975). *The physics, mathematics, and chemical communities in the United States, 1870 to 1915: A preliminary statistical report* (typescript). Pasadena: California Institute of Technology.
Kohlrausch, Frederich (1870). *Leidfaden der praktischen Physik*. Leipzig: Teubner.
Küchler, G.W. (1904). "Physical laboratories in Germany." *Occasional Reports, India, Department of Education* 4: 183–211.
Lardner, Dionysius (1851–54). *Hand-books of natural philosophy and astronomy*. Philadelphia: Blanchard and Lea.
Lodge, Oliver (1890). "An ideal physical laboratory for a college." *The Electrician* 26: 32–33, 66–68.
Lommel, Eugene (1876). *The nature of light, with a general account of physical optics*. New York: D. Appleton and Co.
MacLeod, Roy (1980). "Evolutionism, internationalism, and commercial enterprise in science: The International Scientific Series, 1871–1910," in A. J. Meadows, ed., *Development of science publishing in Europe*. Amsterdam: Elsevier: 63–94.
Mendenhall, T.C. (1882). "Address." *Proceedings of the American Association for the Advancement of Science* 31: 127–138.
Morus, Iwan Rhys (2005). *When phyiscs became king*. Chicago: University of Chicago Press.

Moyer, Albert (1976). “Edwin Hall and the emergence of the laboratory in teaching physics.” *Physics teacher* 14: 96–103.
Moyer, Albert (1985). “History of physics.” *Osiris* 1: 163–182.
Müller, Johann (1847). *Lehrbuch der Physik und Meteorologie als dritte umgearbeitete und vermehrte Auflage der Bearbeitung von Pouillet's Lehrbuch der Physik* (3d ed.). Braunschweig: Vieweg.
“Object teaching and science” (1871). *Scientific American* 24: 39–40.
Olesko, Kathryn M. (1991). *Physics as a calling: Discipline and practice in the Königsberg seminar for physics*. Ithaca: Cornell University Press.
Olesko, Kathryn M. (1995). “German models, American ways: The ‘New Movement’ among American physics teachers, 1905–1909, ” in H. Geitz, J. Heideking & J. Herbst, eds., *German influences on education in the United States to 1917*. Cambridge: Cambridge University Press: 129–53.
Olmsted, Denison (1831–32). *An introduction to natural philosophy, designed as a textbook for the use of students in Yale College*. New Haven: Hezekiah Howe & Co.
Olmsted, Denison (1844). *Rudiments of natural philosophy and astronomy, designed for the younger classes in academies and for common schools, with numerous engravings illustrative of philosophical experiments*. New Haven: S. Babcock.
Olmsted, Denison (1848). *A compendium of natural philosophy ... to which is now added a supplement containing instructions to young experimenters with a copious list of experiments accompanied by minute directions for performing them* (20th ed.). New Haven: S. Babcock.
Pantalony, David, Kremer, Richard L., & Manasek, Francis J. (2005). *Study, measure, experiment: Stories of scientific instruments at Dartmouth College*. Norwich, Vermont: Terra Nova Press.
Parker, Richard Green (1839). *The Boston school compendium of natural and experimental philosophy, embracing the elementary principles of mechanics, hydrostatics, hydraulics, pneumatics, acoustics, pyronomics, optics, electricity, galvanism, magnetism, electro-/magnetism, & astronomy, with a description of steam and locomotive engines* (2d, stereotyped ed.). Boston: Marsh, Capen, Lyon and Webb.
Phillips, Melba (1981). “Early history of physics laboratories for students at the college level.” *American journal of physics* 49: 522–557.
Phillips, Melba (1983). “Laboratories and the rise of the physics profession in the 19th century.” *American journal of physics* 51: 497–503.
Pickering, Edward C. (1871). “Physical laboratories.” *Nature* 3: 241.
Pickering, Edward C. (1873–76). *Elements of physical manipulation*. New York: Hurd & Houghton.
Pickering, Edward C. (1877). *Compilation of the papers on physics*. Cambridge: n.publ.
Pupin, Michael (1926). *From immigrant to inventor*. New York: Scribners.
Reingold, Nathan (1981). “Refugee mathematicians in the United States of America, 1933–41: Reception and reaction.” *Annals of science* 38: 313–338.
Rezneck, Samuel (1960). “The education of an American scientist: H.A. Rowland, 1848–1901.” *American journal of physics* 28: 155–162.
Rezneck, Samuel (1962). “An American physicist's year in Europe: Henry Rowland, 1875–1876.” *American journal of physics* 30: 877–886.
Ricketts, Palmer C. (1933). “The first chemical and physical laboratories for the use of individual students.” *Review of scientific instruments* 4: 571–574.
Rogers, Emma Savage (ed.). (1896). *Life and letters of William Barton Rogers*. Boston: Houghton, Mifflin and Company.
Rogers, William B. (1864). *Scope and plan of the School of Industrial Science of the Massachusetts Institute of Technology*. Boston: John Wilson and Son.
Rosen, Sidney (1954). “A history of the physics laboratory in the American public high school (to 1910).” *American journal of physics* 22: 194–204.

Rosenberg, Ralph P. (1961). "The first American doctor of philosophy degree." *Journal of higher education* 32: 387–394.

Sheppard, Keith, & Horowitz, Gail. (2006). "From Justus von Liebig to Charles W. Eliot: The establishment of laboratory work in U.S. high schools and colleges." *Journal of chemical education* 83: 566–570.

Silliman, Benjamin, Jr. (1859). *First principles of physics, or natural philosophy designed for the use of schools and colleges*. Philadelphia: Peck & Bliss.

Simon, Josep (2004). *Adolphe Ganot (1804–1887) and his textbook of physics*. Unpublished M.Sc., University of Oxford, Oxford.

Simon, Josep (2008). "The Franco-British communication and appropriation of Ganot's Physique (1851–1881)," in J. Simon & N. Herran, eds., *Beyond borders: Fresh perspectives in history of science*. Cambridge: Cambridge Scholars Publishing: 141–68.

Stratton, Julius A., & Mannix, Loretta H. (2005). *Mind and hand: The birth of MIT*. Cambridge: MIT Press.

Thelin, John R. (2004). *A history of American higher education*. Baltimore: Johns Hopkins University Press.

Tobias, Marilyn (1982). *Old Dartmouth on trial: The transformation of the academic community in nineteenth-century America*. New York: New York University Press.

Turner, Steven (2004). "The reluctant instrument maker: A.P. Gage and the introduction of the student laboratory." *Rittenhouse* 18: 40–61.

Wells, James P. (1980). *Annals of a University of Iowa Department: From natural philosophy to physics and astronomy*. Iowa City: Department of Physics and Astronomy, University of Iowa.

Whitman, Frank P. (1898). "The beginnings of laboratory teaching in America." *Science* 8: 201–206.

Wiener, Otto (1909). "Das physikalische Institut," in O. Kirn & E. Friedberg, eds., *Festschrift zur Feier des 500 jährigen Bestehens der Universität Leipzig*. Leipzig: Hirzel: 4/ii: 24–60.

Wightman, Joseph M. (1838). *Illustrated catalogue of philosophical, chemical, astronomical and electrical apparatus*. Boston: G.W. Light.

Zapffe, Carl A. (1969). "Gustavus Hinrichs, precursor of Mendeleev." *Isis* 60: 461–476.

Archives

Cornell University Library, Rare and Manuscript Collections, Ithaca, New York
Ernest Fox Nichols Papers, 14–22–m729.

Dartmouth College Library Special Collections, Hanover, New Hampshire
Charles S. Cook Papers, MS–529.
Edwin B. Frost Papers, MS–1034.
Dartmouth College Registrar's Papers, DA–80.
Ellis, Benjamin Franklin. (1889). Magnetism at Hanover. Unpublished Honors thesis in physics, Dartmouth College, Hanover.
Frost, Edwin B. (1886). On a possible relation between kinetic electricity and the polarization of light by metallic reflection. Unpublished Thesis for Final Honors in Physics, Dartmouth College, Hanover.
Loveland, Frank Olds. (1886). Magnetism at Hanover, N.H. Unpublished Honors thesis in physics, Dartmouth College, Hanover.

Johns Hopkins Library Rare Books and Manuscripts, Baltimore, Maryland
D.C. Gilman Papers, MS 1.
Henry Rowland Papers, MS 6.

Massachusetts Institute of Technology Archives, Cambridge Massachusetts
William Rogers Papers, MC 1.

THE EVOLUTION OF TEACHING INSTRUMENTS AND THEIR USE BETWEEN 1800 AND 1930

Paolo Brenni

TEACHING INSTRUMENTS

Scientific instruments can be roughly divided into various categories depending on their use: (*a*) research and measurement; (*b*) professional and industrial; (*c*) and teaching and didactic instruments (Turner, 1983, 1991). The latter were (and still are) used in schools, colleges and universities for teaching purposes, for demonstrating well known physical laws, for producing and showing various phenomena in a vivid way (Brenni, 1998; Turner, 1996). Didactic instruments hardly allow to increase our knowledge about nature and its laws; they are conceived merely for illustrating well established facts in a clear and convincing way. These instruments were used in the frame of 'normal science', a universally accepted and non-problematic science.[1] No doubt, teaching instruments complement lessons by their illustrative character thereby adding an exciting theatrical touch; the famous scientist Louis Pasteur wrote, "ce serait avec peine que l'on pourrait saisir l'attention des élèves sans les experiences" (quoted in Hulin, 1992, p. 1409).

A large number of scientific instruments—such as the feather and guinea tubes, Mach's pendulum, the wave machines, the mechanical paradox of the rising cone, or the communicating vessels—were conceived as purely didactic apparatuses and were useless for any kind of scientific research. But if we study the birth and evolution of didactic instruments, we can see that many of them were derived from apparatuses which had been originally conceived and used for research and measurement. When these apparatuses had completely run their course and were not capable of producing new data or hitherto unknown phenomena, when they had exhausted their investigating potential, they were used as didactic instruments, often in simplified versions. Some examples would illustrate this point. In the seventeenth century, the French natural philosopher Edme Mariotte introduced the elastic collision apparatus which comprised a wooden frame with a series of pendulums aligned on the same plane. With it, he studied the behaviour of the pendulums and the elastic transmission of movements through the collisions. In the eighteenth century, the Dutch Willem Jacob's Gravesande, the French Jean Antoine Nollet, and other natural philosophers and demonstrators greatly improved the apparatus by adding a series of accessories and used it to study elastic

1 About the concept of 'normal science', see Kuhn (1970).

and inelastic collision. By the end of the century, this kind of apparatus was commonly used in every physics cabinet as a purely didactic device (Figure 1). The collision apparatus, which was produced over time with minor modifications by many instrument makers, remained a classic demonstration apparatus throughout the nineteenth and twentieth centuries. It continues to be sold under the name 'Newton's cradle' as a popular conversation piece (Brenni, 1998). Another more recent example is the cloud chamber developed by the Scottish physicist Charles Wilson at the beginning of the twentieth century which was originally used for detecting particles of ionising radiation. But with the introduction of more sophisticated and powerful detectors (the bubble chamber with superheated liquid, electronic detectors, etc.), the 'classical' Wilson's chamber was superseded in research by the beginning of the 1960s (Galison, 1997). Nevertheless, a simplified form of this instrument continues to be found as a demonstration device in didactic laboratories (Figure 2). In many cases, it can be said that when the utility of an instrument as a research tool decreased, its didactic role increased.

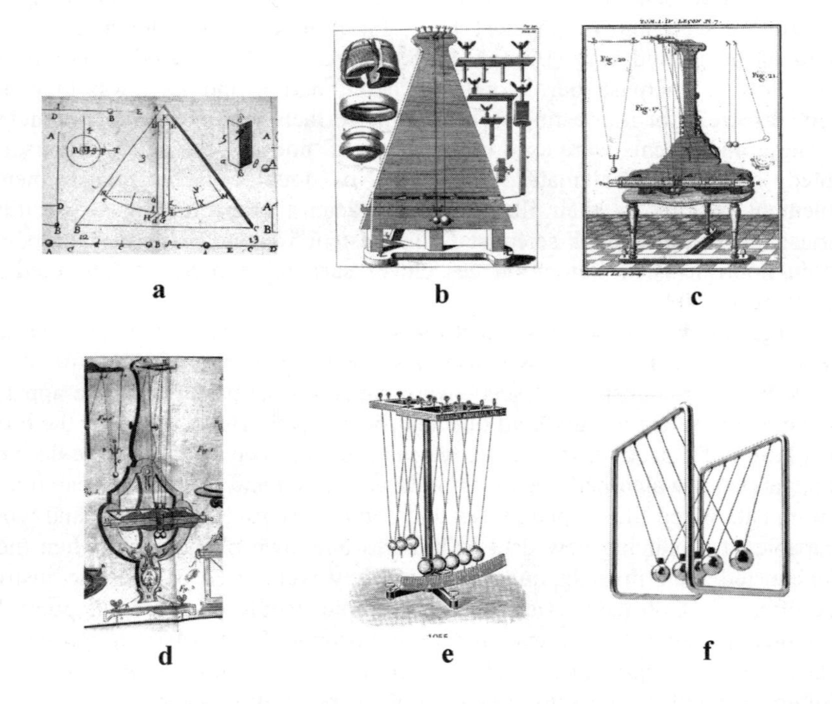

Fig. 1: The evolution of the percussion apparatus: from research apparatus to scientific toy. (a) Mariotte (1673), (b) 's Gravesande (1720), (c) Nollet (1743), (d) Sigaud de la Fond (1774), (e) E. Leybold's Nachfolger (1910 c.), (f) the scientific toy sold today as 'Newton's cradle'

Fig. 299 (1:7) 559 57
Chambre de Wilson d'après Schürholz,
montée sur une pince de table (301 06), raccordée à 2 batteries d'anode (522 34), éclairée par une lampe (450 51) sous carter (450 60 a) avec condenseur à une lentille (460 17) (sans fente réglable) montés sur support (300 11) et transformateur (562 76). Observation horizontale à l'aide du miroir plan (460 28) monté sur pied (300 02), tige (300 42) et noix LEYBOLD (301 01)

Fig. 300 (1 : 3)
Chambre de Wilson d'après Schürholz
avec dispositif d'éclairage constitué par: lampe (450 51) sous carter (450 60 a) et condenseur à une lentille (460 17). Dans l'intérieur de la chambre on distingue à gauche la préparation radioactive piquée sur une aiguille.

Chambre de Wilson d'après Schürholz. Chambre à brouillard avec pompe à commande à main pour rendre visibles les trajectoires des rayons α émis par la préparation fournie avec l'instrument. L'éclairage latéral se fait à l'aide de la lampe (450 51) sous carter (450 60 a, page 53) avec condenseur à une lentille (460 17, page 55), montée sur support à lentilles (300 11, page 108). Pour la décharge, nous recommandons un courant continu de 100 à 200 volts, pris, par exemple, sur une ou deux batteries d'anodes (522 34, page 72) (Fig. 299 et 300) 559 57

Pour d'autres expériences sur la radioactivité,
voir: l'Electroscope de Wulf (Nos. 546 24 - 28, pages 83 - 85)

93

Fig. 2: A simplified and didactic version of the Wilson's chamber sold by Leybold's Nachfolger in 1952.

Other apparatuses belonging to the so called 'physique amusante' are essentially scientific toys which can have some didactic value. They produce amusing, surprising or apparently paradoxical effects, which can attract the attention and arouse the curiosity of an audience. The popular Brewster kaleidoscope illustrates in a fascinating way the multiple reflections of a couple of angular mirrors; the taumatrope is useful for explaining the persistence of images on the retina; the astonishing Hero's fountain is an example of the possible application of pneumatics and hydrostatic phenomena (Turner, 1983, pp. 291–308).

Models can also be considered didactic instruments. A model of a steam engine in reduced scale is useful for explaining the working of its mechanical components; a tri-dimensional representation of a molecule illustrates the distribution and the position of its atoms; and the hydraulic or mechanic model of an electric circuit can help in the better understanding of the behaviour of electric currents (Brenni, 2004). Furthermore, models of machines and technical apparatuses such as telegraphs, telephones, phonographs, photographic cameras, etc., were very common in nineteenth century didactic collections, when the illustration of the practical applications of physical phenomena were an important part of the lecturing of physics.

Instruments can also become didactic because of the context in which they are employed. A pneumatic pump used for showing various phenomena related to atmospheric pressure, in an amphitheatre, for example, can certainly be considered a teaching apparatus. On the other hand, the same device is a research device when part of an instrumental system used for studying the behaviour of gases or electrical discharges in low pressure tubes. The same can be said for electrostatic machines, induction coils, calorimeters, barometers, etc., whose status as instruments for didactic, research or professional use mainly depended on the context in which they were used. So we can say that in many cases the borders between didactic, research, and professional instrument are not well defined and are totally independent from their technical characteristics. Until the beginning of the twentieth century, in fact, many apparatuses (as the mentioned pneumatic pumps) could be used for both teaching and research and measurement. Only after World War I was an increasingly marked separation between the two fields of application visible. On one hand, research and measurement instruments became more and more sophisticated, powerful, and complex and their performances largely exceeded the normal needs of didactic experimentation and demonstrations. On the other hand, didactic apparatuses tended to be cheaper, simpler and user-friendly.

Many didactic instruments reached their maturity rather soon after they were conceived: their function as well as their design was improved and defined in the eighteenth or early nineteenth centuries. Therefore they remained in catalogues for a very long time. Some instruments—like those derived from the original apparatuses of the famous lectures delivered by demonstrators such as Musschenbroek, 's Gravesande, Nollet, Desaguliers, and many others—were produced for more than a two centuries with only a few modifications (Brenni, 1998 and 2002a).

The nineteenth century was characterised by a spectacular multiplication of teaching instruments. During this period, the education system developed greatly both in Europe and America. Thousands of schools and colleges were founded, many new universities were instituted and the old ones were modernised and their scientific collections were remarkably enriched. Newly created polytechnic schools and technical institutes provided the education for engineers and technicians. All these institutions needed an enormous number of instruments and in a very demanding market, British, French, and German makers competed with each other to produce new and more functional didactic apparatuses which, together with the old and well known 'classical' ones, were illustrated and described in the commercial catalogues. In this paper I will trace the evolution of these instruments and their use between the first decades of the nineteenth century and the beginning of the twentieth century.

THE DIFFUSION OF DIDACTIC INSTRUMENTS

The systematic use of scientific instruments for studying nature and for illustrating its phenomena is a result of the Scientific Revolution, an affirmation of the experimental method based on direct observation and experimentation. Instruments used for demonstrations in teaching were well known since the second half of the seventeenth century (see, for example, Turner 1987, especially pp. 190–202). In the eighteenth century, thanks to the activities of lecturer demonstrators, who wrote physics treatises illustrating Newtonian physics in an essentially phenomenological way and conceived (and often manufactured) the apparatus for doing so, the number of available instruments was largely increased (see, for example, Gauvin and Pyenson, 2002). They could be found in amateur private collections, and slowly diffused into the first scientific cabinets of religious schools, colleges and universities, and with itinerant lecturers demonstrators who used them in their lessons. For the courts and the so-called polite society, scientific evenings enriched with spectacular experimental demonstrations became very fashionable and were highly appreciated (Sutton, 1995). In these scientific evenings, a sincere intellectual curiosity was accompanied by a taste for novelties and the pleasure of an exciting social event. In the second half of the eighteenth century, an increasing number of chairs for physics were instituted in many European universities,[2] while physics collections were built up in many schools and colleges of various types and level. While physics was often included in the teaching of philosophy, its teaching remained quite marginal compared to the main and classical curricula. Experiments, performed by itinerant lecturer demonstrators visiting the colleges, were sometimes open to the public. Important physics cabinets were also included in the guides and in the descriptions of the most remarkable sights of a city. However, the size and the characteristics of such cabinets varied. They depended on several factors such as the importance of the institutions, the curricula which they offered, and the personal preferences, scientific interests and wealth of professors, who often acquired the apparatuses on their own.[3]

From the end of the eighteenth century, the educational system in Europe and especially in France developed rapidly (see, for example, Bishop, 1994; Brock, 1996). Even a short overview of the history of educational systems in Europe and in the United States would be far too long and complex for the purpose of this article.[4] Because of the different social, cultural, political, and religious situations in European countries and in the United States, the development of public education, creation of various types of schools, reform of the older universities and the institution of new ones, and the foundation of polytechnic and technical institutes happened in very different ways and at different speeds. However, several characteristics were common more or less everywhere in the western world.

2 See the example of the University of Pavia in Bellodi, et al. (2002).

3 For the case of physics teaching in eighteenth century France, see Balpe (2001), pp. 39–61.

4 For the development of laboratory and technical education see, for example, Fox and Guagnini (1999).

School education became available to an increasing number of people and slowly less elitist; elementary education tended to become compulsory; the number of high school, colleges, and universities rose dramatically. On the other hand, where mechanisation and industrialisation were rapidly progressing, teaching of scientific disciplines and technical subjects became a necessity and was gradually assuming an ever-increasing role in the curricula. The teaching of physics was systematically introduced in an increasing number of schools, colleges, and universities and the needs of didactic instruments rose dramatically.

The case of France is emblematic and particularly well documented (see, for example, Balpe, 2001; Leon and Roche, 2003). In 1795, the French revolutionary government instituted the écoles centrals for secondary education in which the teaching of physics and chemistry was given importance, thus leading to the installation of scientific cabinets. But, in spite of the funding allocated by the government at the end of the eighteenth century, only about half of them were equipped with the proposed cabinets, whose instruments came from various sources. Some were newly acquired but often they came from older private or institutional collections. In 1802 Napoleon eliminated the écoles centrales and substituted to them with the lyceés (high schools)[5] which were financed by the state, and with the secondary schools (collèges) financed by private or by local municipality (Balpe, 2001, pp. 39–65). At the beginning of the nineteenth century, however, the teaching of physics was very heterogeneous and extremely dependent on personal judgement, the taste and attitude of the professors, and of course on the availability of demonstration instruments. In 1821 the installation of a physics cabinet and of a chemical laboratory in the *collèges royaux* became compulsory[6] along with acquiring the *Conseil royal de l'instruction*, the first official and systematic list of didactic instruments for equipping the cabinets of the *collèges* (Belhoste, 1995, pp. 98–101). In the list which included about 100 items, one could find all the classical apparatuses such as the Atwood machine, the two barrelled vacuum pump, a solar microscope or an electrical machine. About 10 of them, including a Gambey's heliostat, a repetition circle, a barometer, and polarising apparatus were denominated 'Machines de precision'. They were not considered necessary to the courses, but useful (if the institution could afford them) for stimulating the private research of professors. The list also indicated the name of the makers (Pixii, Gambey, Fortin, Soleil, and a few others) who could provide the apparatuses, as well as the suggested price for every item (Figure 3).

In 1842, when a larger number of colleges were equipped with instruments and with the introduction of a more ambitious programme for the scientific curricula, the chemist L. J. Thénard, who was responsible for the physics teaching at the Conseil royal de l'instruction publique since 1830, published a new list for the colleges with 169 apparatuses (ibid., pp. 181–85). It included 29 apparatuses for

5 It has to be pointed out that often the *lycées* had various denomination in the nineteenth century. Following the alternations of various regimes, they were also called *lycées impériaux* (with Napoleon I and again with Napoleon III) or *colleges royaux* (under the Restoration).

6 It was compulsory for the *écoles centrales* but not for the Napoleonic *lycées.*

the study of mechanics and gravitation, five for molecular actions (compressibility of liquid, capillarity), 30 for the thermology, four for magnetism, 23 for static electricity, 17 for galvanism, 25 for acoustics, and 34 for optics. Three more instruments denominated 'de précision' were a dividing engine, a Breguet's pointing clock, and a spherometer. The total cost of the proposed collection was 10,000 Francs. The maker's name was not provided, but in 1833, 1837, and 1842 Thénard sent to the physics professors the copies of the catalogues of Pixii, Lecomte and Deleuil.

CATALOGUE

DES PRINCIPAUX

INSTRUMENTS

DE PHYSIQUE,

CHIMIE, OPTIQUE, MATHÉMATIQUES,

ET AUTRES A L'USAGE DES SCIENCES,

qui se fabriquent

CHEZ PIXII, PÈRE ET FILS,

SUCCESSEURS DE DUMOTIEZ,

Ingénieurs-Constructeurs de l'École Polytechnique, du Collége de France, de la Faculté des Sciences de Paris et des autres Facultés, des Lycées, Colléges, Séminaires, etc.

Maison fondée en 1780.

A PARIS,

RUE DE GRENELLE SAINT-GERMAIN, N° 18,

ci-devant rue du Jardinet.

1849

Fig. 3: The front page of Pixii's trade catalogue of 1849. Between 1830 and 1850, Pixii was one of the most important and renowned French maker of physics instruments.

The publication of such lists continued for the rest of the century in France and in other countries. At the time the collections of important lycées had an average of about 250 instruments. And one can consider that by around 1850 most of these institutions possessed fairly equipped cabinets.

The example of France in the first half of the nineteenth century cannot be generalised. As a strong, well organized, and very centralised state, France experienced a pioneering and well structured model for public education, which certainly was not automatically followed in the rest of Europe. Nevertheless, because of the important role of the French language and culture and the Napoleonic conquests, several European counties (such as Italy) copied or at least were strongly influenced by the French system (see, for example, Ruiz-Castell, 2008 and the papers by Cuenca Lorente and Simon Castel in this volume). The situation in England was quite different. There the role of the state in education was less important, and an attitude of laissez faire prevailed in the public educational system (Brock, 1996). Nevertheless, in spite of all the different national situations, in the first half of the nineteenth century the role of physics in secondary education was growing and, consequently, also the number of physics cabinets.

THE USE OF DIDACTIC INSTRUMENTS

In the first decades of the nineteenth century the separation between general and particular physics (physique générale and physique particulière) disappeared.[7] Physics began to be subdivided in Mechanics and Gravitation (including matter, forces, movement, gravitation, etc.), Hydrostatics, Hydrodynamics and Pneumatics (which were often included in the Mechanics), Acoustics, Thermology and Meteorology, Optics, Magnetism, Electricity (static and dynamic, and often divided in Electrostatics, Electrodynamics, and Electromagnetism). Until the beginning of the twentieth century, most of the physic treatises and trade catalogues were compiled following these subdivisions and physics cabinets were ordered in the same way.

By 1800 instrument makers included in their catalogues the didactic apparatuses which had been designed during the eighteenth century by the most famous lecturer demonstrators. But in the first half of the nineteenth century, because of the rapid and spectacular progress of physics and the discovery of many new phenomena, the number of available instruments increased very quickly. Just a few examples will show how the horizons of physics were greatly expanding. Volta's

7 D'Alembert considered that 'general physics' was the science studying the general proprieties and laws common to all bodies, while 'particular physics' considered exclusively the bodies themselves. However, it is possible to find various definitions of the two terms between the late eighteenth century and circa 1850.

invention of the electric cell in 1800 led in a few years to the foundation of electrodynamics and the studies and discoveries by Oersted, Ampère, and Faraday started the birth of electromagnetism. The studies by Young, Fresnel, Malus, and others opened the way to physical optics, which investigated the phenomena of interference, diffraction, double refraction, polarisation, etc. In the field of thermology, where the old caloric theory was losing ground and the concepts of work and energy were beginning to play an important role, a series of systematic researches allowed the determination of the specific and latent heats of bodies, as well as the proprieties and the behaviour of gases. The study of sounds, which for centuries had been almost entirely confined to the realm of music, harmony, and arithmetical proportions, was becoming an important part of physics and many new instruments were invented for investigating the nature of sounds.[8] A natural consequence of such progress was also a massive increase in the number of teaching apparatuses available, a trend which continued for the rest of the century. To keep pace, instrument makers had to constantly enlarge and update their trade catalogues. Pixii, who was one of the most appreciated physics instruments maker of the time, published at least six different editions of his catalogue between 1821 and 1852. The number of items in them grew from about 600 to almost 900 (see Pixii, 1821, 1842, and 1852).

But what kind of physics was taught in lycées and high schools around 1850? Again, we take the example of France.[9] If we examine instructions from the government and analyse the most popular treatises, we can see that they present essentially a dogmatic, phenomenological, and descriptive science, which was explained by an inductive approach. In the so called physique élémentaire, the use of mathematical formalisms was extremely limited if not absent. The descriptions and the replication (at least in a simplified form) of historical and fundamental experiments were essential. Mentioning the famous scientists and their great achievements also had a moral and rhetoric role. In this kind of teaching, the place of instruments, which were necessary for demonstrating the laws and the principles of physics and for showing the related phenomena, was absolutely central. Therefore, a lot of space was dedicated to them both in the treatises and in the lessons (Balpe, 2001). Finally, in an era when new machines and new processes seemed to derive almost automatically from the conquests of science, it was necessary to illustrate the useful practical applications of physics—the didactic collections were enriched with models of hot air balloons, hydraulic pumps and turbines, steam engines and locomotives, telegraphs (and later telephones), dynamo machines and electric motors, electroplating apparatus, and so on.

8 See, for example, Turner (1983 and 1996). As far as electrical instruments are concerned, see Brenni (2000a). For a short history of optical instruments in the nineteenth century, see Brenni (2000b). For the acoustical apparatuses of the nineteenth century, see Pantalony (2009).

9 With the reform of 1852 and the so called bifurcation, a scientific baccalauréat was introduced for the first time. With this reform, the role of sciences and experimentation in the curricula increased.

Today when we browse the inventories of nineteenth century collections, when we analyse the endless lists of apparatuses proposed for equipping the cabinets, when we visit the extant cabinets preserved in several schools and institutes, a question arises spontaneously: were all these instruments really used? As I far as I can say, for several reasons the answer is no. As Blondel rightly pointed out: "En effet au XIXè siècle, les appareils composants les cabinet de physique de lycées et décrits dans les manuels scolaires sont des appareils de démonstration, le plus souvent copies d'instruments historiques. Ils étaient montrés, décrits dans leur fonctionnement mais l'expérience était rarement effectuée" (Hulin 2008, p. 96).

First, in high schools the number of hours dedicated to the physics courses was far too limited for performing time consuming and complicated demonstrations. Only by repeating some of the optical, acoustical or electrical experiments using the classical nineteenth century experimental apparatus can one fully understand the difficulties that could have been encountered.[10] These instruments were certainly elegant and cleverly conceived, but they were not user-friendly. Even the apparently banal projection on a wall of a clean and defined solar spectrum required careful, lengthy preparations and several trials. A good physics demonstration is like the trick of a magician. It works smoothly and is convincing only if it is accurately and carefully prepared. In spite of the fact that many physics teachers and professor could count on the help of skilled assistants and préparateurs, it is hard to believe that certain instruments had really been employed. Many demonstrations (especially the ones related to calorimetry where the achievement of thermal equilibrium was essential) would have certainly taken too long to be performed during the course of a normal lesson. Others were probably too complicated. For example, the liquefaction of gases with the Natterer or with the Cailletet apparatus required (before manipulating the special compression pump) not only the preparation of a suitable gas (for example, carbon dioxide or nitrous oxide), but also the production of ice for the freezing mixture which was necessary to cool it down. Even simpler demonstrations were problematic as stated by an Italian professor:

> A spiegare la costruzione della macchina d'Atwood con tutti i perfezionamenti introdottivi, ed a verificare le leggi sulla caduta dei gravi non basterebbe certamente una lezione intera: ad eseguire con discreta precisione tutte le operazioni necessarie per determinare il peso di un litro di aria, o di un gaz qualunque bisognerebbe spendere più ore; e non men lunghi riuscirebbero gli sperimenti per verificare la relazione che esiste tra il grado di acutezza o di gravità di un suono e il numero di vibrazioni eseguite in un minuto secondo; o per determinare il coefficiente di dilatazione di un corpo, o il suo calorico specifico e via discorrendo. Ed è chiaro che a procedere in un siffatto modo nell'insegnamento degli elementi di fisica bisognerebbe poter disporre non di poche ore per settimana, ma di tutte le ore di insegnamento… (Boltshauser, 1866, p. 275)

10 For the problems related to replication of historical experiments to use of historical instruments, see Heering (1989), and Heering, et al. (2000).

Many teaching instruments were probably just shown and described, and the experiments which could be performed with them were explained. They were used as tri-dimensional illustrations but were not operated. The same can be said for several iconic instruments, which were always proposed by the makers and included in the collections because of their historical role in the evolution of science and technology. A fairly complete collection would also include the apparatuses which had marked the most important progresses of physics, notwithstanding the fact that from a practical point of view they were quite obsolete. For example, probably after 1850, nobody would have really used Laplace-Lavoisier's calorimeter for measurements (other more efficient instruments were available); nevertheless, it was shown as one of the fundamental apparatuses of thermology. The temperamental Coulomb's torsion balance, a key instrument in the development of electrostatics, hardly allowed for the measure of electrostatic forces during a physics lesson (Blondel and Dörries, 1994). But with the torsion balance, it was possible to illustrate how such forces could be measured. Several apparatuses, which had been intended for measurements, were probably only used in an essentially qualitative way. The classic (and very inefficient) Voltaic cell, which remained in physics cabinets even in the twentieth century, was not used as a source of continuous current, but simply as a landmark in the history of electricity and for illustrating the principles of construction of the first electric piles.

Some other instruments, which in certain cases were too sophisticated for teaching, were acquired by wealthy schools as status symbols. Owning a Hipp's chronograph, a Koenig's sound analyser or a Melloni's bench with dozens of accessories for experimenting with radiant heat, automatically increased the reputation of a scientific cabinet and also of the institution which possessed it. And that happened independent of the fact whether these apparatuses were really used or simply remained in their cupboards. Finally, we have to remember that sometimes well equipped collections remained underused simply because of the lack of teachers who were able to perform the experiments.

INSTRUMENT MAKERS AND THE MARKET FOR TEACHING INSTRUMENTS

Around 1850 England and France dominated the European and American markets of physical demonstration apparatuses (see, for example, de Clercq, 1985). Since the mid-eighteenth century, the instrument makers from London were known for their excellence. Their optical instruments for astronomy equipped the best observatories and their navigational, surveying, and drawing apparatuses were highly appreciated by practitioners. English philosophical instruments were praised everywhere; the made-in-London orreries, vacuum pumps, and electrical machines were acquired for furnishing the most prestigious private and institutional physics cabinets (see Clifton,1991, pp. 338–49; Turner,1987, especially pp. 203–230). For many reasons, until the end of the eighteenth century French instrument makers could not compete with their English colleagues and that was particularly true as far as precision instruments were concerned. But a series of changes and transformations, which accompanied the Revolution and the following Napoleonic regime, boosted the French instrument industry which was concentrated in Paris (Brenni, 2006). In a few decades, the makers of Paris developed and improved their production, increased their exports, and started to seriously challenge the London supremacy. The apparatus signed by Gambey, Fortin, Pixii, Lerebours, Chevalier, to mention a few of them, started to be universally appreciated. This situation appeared evident during the 1851 exhibition, when for the first time products of different nations could be compared side by side (Brenni, 2010). The reports of the jurors of the exhibition and the comments of the specialists clearly agreed on the fact that not only had the quality of French instruments improved remarkably, but also that in some branches of physics (such as optics) they were comparable to or even better than the English ones. In fact, the period between about 1830 and 1890 can be considered the gilded age of French philosophical instruments. Sober but elegant, well made and functional, the electrical apparatus of Ruhmkorff, the optical ones of Soleil, Duboscq, and of Lerebours & Secretan, the acoustical instruments by Marloye and Koenig, the apparatus for studying the phenomena related to heat of Golaz, the physics didactic devices and models of Salleron (Figure 4) and Deleuil, and the Molteni projection lanterns and didactic slides were the standard equipments of a physics cabinet for almost half a century. These instruments were constantly reproduced in some of the most popular physics treatises of the time (see below) and could be found in almost every European and American school, college, and university.

In September 1870, the French empire of Napoleon III collapsed in the Franco-Prussian war. The German states were unified under the Prussian crown a few months later. In the following decades the development of German industry as well as the successes of German science were impressive and, by 1900, Germany was the most important continental power and was seriously challenging British supremacy. The educational system was developed, the collaboration between scientists and industrialists, which was often supported by the state, proved to be very fruitful as demonstrated by the emblematic foundation in 1887 of the Physikalisch-Technische Reichsanstalt, one of the most advanced metrological and research institute of the time (see, for example, Cahan, 1989). German precision and instrument industry fully profited from the favourable situation. In the last three decades of the nineteenth century, German instrument makers, who since the early 1800s were especially appreciated for their astronomical, geodetic, and surveying instruments, acquired a new and excellent reputation in the fields of physical, optical, and electrical apparatuses also. Among the main reasons for the success of German manufacturers were the high quality and originality of their instruments, as well as a more modern and efficient design. But their increasing role in the instrument market can be also ascribed to a clever and sometimes aggressive marketing strategy (Brenni, 2002b). For example, they participated en masse at the universal exhibitions of Chicago (1893), Paris (1900), and St Louis (1904) with large collective displays where multilingual representatives were constantly available for informing the potential clients. By the beginning of the twentieth century the firms which specialised in didactic and demonstration instruments, such as Max Kohl of Chemnitz, E. Leybold's Nachfolger of Cologne or Ferdinand Ernecke of Berlin, freely offered ponderous trade catalogues, illustrating thousands of instruments and which were often available in several languages.[11]

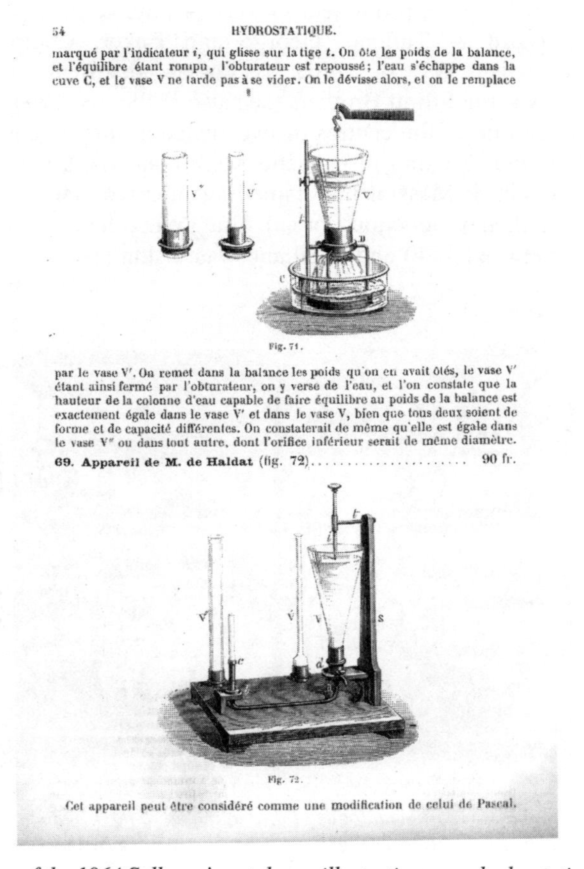

54 HYDROSTATIQUE.

marqué par l'indicateur *i*, qui glisse sur la tige *t*. On ôte les poids de la balance, et l'équilibre étant rompu, l'obturateur est repoussé; l'eau s'échappe dans la cuve C, et le vase V ne tarde pas à se vider. On le dévisse alors, et on le remplace

Fig. 71.

par le vase V'. On remet dans la balance les poids qu'on en avait ôtés, le vase V' étant ainsi fermé par l'obturateur, on y verse de l'eau, et l'on constate que la hauteur de la colonne d'eau capable de faire équilibre au poids de la balance est exactement égale dans le vase V' et dans le vase V, bien que tous deux soient de forme et de capacité différentes. On constaterait de même qu'elle est égale dans le vase V" ou dans tout autre, dont l'orifice inférieur serait de même diamètre.

69. Appareil de M. de Haldat (fig. 72) .... 90 fr.

Fig. 72.

Cet appareil peut être considéré comme une modification de celui de Pascal.

Fig. 4: A page of the 1864 Salleron's catalogue illustrating some hydrostatic apparatus.

Exhibition medals and several letters of appreciation written by teachers and professors were proudly reproduced in the first pages of these publications. Some of the most important German makers of teaching and demonstration instruments manufactured (or at least they could provide) the technical equipment and all the furniture necessary for modern lecture rooms, auditoriums, and collection

11 It is beyond the scope of this paper to mention the many trade catalogues. A large number of them can be viewed at the reference page of the Scientific Instrument Commission, http://www.sic.iuhps.org/refertxt/catalogs.htm.

storerooms (Figure 5). These makers could also provide complete collections of instruments for educational institutions of different types and levels. At the beginning of the twentieth century, the situation of the manufacture and trade of didactic instruments could be summarised as follows: England remained one of the most important instrument producers and could count on a large domestic and colonial market. Among the prominent makers of physics instruments there were firms such as Baird & Tatlock, George and Baker, Cambridge Scientific Instruments or Griffin & Co, while as far as electrical instruments were concerned, there was the Elliott Brothers (see also Williams, 1994).

Paris remained one of the centres of excellence in instrument making. As far as demonstration and teaching instruments were concerned, some firms such as Ducretet or Radiguet & Massiot maintained an international appeal and sold an important part of their production abroad. But France had lost the predominant position it held between 1840 and 1880 and was losing ground in the world market.

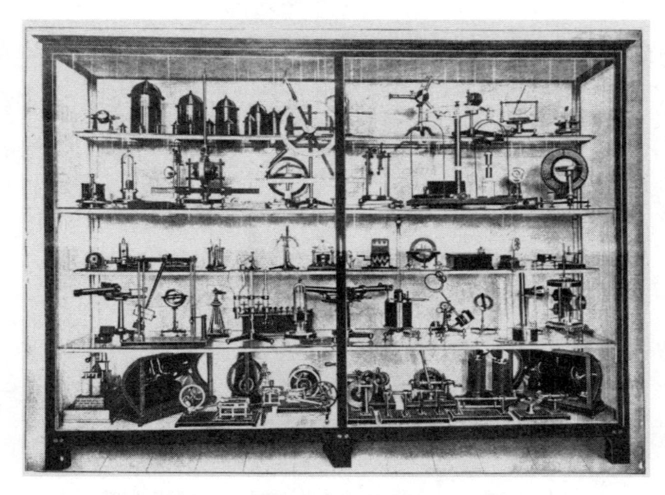

Fig. 5: An image from a Max Kohl catalogue of the beginning of the twentieth century illustrating a typical cupboard with several physics instruments made by this German firm.

The rising star in the firmament of instruments was the German industry which was supplying a large portion of the apparatuses required by teaching institutions of every order and level both in Europe and in America. As far as the Austro-Hungarian Empire was concerned, we can mention at least two firms of Vienna: Lenoir & Foster and W. J. Rohrbeck's Nachfolger.

The United States, while still importing a large number of instruments, was becoming less dependent on European manufacturers. In Italy, Holland, Spain, Portugal, Switzerland, the Scandinavian countries and other countries on the periphery of the European industrial and political powers there were some good makers. But their production generally supplied only the local market and, apart from a very few exceptions, did not play an important role on the international scene.[12]

INSTRUMENTS IN PHYSICS TREATISES

The role of physics treatises and specialised literature in the diffusion of teaching instruments and in the formation of nineteenth century physics cabinets cannot be overestimated. A detailed analysis of these topics is not the aim of this paper, nevertheless we have to consider at least the most influential texts.

In the first decades of the century several authors such as R. J. Hauy, C. M. S. Pouillet, J. B. Biot, and others contributed to renew the physics literature in France. But it was only after 1850 that several secondary school physics treatises became phenomenal bestsellers, were translated and used abroad, and inspired many similar books. Among the most famous authors, we can mention the names of A. Ganot, J.C. D'Almeida, E. Fernet, A. Daguin, A. Privat-Deschanel, F. Marcet, etc. Probably the most famous of them was the Adolphe Ganot, professor of physics in Paris, who published in 1851 the first edition of his *Traité élémentaire de physique expérimentale et appliqué.*[13] Ganot's *Traité* was popular for more than half a century and hundreds of thousands of copies were sold. (In 1881, Ganot himself affirmed that in about 30 years roughly 204,000 copies of the *Traité* had been sold! See Figure 6.) It was constantly updated with new subjects and new instruments and in 1913 appeared its 25th French edition. Furthermore, it was translated into English, Italian, Spanish, Dutch, German, Swedish, Spanish, Polish, Bulgarian, Turkish, and Russian.[14] As it was rightly remarked: "...Ganot's textbooks were certainly amongst the most widely translated and read, and as such made a major contribution at an international level to the configuration of physics as a discipline. By the 1880s, they were considered standard works of physics by a wide range of readers across the social and cultural spectrum in France, as well as in countries such as Britain (Simon, 2008, p. 141)." Ganot's *Traité* also largely contributed to the establishment of standard instrumental equipment for colleges and secondary schools in Europe, America as well as in India and Japan, where the English versions were well known. As observed by Simon[15], Ganot's *Traité* was often neglected by historians because it did not belong to the hegemonic period of French science, which is normally situated between the end of the eighteenth century and the 1830s. Nevertheless, we can affirm that with his emphasis on demonstration, instrument design, and experimental manipulations, it reflected

12 About the evolution of the instrument industry in the nineteenth century, see de Clercq (1985) and Brenni (2002c).

13 See, for example, Ganot 1853, 1894, and other editions. In 1859 Ganot also published *Cours de physique purement expérimentale.*

14 See Gires (2006), pp.19–24 and Simon (2008). From the 19th edition onward the treatise was re-edited and updated by G. Maneuvrier, professor of physics and natural sciences and under-director of the physics laboratory of la Sorbonne.

15 See note 14.

for several decades the hegemonic era of French physics instruments and instrument makers. It is important to point out that Ganot not only contributed to the teaching of physics in countries such as France with a strong state directed educational system, but (translated and adapted) also in the much more heterogeneous British system. Ganot's treatise became a landmark in the scientific educational literature and was copied and taken as a model by several other French and non-French authors. One of the reasons for its success was the large number and outstanding quality of its engravings illustrating hundreds of instruments produced by the best French makers. Instead of the schematic diagrams used in several textbooks of the first half of the nineteenth century such as the Pouillet, or the poorer illustrations which could be found in similar books published elsewhere, the wood engravings of Ganot's *Traité* reproduced the apparatuses in a vivid, clear, and very readable way. Even compared to contemporary photographs, these engravings with their perfect shadows and perspectives give a sensation of tri-dimensionality and produce a hyper-realistic rendering of the objects they represent. These illustrations were reproduced and copied for decades in many other treatises, articles, and trade catalogues. Therefore, the diffusion of Ganot's treatises acted as a powerful international advertising tool for the French (or French-style) apparatuses. Professors and teachers became used to them and included them among the items in their cabinets. This fact is still evident today: historians of instruments find in Ganot's textbooks a detailed and accurate catalogue of a large number of apparatuses which are nowadays studied as scientific and historical artefacts. Other French treatises, such as the ones published by Drion and Fernet (1861) or by Privat-Deschanel (1869), also had a wide reach and contributed to diffusion of the French style in teaching physics.

In Germany, one of the most popular and long-lived treatise was the one by Müller and Pouillet. In the early 1840s the physicist and mathematician Johannes Heinrich Müller undertook the task of translating and adapting in German the famous French physics treatise of Claude Servais Mathias Pouillet (1791–1868). Müller's objective was to produce a good and accessible physics textbook for the largest possible audience: "Es ist deshalb von der grösseren Wichtigkeit, dass die Naturwissenschaften durch zweckmässige Lehrbücher möglichst zugänglich gemacht werden" (Müller, 1844, p. vi). Since its first edition the Müller-Pouillet treatise presented several differences from the original Pouillet; in the subsequent editions (Müller died during the compilation of the eighth one) the traces of the French model tend to disappear. The 10th edition published by Leopold Pfaundler between 1905 and 1915 was in four volumes totalling more than 4,400 pages with over 3,000 illustrations (Figure 7).[16]

16 See Pfaundler (1905–1914). It is curious to note that in spite of the fact that the last editions were completely different from the original Pouillet textbook, the title of treatise always remained *Müller-Pouillet's Lehrbuch der Physik* and was commonly known as Müller-Pouillet.

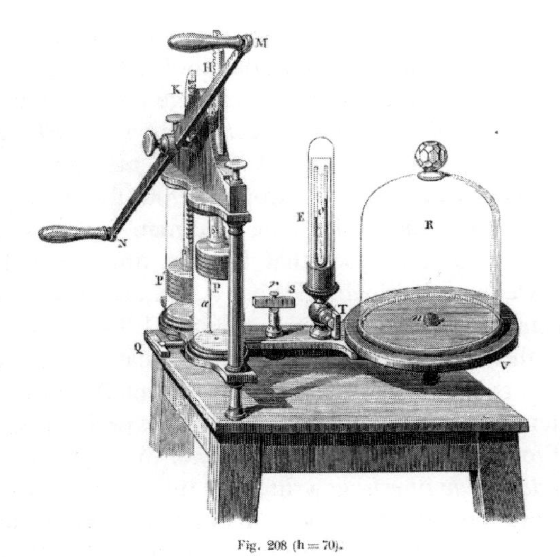

Fig. 6: The page of Ganot's Traité élementaire de physique of 1884 (19th edition) illustrating a vacuum pump. The success of this treatise was also due to a large number of engravings of excellent quality.

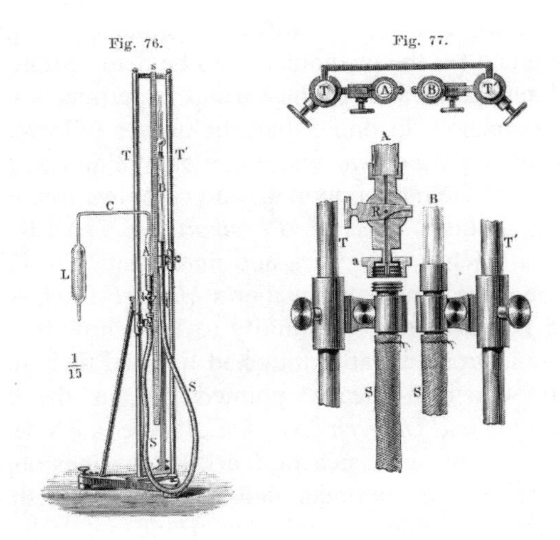

Fig. 7: A page of the Müller-Pouillet's Lehrbuch der Physik of 1898 (ninth edition). It was probably the most influential German physics treatise.

As in the case of the Ganot, one of the reasons of the success of Müller's treatise was the presence of a large number of illustrations (wood engravings, tables, etc.) carefully chosen, selected, and improved by Müller himself. In the first editions

Müller, like Ganot, presented the best instruments produced by the Parisian makers. But after the 1870s, with the introduction of new acoustical, optical and electrical apparatuses and with the great development of the German science and precision industry, the later editions of the *Lehrbuch der Physik* described and illustrated an ever increasing number of instruments made in Berlin, Köln, Chemnitz, Jena, München or Dresden. Around 1900, the treatise became one of the most important references for the German makers of instruments. Nevertheless, we have to remember that by that time several other excellent physics textbooks were available in Germany (see, for example, Donle, 1907; Lommel, 1895; Weinhold, 1881; Wiedemann, Ebert 1904).

There was another type of technical treatise that was meant not only for teachers and professors who taught physics, but especially for préparateurs[17] who had to set up the demonstration and take care of the experiments. Among them we can mention the French *Nouveau manuel complet du physicien-préparateur; ou Description d'un cabinet de physique* written by Julien Fau,[18] who was a doctor of medicine, in collaboration with the famous instrument maker Charles Chevalier. It was not a physics treatise but a manual describing all the instruments of an ideal physics cabinet, their functions, uses and how to operate them for the experiments and demonstrations. In the foreword the author, who mentioned as a model the old fashioned but still used Sigaud de Lafond's *Description d'un cabinet de physique*, said, "...la seule prétention est de faire connaître les instruments nécessaires aux physiciens, abstraction faite de toute considération théorique" (Fau and Chevalier, 1853, p. 2 ; Figure 8). For all the different branches of physics, the book systematically presented (*a*) the phenomenon to be demonstrated; (*b*) the description of the related apparatus; and (*c*) the various experiments useful for showing the mentioned phenomenon. In doing that, the author followed the earlier mentioned *Eléments de physique expérimentale* written by Pouillet.

In Germany, one of the most important and complete treatises concerning the use of physical apparatuses was the *Physikalische Technik* by Joseph Frick (second ed. 1856), a teacher of physics and mathematics in Freiburg. The book appeared around the same time as Chevalier's *Manuel.* Frick was convinced that many teachers did not have the opportunity to learn how to experiment and to properly manipulate instruments after they had finished their studies. The lack of experimental skill was also clearly pointed out in the *Encyklopädie des gesammten Erziehungs- und Unterrichtswesens*: "Aber auch der beste Apparat ist für die Schule werthlos, wenn er sich in den Händen eines unpraktischen, wenn auch noch so gelehrten Lehrers befindet, und manches gute Instrumentist schon zu Grunde gegangen, weil es von ungeschicken Händen falsch behandelt wurde" (Schmid, 1867, p. 61).

17 "Préparateurs" were skilled laboratory assistants who had great knowledge of the instruments and their functions. They prepared the experiments; were responsible for the scientific collections; maintained the instruments in good condition; and they made the necessary repairs. Sometimes they also made simple pieces of apparatus on demand.

18 Fau was also the author of a famous illustrated book of anatomy. See Fau (1845).

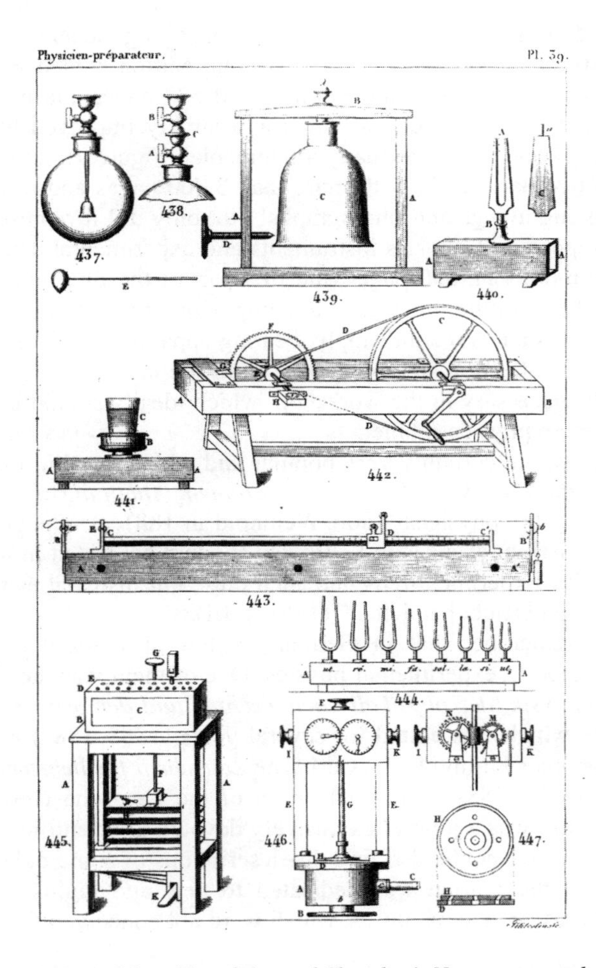

Fig. 8: One of the tables of Fau and Chevalier's Nouveau manuel complet du physicien préparateur of 1853. This treatise gave the instructions for using physics instruments and performing experiments.

Therefore Frick's treatise had to provide an introduction to experimental physics; to describe the best way to perform demonstrations successfully; to help teachers and professors choose and acquire the right apparatus from the best makers; and to give instructions for the construction of some simple and efficient instruments. In presenting his subject, Frick followed Müller–Pouillet's physics textbook. Even more than the physics textbooks, *Physikalische Technik* showed the progress of the technology of science teaching in half a century. It was also a powerful herald for the German instrument industry, which was booming around 1900. In fact, the name and addresses of the makers of most of the illustrated instruments were carefully reported in the extensive footnotes accompanying the text.

Frick's treatise was very successful. Its second edition, which was published in 1856 and was also translated into English, had 588 pages and more than 800 illustrations. But the seventh edition edited and largely improved by Otto Lehmann between 1904 and 1909 was really monumental (Figure 9). It was divided in two volumes of two parts each, with more than 3,700 pages and illustrated with more than 7,600 engravings and photographs! Not only did it represent the most complete encyclopaedia of physics instruments and experimental demonstrations, it also dedicated more than 600 pages (the first part of the first volume) to carefully describing the furniture, the technical equipments (projections, light sources, electricity, water, gas and vacuum supply, heating apparatus, etc.) of an ideal lecture room and of a collection storeroom as well as the installations, the tools and the machine tools necessary in the workshop, which ideally should have been installed near a modern physics cabinet.

Less ponderous, but certainly very popular and widely used in the secondary schools was also the *Physikalische Demonstrationen: Anleitung zum Experimentieren*, published by the physicist Adolf Weinhold in 1881. As the author clearly explained "… [es ist] durchaus kein Lehrbuch, sondern behandelt in wesentlichen nur die Technik des Experimentirens und auch diese nur insoweit es sich um Demonstrationen im Unterricht handelt…" (1881, p.III).

We have to remember that in Germany a few specialised journals were specifically dedicated to experimental physics. One of them was the *Repertorium der Experimentalphysik, der physikalischen Technik und der astronomischen Instrumentenkunde*, which was edited for several years (1865–1882) by the physicist and the astronomer Philipp Franz Carl. The *Zeitschrift für Instrumentenkunde*, which was founded in 1881 as official organ of the Deutsche Gesellschaft für Mechanik und Optik, was a journal exclusively devoted to scientific instrumentation. In this journal, one could also find the descriptions of some didactic apparatuses, but most of the journal was dedicated to new and highly sophisticated measurement and research apparatuses, which were not conceived for schools and colleges.

Finally, the "*Zeitschrift für den physikalischen und chemischen Unterricht*" appeared in 1887, edited by the teacher Friedrich Poske in collaboration with the school director Bernhard Schwalbe and Ernst Mach, who would become a famous physicist and philosopher. Contrary to the *Zeitschrift für Instrumentenkunde*, the latter exclusively presented and illustrated new teaching and demonstration instruments as well as didactic experiments. The *Zeitschrift für den physikalischen und chemischen Unterricht* was published every two months and clearly showed how the art of science teaching was advanced and important in Germany. No similar publication could be found elsewhere before 1900. The English *Journal of Scientific Instruments,* for example, was launched only in 1923.

1330 IX. Kapitel: Dynamik.

Die Wellenmaschine von Mach gleicht einigermaßen dem Stoßapparat, insofern in horizontaler Reihe nebeneinander eine Anzahl kleiner Kugeln an je zwei divergierende Fäden angehängt sind, doch nicht so, daß sie sich berühren, sondern relativ große Zwischenräume zwischen sich lassen. Die beiden Leisten, an welchen die Fäden befestigt sind, sind durch Querstäbe mit Scharnieren miteinander verbunden, so daß man sie leicht einander nähern oder wieder trennen kann. Bei der Näherung verschiebt sich gleichzeitig die bewegliche etwas in ihrer Richtung, so daß die Aufhängepunkte der Fäden doch in gleichem Abstande bleiben. In dieser Anordnung (Fig. 3408) ist der Apparat für Demonstration transversaler Wellen geeignet, in der ursprünglichen dagegen für longitudinale. Da die Umstellung sich leicht auch während der Bewegung der Kugeln bewirken läßt, so kann man in sehr hübscher Weise momentan eine transversale Wellenbewegung in eine longitudinale übergehen lassen und umgekehrt.

Fig. 3404.

Fig. 3405. Fig. 3406.

Will man zunächst transversale Wellen demonstrieren, so werden die beiden Leisten dicht aneinander gelegt, die Kugeln durch ein Lineal *ik* alle gleichzeitig zur Seite geschoben und dieses in eine Nut eines zweiten Lineals eingelegt. Zieht man nun *ik* längs seiner Achse weg, so fällt eine Kugel nach der anderen zurück, sie

Fig. 9: A page of Frick's Physikalische Technik (seventh edition) of the beginning of the twentieth century. It certainly was the most comprehensive treatise specifically dedicated to physics instrumentation.

THE BELLE EPOQUE

The years between the end of the nineteenth century and World War I certainly marked the gilded age of teaching instruments and of physics cabinets. In the last decades of the century, many new apparatuses were added to the armamentarium of physics teaching: vacuums tubes for spectroscopy and for showing the proprieties of electrical discharges in gases, instruments for generating alternating currents, electrical oscillations and electromagnetic waves, as well as new optical and acoustic instruments, etc. In the same period, projection technology made remarkable progress and its use in classrooms became more and more common. Weinhold wrote, "Für die Ermöglichung gleichzeitiger Beobachtung durch eine grössere Zahl von Personen findet in neuerer Zeit die optische Projektion immer häufigere Anwendung und vielfach mit vollem Rechte und grossem Nutzen" (1881, p. iv). Better projections apparatus (scioptikon), more powerful light sources (lime light, electric arc, oxy-etheric burner, etc.), and the progress of photography contributed to the diffusions of lectures enriched with slides. Specialised firms, like the French Molteni, proposed in their catalogues many types of lanterns and projectors as well as thousands of slides illustrating every possible subject from art to history, from archaeology to geography, from natural science to astronomy (Molteni, no date; Figure 10).

And as far as physics was concerned, the slides could reproduce portraits of famous scientists, engravings and illustrations taken from the most popular physics treatises, pictures of instruments or schemes of experiments. Projections also allowed presentations of several experiments in an objective rather than a subjective way. Special transparent apparatus could be placed in front of the projector so that on the large screen the audience could see the silhouette of a galvanometer with a scale and the moving needle, the bubble of gas produced in an electrolytic cell, the line of a magnetic field visualised with iron filing, etc. Projection microscopes (solar or lucernal depending on whether they were illuminated by the sun or artificial light) were well known since the eighteenth century (Heering, 2008). In the nineteenth century these instruments underwent great improvements, while several other optical projection apparatuses were conceived of in order to show on a screen phenomena like diffraction, interference and polarization.

As far as possible, physics cabinets had to be constantly kept up-to-date and follow the evolution of teaching material. In France, the catalogues published since 1821 by the Ministry of Public Education which listed the apparatuses to be used in high schools were periodically updated. Two new versions appeared in 1884 and 1900 (Figure 11). The former counted 398 physics instruments, the latter 378 (Ministère, 1884, 1900). Both contained two parallel lists, A and B. List A addressed the most important high schools (*lycées*),[19] list B the smaller ones, which did not have great financial resources. Besides the old instruments, the

19 The physics collections of two important French high schools are well documented in Gires (2004 and 2006).

catalogues included several new apparatuses and devices such as the phonograph, the telephone and (in 1900) an X-ray apparatus.

RADIGUET & MASSIOT, 44, rue du Château-d'Eau, PARIS. 9

Modèle du Ministère de l'Instruction Publique

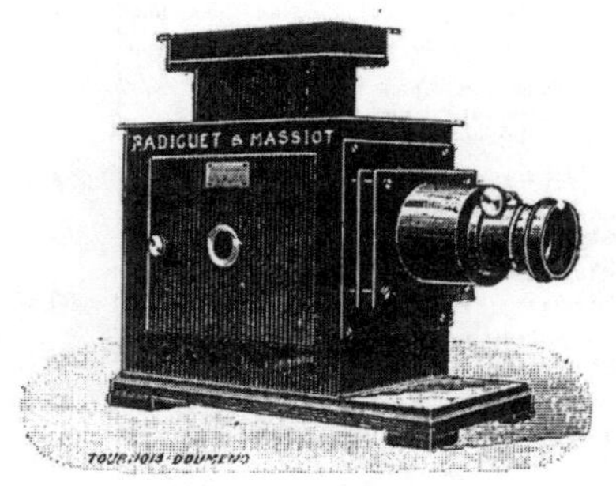

N° 5. — Appareil du Ministère.

Appareil adopté par le Ministère de l'Instruction Publique. D'une construction très soignée et en même temps simple pour qu'il puisse être vendu bon marché, cet appareil se recommande tout spécialement aux instituteurs soucieux d'illustrer par des projections leurs cours ou conférences.

Le corps de la lanterne est en tôle russe extra-forte. La devanture entièrement en cuivre poli et verni reçoit l'objectif double achromatique monté à crémaillère. L'appareil est monté sur un socle en acajou verni qui donne à l'ensemble l'aspect d'un instrument de précision et en assure la stabilité parfaite.

Code télégraphique	Nos de référence		Prix	Poids approximatifs
Adeni	5	**L'appareil** est livré en boîte formant socle, avec condensateur de 103 %, châssis double va-et-vient à hausse, objectif double achromatique donnant une image de 2m50 à la distance de 5 mètres, lampe R et M n° 528, à incandescence par l'alcool, complète avec support à vis calantes	105 fr.	9 kil
Adeo		**Le même appareil**, avec lampe R et M n° 733, à récipient indépendant......	115. »	9 kil.
Adeptio		**Le même appareil**, sans éclairage, mais avec cheminée et support à tige permettant de recevoir tous les systèmes d'éclairage.	80. »	7.900
		(Voir Chapitre spécial : **Eclairage des Appareils, 6e partie. Chapitre I du catalogue général**).		

Fig. 10: A page of Molteni's catalogue of 1900 circa illustrating a slide projector to be used for educational purposes.

But in spite of the fact that in the introductory remark of the 1900 catalogue one could find the following statement: "Il convenait aussi de faire disparaître des instruments anciens et imparfaits que se trouvent actuellement remplacés par d'autres plus conformes au développement de la physique", the list still included old fashioned items such as the culbuteurs chinois, Hero's fountain, or various apparatuses for the electrostatic dance. These were more old fashioned curiosities and amusing toys rather than useful didactic apparatuses. Furthermore, one could find in it nine different types of barometers, without counting their accessories!

— 18 —

DÉSIGNATION DES OBJETS.	A.	B.	OBSERVATIONS.
	fr. c.	fr. c.	
Physique. (Suite.)			
Calorimètre pour la méthode des mélanges..	45 00	45 00	
Appareil de Regnault pour la détermination des chaleurs spécifiques.............	350 à 600	//	(*)
Calorimètre de Favre et Silbermann.......	125 à 400	//	(*)
Appareil de Dulong et Petit pour la détermination des chaleurs spécifiques par la méthode du refroidissement............	80 00	//	(*)
Appareil de Clément et Desormes.........	140 00	//	M. S.
Appareil de Despretz pour la mesure de la chaleur latente de la vapeur d'eau.......	155 00	//	
Appareil d'Ingenhousz pour la conductibilité des solides.........................	40 00	40 00	
Appareil d'Ingenhousz pour la conductibilité des liquides.......................	10 00	10 00	
Miroirs conjugués.....................	180 00	85 00	
Cube de Leslie avec miroir et thermomètre différentiel.........................	65 00	65 00	
Thermomètre différentiel seul...........	12 00	12 00	
Appareil de Melloni pour l'étude de la chaleur rayonnante....................	800 00	//	
Électricité statique.			
Pendules électriques...................	10 00	6 00	
Bâton de verre et bâton de résine.........	8 00	8 00	
Bâton et plaque de caoutchouc durci.......	10 00	10 00	
Bâton de verre dépoli à l'extrémité.......	3 50	3 50	
Bâton de cuivre à manche isolant.........	8 00	8 00	
Disque de cuivre et disque de bois avec manches isolants....................	20 00	20 00	
Deux disques en glace, l'un poli et l'autre dépoli, avec manches isolants...........	20 00	20 00	
Tabouret isolant.......................	20 00	20 00	
Isoloir à acide sulfurique de M. Mascart....	18 00	//	
Balance de Coulomb....................	110 00	//	

Fig. 11: A page of the official list of scientific instruments for high schools published in 1884 by the French Ministry of Public Education.

Other countries proposed similar official or semi-official lists. In Germany, for example, a *Normalverzeichniss* (standard catalogue) of instruments for higher education institutions was proposed and accepted at the fifth general assembly of the Verein zur Förderung des Unterrichts in der Mathematik und den Naturwissenschaften in 1896. At the same time, an analogous list was published in Austria by the Wiener Verein zur Förderung des physikalischen und chemischen Unterrichts (see Zwei Normalverzeichnisse physikalischer Apparate, 1896). The German list, which remained open to improvements, was only a suggestion and not a necessity. It appeared very difficult to propose different lists for every type of school and therefore, like the French catalogue, it was simply divided into two groups, A and B, whose boundaries were not well defined.

The first group included all the teaching apparatuses which were generally considered necessary (notwendig), the second had the desirable (wünschenswert) ones. If we compare the French, German, and Austrian lists, we can see that the differences were relatively small. Most of the instruments were the same and they only differed from the point of view of the design following the national styles. The most important German manufacturers of teaching apparatuses such as Max Kohl or Ferdinand Ernecke carefully listed the Normalverzeichniss at the end of their catalogues, indicating for each items their own catalogue numbers and prices. But the former also divided the two groups A and B in two subgroups of objects in case the purchaser for economic reasons would have preferred to make the acquisition in two phases.

Between 1900 and 1915, the catalogues of Max Kohl, E. Leybold's Nachfolger, and those of a few other manufacturers, seemed to be attained by elephantiasis (see, for example, Kohl AG, [1911]; E. Leybold's Nachfolger, [1913]; and Zoller, [2009]). Thousands of pages of illustrations proposed every kind of instrument to showing every possible phenomenon and demonstrate all the laws of physics and their corollaries. Several different apparatuses were available for performing the same experiment or the same demonstration. To old ones, often conceived in the eighteenth century, were added new ones. Accordingly, at the beginning of the twentieth century one could see the coexistence (both in the trade catalogues as well as in the cabinets) of instruments of different ages. The eighteenth century's double barrelled vacuum pump survived next to the most recent Gaede's rotary pump and the old fashioned astatic galvanometers with the direct reading ammeter. In fact, instrument makers also tended to maintain in their catalogues (but probably not in their stocks) instruments which were old fashioned, obsolete or superseded by more modern and efficient devices. In doing so, they could inflate their catalogues and impress their clients with ponderous volumes, even though many instruments were made on demand only. Today, in an age when production has to be rationalised, standardised, and simplified, it seems quite extraordinary that the firm Max Kohl in its catalogue Nr. 50 of 1911 could offer a dozen different apparatuses just for showing the hydrostatic paradox, about 90 electrostatic machines (Figure 12), about 100 vacuum pumps, and about 100 induction coils of different sizes and types!

THE FIRST CRITICISMS

In his *Didaktik und Methodik der Physik*, the physicist and pedagogue E. Grimsehl wrote:

"Physikalischer Unterricht kann überhaupt nur dann erteilt werden, wenn eine ausreichende Apparatensammlung und ein besonderes Unterrichtszimmer vorhanden sind (1911, p. X, 39)." This peremptory statement showed how, at the beginning of the twentieth century, the teaching of physics was not even conceivable without a collection of instruments and a purpose made lecture room.

818 Elektrostatik. Nr. 60 397 —

60 397 — 60 406. 1 : 6 — 1 : 24.

60 417 — 60 424. 1 : 11 — 1 : 27.

60 425 — 60 430. 2 : 7 — 1 : 7.

60 407 — 60 416. 1 : 8 — 1 : 27.

Selbsterregende Influenzmaschine nach Töpler, mit 1 festen und 1 rotierenden Scheibe, Figur, mit massiven Hartgummisäulen, auf welchen die Kämme sitzen, um ohne weiteres die Scheiben abnehmen zu können; Gestell aus poliertem Mahagoniholz. — Mark

Listen-Nr.	60 397	60 398	60 399	60 400	60 401	60 402	60 403	60 404	60 405	60 406
Durchmesser der rotier. Scheibe	26	31	36	41	47	52	57	62	75	90 cm
ℳ	70.—	90.—	120.—	145.—	185.—	210.—	315.—	400.—	540.—	600.—

Selbsterregende Influenzmaschine nach Töpler, Figur, mit 2 festen und 2 rotierenden Scheiben, Grundbrett und Säulen von Mahagoni, Kämme auf Hartgummisäulen sitzend, hochelegant ausgeführt und die doppelte Menge Elektrizität liefernd als solche mit nur 2 Scheiben.

Listen-Nr.	60 407	60 408	60 409	60 410	60 411	60 412	60 413	60 414	60 415	60 416
Durchmesser der rotier. Scheiben	26	31	36	41	47	52	57	62	75	90 cm
ℳ	135.—	170.—	220.—	270.—	345.—	380.—	500.—	585.—	650.—	825.—

Kl. 5707, 5598. 2134, 5709.

Fig. 12: One of the several pages of the Max Kohl catalogue N.50 which were dedicated to the electrostatic machines.

Grimsehl continued on for pages to describe in detail the ideal spaces, appropriate electricity supply, well-furnished workshop, rules for managing and maintaining an inventory of the collection, the role of apparatuses, the way of experimenting, and so on. In the first decade of the twentieth century, while the number of available teaching instruments reached a peak and most of the European secondary schools and colleges were equipped with scientific cabinets, physics was changing and its teaching had to change too (Figure 13). In a few years, the discovery of the X-ray, of radioactivity, of the electron and new theories of quanta and of special relativity were violently shaking the apparently indestructible temple of classical physics.

Fig. 13: The main room of the physics cabinet of the Istituto Tecnico of Florence around 1900. The instrument collections (which is preserved) is typical for a large educational institution at the beginning of the twentieth century. (© Fondazione Scienza e Tecnica, Firenze)

The Second Industrial Revolution, which started towards the last decades of the nineteenth century, transformed western society with the successes of the chemical industry, massive steel production, the diffusion of electric light and power, the increasing importance of oil extraction and the birth of the automobile industry. A larger proportion of the population had access to education, the number of workers employed in industry grew, and technical activities and professions attracted more and more people. In this changing world, science teaching had to be modernised too. In France, school programmes and curricula were modified several times during the second half of the nineteenth century, but in 1902 a radical and general reform of the educational system was introduced (Balpe, 2001; Belhoste, Gispert, and Hulin, 1996; Hulin, 2000 and 2007). As far as experimental sciences were concerned, the 1902 reform marked a turning point. Sciences had to be an aspect of a renewed humanism and had to be developed within the frame of a positivistic culture: "Les études scientifiques doivent comme les autres, contribuer à la formation de l'homme."[20] Without entering into a detailed description of the various aspects of the reform, we can see how it was supposed to practically change the teaching of physics.

The deductive, dogmatic, and historical approach had to be abandoned in favour of an inductive method. Very important was the introduction of practical experimental exercises and graphic representations of the result of the experiments.[21] The French chemist and metallurgist Henri le Chatelier affirmed: "Il ne suffit même pas de regarder les expériences faites au cours par le prosefesseur. Il faut les avoir répétés soi-même" (quoted in Hulin, 2007, p. 66). The instruments had to be simplified, the outdated ones had to be abandoned, and many old fashioned experiments and demonstrations, which were only interesting from a historical point of view, had to be banned from the lessons. A physics professor commented on the practical exercises introduced in 1902: "...les anciens appareils ne conviennent pas en général au nouvel enseignement. Crées pour être exposés sur la table, pendant la leçon du professeur, ces appareils décoratifs n'avaient pas été prévu comme devant servir à l'expérience" (Lemoine, 1907). The endless descriptions of the apparatuses and the detailed presentations of experiments were to be eliminated. It is very interesting to read some of the comments made the physicist Henri Bouasse, author of many books and treatises, who actively participated to the debates related to the reform, and contributed to elaborate the new programmes. In 1901 in the article 'Enseignement des sciences physiques dans l'enseignement secondaire' (Bouasse, 1901 and 1913; also in Hulin 2000 and 2001, pp. 135–54), he heralded a renewal of physics teaching and the modernisation of physics treatises and laboratory equipment which he called "bazaar instrumental".

> Les traité élémentaires sont d'étranges recueils d'instruments démodés; on y trouve comme dans un musée d'antiquités où l'on a réuni pêle-mêle les legs des temps passés, sans avoir le courage de choisir. Tout cela tome sur la tête de élèves qui sont abruti sous cette avalanche et s'en relève difficilement.
>
> Il y a d'abord les instruments qui ne sont justifiés ni par la pratique ni par la théorie. .... Ces appareils sont absurdes. Ils constituèrent autrefois un progrès, en ce sens qu'ils ont remplacé des appareils plus absurdes encore; ils font aujourd'hui scandale. Il y a les appareils qui ont pu donner des résultats exacts, mais qui sont incommodes, définitivement abandonnés et qui ont même disparu du catalogue des constructeurs. ...
>
> Il y a les appareils qui font double emploi, l'un n'étant que le perfectionnement de l'autre, ou même une modification sans importance. ...
>
> Il y a les appareils utiles dans la pratique, mais qui n'ont aucun intérêt pédagogique, parce que leur théorie est complexe et leur graduation peur rationnelle et empirique. ...
>
> Il y a les appareils qui conservent un intérêt historique douteux; *la machine pneumatique*

20 A statement by Louis Liard quoted in Hulin (2000), p. 248. Louis Liard (1846 1917) was a French philosopher and administrator. He founded the École pratique des hautes etudes in 1886 and became vice-rector of the Académie de Paris in 1902.

21 In France, the idea of introducing experimental exercises and manipulation in physics and chemistry for students was not new, but in the second half of the nineteenth century, they could be realised only after some special advanced curricula.

> *ordinaire* qui n'est jamais employée et dont on assomme les élèves, en leur infligeant jusqu'au mémorable perfectionnement de Babinet, *l'appareil du général Morin*, de construction antédiluvienne et qui se refuse obstinément de fonctionner, etc. etc. Il y a les appareils de pure démonstration que la routine fige dans une forme archaïque, incommode et souvent absurde. Il ne fonctionnent pas, le professeur n'a pas l'idée de s'en servir. …
> Je ne plaisante pas. Tous les professeurs savent si bien que ces appareils ne valent rien, qu'ils passent trois quart de leur temps à le démontrer aux élèves; de sorte qu'on pourrait dire, sans paradoxe, qu'un cours de physique élémentaire est le catalogue méthodique et expliqué des appareil qu'ils ne faut pas employer.[22]

The colourful and caustic comments of Bouasse were certainly exaggerated (he was used to these kind of language), but nevertheless he touched a very delicate point. Too many physics teaching instruments at the beginning of the twentieth century did not correspond anymore to the need of a modern science. Rightly did he point out the necessity of reducing the number of simple and perfected apparatuses and abandoning the historical ones. In a more elegant an less vehement way, Grimsel in Germany expressed the need to get rid of obsolete instruments:

> Ich halte es für falsch, wenn wir heute noch Spiegelgalvanometer im Unterricht zu verwenden, die sich der alten Form des SCHWEIGGERschen Multiplikators anschliessen, (…). So gehet es mit manchen anderen Apparaten, die nur deshalb vielleicht noch ihr unsterbliches Dasein in der Schulensammlungen haben, weil ein um die Schulphysik verdienter Pädagoge sie vor vielen Jahren einmal konstruirt und empfohlen hat, und weil dessen Name daran klebt. Ich bin überzeugt, dass derselbe praktische Mann diese Apparate heute selbst nicht mehr gebraucht oder gebrauchen würde, wenn er die neueren Apparate kennt (Grimsehl 1911, pp. X, 70).

But, as it always happens in the world of public education, things changed very slowly. The introduction of new instruments for practical student experiments in high schools was hindered by the difficulties in finding the necessary finances. The use of classical teaching instruments was often so firmly rooted in the habits of teachers and professors, that there was resistance in abandoning them.[23] The same could be said about the physics treatises which had accompanied generations of teachers and students. On the other hand, instrument makers, while producing new apparatuses, were not keen to leave out the old products which fattened their catalogues and their balance sheets for years. If the seeds of a profound change had been dispersed at the beginning of the century, we can consider that World War I, which dramatically closed the belle époque with a bloodbath, also marked the end of the glorious 'brass and glass' era.

22 I have shortened the text. Where I inserted ellipsis points, he mentions several apparatuses which he considered useless and absurd. Among them there were, for example, the Laplace and Lavoisier calorimeter, the hygrometers of Daniel, Régnault and Alluard, Armstrong's electrical machine, the Wheatstone photometer, Gambey' magnetic compass, the magneto-electric machines of Pixii and Clarke, Morin's fall machine, Bourbouze's galvanometer, Wedgwood's pyrometer, and many others. Emphasis in the original.

23 In the past few years I visited many schools, saw their collections and met teachers who still like to use some of the apparatus which had been acquired a century ago!

THE BEGINNING OF A NEW ERA

The 1920s and 1930s saw several changes. Many ideas of renewal and reform, which had been proposed before World War I, were brought to fruition. In Germany, Robert Wichard Pohl developed a new way of teaching physics, conceived a new generation of didactic teaching apparatuses, and wrote a series of entirely new textbooks (see R. Wittje's paper in this volume). Everywhere in Europe, several die-hard physics treatises were abandoned or largely modernised. In the new textbooks, the realistic wood engravings illustrating the instruments in every detail were often substituted by simpler technical schemes. Large parts of the old school collections, which were saturated with hundreds of instruments accumulated over decades, were slowly neglected in their cupboards and became silent museums. Instruments as well as their material were changing. An increasing number of modular apparatuses were introduced on the market (Figure 14). They allowed a series of demonstrations and experiments to be performed with a limited number of elements, and in a period of crisis and lack of funding, they were more affordable than the older instruments. High voltage transformers slowly took the place of electrostatic machines and induction coils. The old fashioned piston or mercury vacuum pumps were substituted by simpler and more efficient industrial pumps. The first electronic instruments (lamp oscillators, oscilloscopes, amplifiers, etc.) replaced many mechanical or electromechanical apparatuses which had been used for a long time for experimenting with electricity or acoustics. Compact and user-friendly direct reading electrical measuring apparatuses were finally adopted for demonstrations and practical exercises (Brenni, 1997; Sydenham, 1979).

Several instrument makers in France, England, and Germany who had to contribute to the war effort in producing optical apparatuses for the military or precision elements for weapons changed their activities. Producing hundreds of different apparatuses which sometime had a different design but had the same use was not possible anymore. Rationalising and simplifying production was a must. In 1926, A. Schmidt the director of the firm Max Kohl wrote: "Die Entwicklung der Firma ist seit dem Krieg in eine dritte Phase getreten. Die im Krieg gemachten Erfahrungen haben neue Wege für die Herstellung und Kontrolle der Apparate gezeigt, eine ganze Reihe neuer Apparate und Instrumente ist zu den vorhandenen hinzugekommen… (1926, p. 24)".

In the years 1925 and 1928, the firm Max Kohl still printed a large three volume catalogue of about 1,000 pages and some other small lists illustrating the usual apparatuses, while E. Leybold's Nachfolger produced in 1929 a 461 page catalogue with around 5,600 items (but the catalogue of 1913 had 970 pages and listed more than 13,000 items!; Zoller, 2009). These publications were their swan songs: the untenable race for the largest catalogue was over. Kohl never produced any list before World War II and E. Leybold's Nachfolger launched a reduced catalogue of about 230 pages in 1938.

E. LEYBOLD'S NACHFOLGER, KÖLN-BAYENTAL

Electricité
Electromagnétisme
Expériences fondamentales

Dispositif expérimental pour la démonstration de l'action réciproque entre le champ magnétique et le courant électrique d'après le professeur R. W. Pohl.
Le fil parcouru par un courant engendre un champ magnétique, qui est orienté perpendiculairement à l'aiguille aimantée. Celle-ci prend une position d'équilibre perpendiculaire au fil. (Expérience d'Oerstedt; voir fig. à droite.) (Fig. 1/6 grand. nat.)

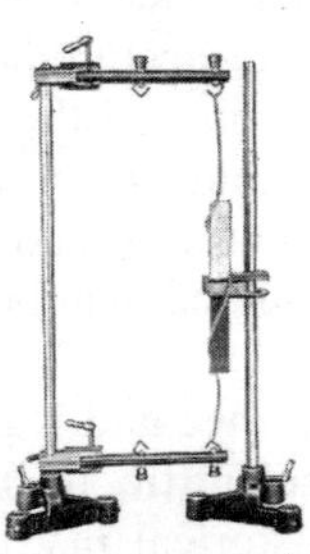

Si on envoie un courant dans le conducteur déformable, celui-ci s'enroule autour de l'aimant placé tout contre lui. Les lignes de force circulaires se développant autour du conducteur mobile, cherchent à prendre la même direction que celles de l'aimant.

Si on envoie un courant dans un conducteur, celui-ci est dévié dans la direction indiquée, ou dans la direction inverse, selon le sens du courant.

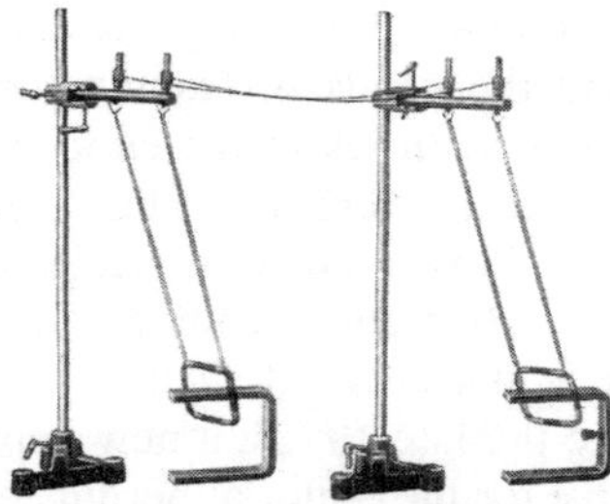

Si on relie les deux pendules l'un avec l'autre, et qu'on laisse osciller l'un, dans le champ magnétique de l'aimant, le courant alternatif induit provoque une oscillation égale de l'autre. Ceci à condition que les deux pendules aient les mêmes caractéristiques et la même longueur.

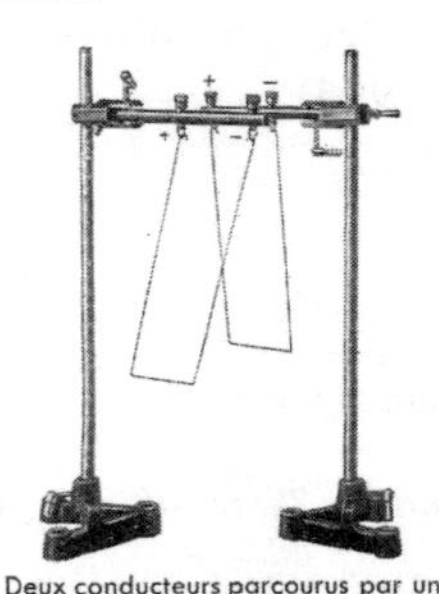

Deux conducteurs parcourus par un courant s'attirent ou se repoussent.

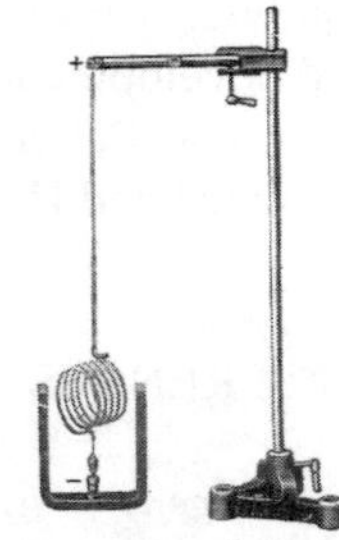

Les lignes de force du champ magnétique de la bobine parcourue par un courant cherchent à se superposer avec celles de l'aimant en fer à cheval. Si les deux directions sont perpendiculaires, la bobine tourne quand le courant passe.

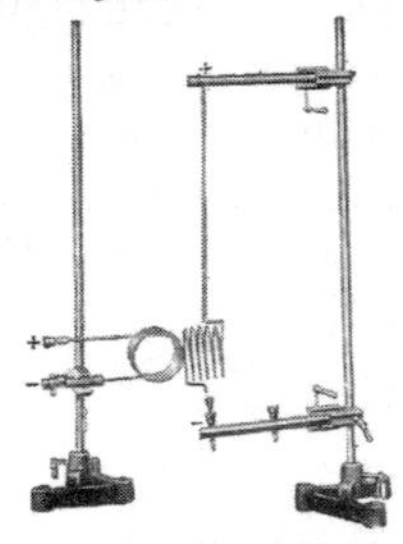

Les lignes de force du champ magnétique de la bobine (mobile) parcourue par un courant cherchent à se superposer avec celles de la bobine fixe. Si au commencement, ces lignes de force sont orientées perpendiculairement les unes par rapport aux autres, il en résulte une rotation.

169

Fig. 14: A page of E. Leybold's Nachfolger's catalogue of 1938 illustrating several experiments of electrodynamics done with the modular and simplified apparatus proposed by R. W. Pohl.

The introduction to the French version reads: "En comparant ce catalogue avec les édition précédentes, il sera facile de remarquer que de nombreux appareils ont été perfectionné et que nous en avons créé de nouveaux. Nous avons, par contre, supprimé tous les appareils ne présentant plus qu'un intérêt historique ou qui ne se prêtent pas à un enseignement vivant " (E. Leybold's Nachfolger, 1938).

World War II was a coup de grace for the old style instruments still available in the market. In the two decades after the war, most of the European and American teaching collections were completely renewed. In the late twentieth century, physics teaching in high schools and in colleges had hardly anything to do with the phenomenological and descriptive science presented in Ganot's treatises. The so called classical physics is nowadays only a small aspect of contemporary physics, whose place in a positivistic (but also today ridiculous) hierarchy of sciences is certainly not as important as it used to be. But what happened to the thousands and thousands of teaching apparatuses, which had been accumulated for almost two centuries in the cupboards of physics cabinets all around the world? Several of them were lost. They were destroyed by the wars, sold as scrap metal, or simply thrown away in occasion of renovations of laboratories and school buildings, when nineteenth century instruments were considered cumbersome, obsolete, and useless artefacts.

Since the late 1970s, a new approach in the history of science and a growing sensibility for the material heritage of the past (also for the scientific, technological, and industrial ones) triggered a renewed interest for historical instruments. Since then, a large number of didactic physics collections around the world have been restored and catalogues reordered, which are now visible through the internet and sometimes accessible to the public. These collections are now fascinating material witnesses of an era of great development and ingenuity in science teaching (see, for example, Bud, Cozzens, and Potter, 1992; Bud and Warner, 1998; Grob and Hooijmaijers, 2006; Van Helden and Hankins, 1994).

REFERENCES

Balpe, C. 2001. *Enseigner la physique au collège et au lycée.* Rennes: Presses Universitaires de Rennes.

Belhoste, B. 1995. *Les sciences dans l'enseignement secondaire français Textes officiels*, Tome I: 1789–1914. Paris: Inst. national de recherche pédagogique.

Belhoste, B., H. Gispert, and N. Hulin (eds). 1996. Les sciences au lycée. Un siècle de réformes des mathématiques et de la physique en France et à l'étranger. Paris: Vuibert, INRP.

Bellodi, G., F. Bevilacqua, G. Bonera, and L. Falomo (eds). 2002. *Gli strumenti di Volta. Il Gabinetto di Fisica del'Università di Pavia.* Milan: Università Degli Studi Pavia, Hoepli.

Bishop, G. 1994. *Eight Hundred Years of Physics Teaching.* Basingstoke: Fisher Miller Publishing.

Blondel, C. and M. Dörries (eds). 1994. *Restaging Coulomb: Usages, controverses et réplications autour de la balance de torsion.* Firenze: Olschki.

Blondel, C. 2000. 'L'impact d la réforme de 1902 sur l'enseignement de l'électricité', in N. Hulin (ed.), *Physique et « humanité scientifique »s. Autour de la réforme de l'enseignement de 1902. Etudes et documents.* Villeneuve d'Ascq: Presse Universitaires du Septentrion.

Boltshauser, G. A. 1866. 'Riflessioni sull'ordinamento dei Gabinetti di Fisica delle Scuole Secondarie', *Rivista contemporanea nazionale Italiana*, XLV, pp. 272–85.

Bouasse, H. 1901. 'De l'enseignement des sciences expérimentales dans les lycée', in *L'Enseignement secondaire*, 11, pp. 183–183 and 12, pp. 203–206.

Bouasse, H. 1913. 'Du rôle pédagogique des expériences et des manipulations', Introduction to *Cours de thérmodynamique, Deuxième partie, Tome deuxième du Cours de Physique*, Second Edition. Paris: Ch. Delagrave, pp. xvii–xix.

Brenni, P. 1997. 'Physics Instruments in the 20th Century', in J. Kriege & D.Pestre (eds), *Science in the 20th Century*, Amsterdam: Harwood Academic Publisher, pp. 741–757.

———. 1998. 'La funzione degli strumenti scientifici nella didattica fra Settecento e Ottocento', *Studi Settecenteschi*, 18, pp. 421–31.

———. 2002a. 'Jean Antoine Nollet et les instruments de physique expérimentale', in

J. F. Gauvin and L. Pyenson (eds), L'art D'enseigner la Physique Les Appareils de Démonstration de Jean-Antoine Nollet. Sillery, Québec: Septentrion, pp. 11–27.

———. 2002b. '19th Century Scientific Instrument Advertising', *Nuncius, Annali di Storia della Scienza*, 2, pp. 497–514.

———. 2002c. 'La industria de precisión en el siglo XIX Una panorámica de los instrumentos, los constructores y el mercato en diferentes contextos nacionales', in Bertomeu Sanchez, Garcia Belmar (eds), *Obrint les caixes negres. Collecció d'instruments científics de la Universitat de València*, València: pp. 53–72 (and translated in Enghlish pp. 425–433).

———.2004. 'Mechanical and Hydraulic Models for Illustrating Electromagnetic Phenomena', *Nuncius Annali di Storia della Scienza*, Anno XIX, fasc. 2, pp. 629–57.

———. 2006. 'Artist and Engineer: The Saga of 19th Century French Precision Industry', The Annual Invitation Lecture, *Bulletin of the Scientific Instrument Society*, 91, pp. 2–11.

———. forthcoming. 'La science Française au Crystal Palace', in P. Bret, I. Gouzévitch, and L. Pérez (eds), *Les techniques et la technologie entre France et Grande-Bretagne (XVIe–XIXe siècle), Documents pour l'histoire des techniques*, n° spécial, to be pulished in 2010.

Brock, W. H. 1996. Science for all: Studies in the History of Victorian Science Education. Aldershot: Variorum.

Bud, R., Cozzens S.E. and Potter R.F. 1992, *Invisible Connections: Instruments, institutions and science*, Bellingham (Wash.): SPIE Optical Engineering Press.

Bud, R., and D. Warner (eds). 1998. *Instruments of Science: An Historical Encyclopedia*. New York: Garland.

Cahan, D. 1989. *An Institute for an Empire: the Physikalisch-Technische Reichsanstalt, 1871–1918*. Cambridge: Cambridge University Press.

Clifton, G. 1991. 'La produzione di scientifici in Inghilterra', in G.L'E. Turner (ed.), *Gli Strumenti*. Torino: Einaudi.

de Clercq, P. 1985. *'Nineteenth-Century Scientific Instruments and their Makers' paper presented at the Fourth Scientific Instrument Symposium, Amsterdam, 23–26 October*. Amsterdam : Rodopi.

Donle, W. 1907. *Lehrbuch der Experimentalphysik für den Unterricht an höheren Lehranstalten*, Fourth Edition. Stuttgart: Fr.Grub Verlag.

Drion, C., and M. Fernet. 1861. *Traité de physique élémentaire*. Paris: Masson et fils.

Fau, J. 1845, *Anatomie des formes extérieures du corps humain, à l'usage des peintres et des sculpteurs*, Paris: Méquignon-Marvis.

Fau, J. and C. Chevalier. 1853. *Nouveau manuel complet du physicien-préparateur* Paris: Roret.

Fournier-Balpe C. 1994, *Histoire de l'enseignement de la physique dans l'enseignement secondaire en France au XIXe siècle.* Thèse de Doctorat en didactique des sciences soutenue sous la direction de D.Pestre, Université Paris XI, Paris.

Fox, R., and A. Guagnini (eds) 1999. *Laboratories, workshops, and sites – Concepts and practices of research in industrial Europe, 1800–1914.* Berkeley: Office for History of Science and Technology.

Frick, J. 1856. *Die physikalische Technik oder Anleitung zur Anstellung von physikalischen Versuchen und zur Herstellung von physikalischen Apparaten*, Second Edition. Braunschweig: Friedrich Vieweg und Sohn.

Ganot, A. 1853. *Traité élémentaire de physique expérimentale et appliquée*, Second Edition. Paris: chez l'auteur.

———. 1859. *Cours de physique purement expérimentale: à l'usage des personnes étrangères aux connaissances mathématiques*. Paris: chez l'auteur.

Ganot, A. and G. Maneuvrier. 1894. *Traité de physique élémentaire*, 21st Edition. Paris: Hachette.

Gauvin, J. F. and L. Pyenson (eds). 2002. *L'art d'enseigner la physique: Les appareils de démonstration de Jean-Antoine Nollet*. Sillery, Québec: Septentrion.

Gires, F. (ed.). 2004. *Physique impériale. Cabinet de physique du Lycée impérial de Périgueux*. Niort: ASEISTE.

———. (ed.). 2006. *L'empire de la physique. Cabinet de physique du lycée Guez de Balzac d'Angoulême*. Niort: ASEISTE.

Grimsehl, E. 1911. *Didaktik und Methodik der Physik*. München: C. H. Beck.

Grob B. and Hooijmaijers H. (eds.) 2006. *Who Needs Scientific Instruments? Conference on Scientific Instruments and their Users, 20–22 October 2005*. Leiden: Museum Boerhaave.

Heering, P. 2008. 'The enlightened microscope: re-enactment and analysis of projections with eighteenth-century solar microscopes', *The British Journal for the History of Science*, 41, pp. 345–367.

Hulin, N. 1992, 'Caractère expérimental de l'enseignement de la physique XIXème–XXème siècles', *Bulletin de l'Union de Physiciens*, 86, pp. 1403–15 and 1565–1530.

———. (ed.). 2000. *Physique et «humanité scientifique»s. Autour de la réforme de l'enseignement de 1902. Etudes et documents*. Villeneuve d'Ascq: Presse Universitaires du Septentrion.

———. (ed.). 2001. *Études sur l'histoire des sciences physiques et naturelles*. Lyon: ENS.

———. 2007. *L'enseignement secondaire en France d'un siècle à l'autre 1802–1980*. Paris: Institut National de Recherche Pédagogiques.

Max Kohl A.G. 1911 (no date in catalogue). *Physikalische Apparate Preisliste Nr.50 Band II und III*. Chemnitz: Max Kohl A.G.

Kuhn T. 1970. *The structure of scientific revolutions*, Second Edition. Chicago: University of Chicago Press.

Lehmann, O. (ed.). 1904–1909. *Dr. Frick's Physikalische Technik oder Anleitung zu Experimentalvorträgen sowie zur Selbstherstellung einfacher Demonstrationsapparate*, Seventh Edition. Braunschweig: Friedrich Vieweg und Sohn.

Lemoine, J. 1907. 'Étude critique des collections de Physique, Chimie, Histoire Naturelle', *Bulletin de l'Union de Physiciens*, 2, pp. 23–25.

E. Leybold's Nachfolger. 1913 (no date in catalogue). *Einrichtungen und Apparate für den physikalischen Unterricht sowie Übungen in Praktikum nebst Literaturangaben*. Cologne: Paul Gehly.

———. 1938. *Appareil de physique*. Köln-Bayental: E. Leybold's Nachfolger.

Leon, A. and P. Roche. 2003. *Histoire de l'enseignement en France*. Paris: PUF.

Lommel, E. 1895. *Lehrbuch der Experimentalphysik*, Second Edition. Leipzig: Johann Ambrosius Barth.

Ministère de l'instruction publique et des beaux-arts. 1884. *Catalogue du matériel scientifique des lycées et des collèges de garçon*. Paris: Imprimerie Nationale.

———. 1900. *Catalogue du matériel scientifique des lycées et des collèges de garçon*. Paris: Imprimerie Nationale.

Molteni, A. no date. *Catalogue des tableaux sur verre en noir et en couleur pour l'enseignement par projections*. Paris: Molteni.

———. no date. *Les projections lumineuses à l'école et chez l'amateur*. Paris: Molteni.

Müller, J. 1844. *Lehrbuch der Physik und Meteorologie*, Second Edition. Braunschweig: F. Vieweg und Sohn.

Pantalony, D. 2009. *Altered Sensations: Rudolph Koenig's Acoustical Workshop in Nineteenth-Century Paris*. Dordrecht, Heidelberg, London, New York: Springer.

Pfaundler, L. (ed.). 1905–1914. *Müller-Pouillet's Lehrbuch der Physik und Meteorologie*, 10th Edition. Braunschweig: F.Vieweg & Sohn.

Pixii. 1821. *Notice de différents instruments de physique, d'optique de mathématique et autres à l'usage des sciences*. Paris: imprimerie de P.N. Rougeron.

Pixii père et fils. 1842. *Catalogue des principaux instruments de physique, chimie, optique, mathématique et autres à l'usage des sciences*. Paris: chez l'auteur.

———. 1852. *Catalogue des principaux instruments de physique, chimie, optique, mathématique et autres à l'usage des sciences*. Paris: chez l'auteur.

Privat-Deschanel, A. 1869. *Traité de physique*. Paris: Hachette.

Ruiz-Castell, P. 2008. 'Scientific Instruments for Education in Early Twentieth-Century Spain', *in Annals of Science*, Volume 65, pp. 519 – 527.

Schmid. K. A. 1867. *Encyklopädie des gesammten Erziehungs- und Unterrichtswesens*, Volume VI. Gotha: R. Besser.

Schmidt, A. 1926. *Geschichte der Firma E. Leybold's Nachfolger 1850–1925*. Köln: Paul Gehly.

Simon, J. 2008. 'The Franco-British communication and appropriation of Ganot's physique (1851–1881)', in J. Simon, N. Herran, T. Lanuza-Navarro, P. Ruiz-Castell, and X Guillem-Llobat (eds), *Beyond Borders: Fresh Perspective in History of Science*. Newcastle: Cambridge Scholars Publishing, pp. 141–68.

Sutton, G.V. 1995. *Science for a Polite Society Gender, Culture and the Demonstration of Enlightment*. Oxford: Westview Press.

Sydenham, P.H. 1979, *Measuring Instruments: tools of knowledge and control*. New York: Peter Peregrinus ldt. in association with the Science Museum London.

Turner, A. 1987. *Early Scientific Instruments*, London: Sotheby's Publications

Turner, G. L'E. 1983. *Nineteenth-Century Scientific Instruments*. London and Berkeley: Sotheby Publications and University of California Press.

———. (ed.). 1991. *Gli Strumenti*. Torino: Einaudi.

———. 1996. *The Practice of Science in the Nineteenth Century Teaching and Research Apparatus in the Teyler Museum*. Haarlem: Teyler Museum.

Van Helden A. and Hankins T.L. (eds). 1994. 'Instruments' special issue of *Osiris*, 2nd series, vol. 9.

Weinhold, A. F. 1881. *Physikalische Demonstrationen. Anleitung zum Experimentieren*. Leipzig: Von Quandt & Händel.

Wiedemann E., Ebert H. 1904. Fifth Edition, *Physikalisches Praktikum*, Braunschweig: F. Vieweg und Sohn.

Williams, M. E. W. 1994. *The Precision Makers. A history of the instruments industry in Britain and France, 1870–1939*. London: Routledge.

Zoller, P. 2009. 'Physics Experiments for Everyone: German Makers', *Bulletin of the Scientific Instrument Society*, 102, pp. 21–28.

Zwei Normalverzeichnisse physikalischer Apparate. 1896. *Zeitschrift für den physikalischen und chemischen Unterricht*, Heft IV, pp. 175–83.

"SIMPLEX SIGILLUM VERI": ROBERT POHL AND DEMONSTRATION EXPERIMENTS IN PHYSICS AFTER THE GREAT WAR

Roland Wittje

ABSTRACT

Lecture demonstrations, the hallmark of experimental physics lectures in the nineteenth century, have given way to the once-mocked chalk and blackboard physics in the second half of the twentieth century. This paper examines the nature and purpose of demonstration experiments and the path of their transformation. What is it that was actually demonstrated? How and why did lecture demonstrations change and finally loose much of their importance? With these questions in mind, I will trace the trajectory of lecture demonstrations in physics teaching from around 1900 up until the interwar period. I especially want to draw upon the conceptualisation of lecture demonstrations and physics teaching, marketed by the dominating German instrument companies Max Kohl and E. Leybold's Nachfolger, followed by the system of lecture demonstrations developed by Robert Wichard Pohl, Professor of Experimental Physics at Göttingen, after the First World War.

In 1924 Pohl and the instrument company Spindler & Hoyer started a cooperation to produce a system of demonstration instruments which were organised along the lines of Pohl's series of physics textbooks. Vividness, simplicity and the reject of historism were the main features of the teaching system. Pohl's ideas for the re-organisation of lecture demonstration fitted well in a time of rapid changes in physics, shortage of money and German postwar society. His elaborated system of textbooks and demonstration technology turned out to renew and soon to dominate lectures on experimental physics at universities and other higher education institutions in Germany and other countries.

INTRODUCTION

The interwar years have generally been characterised as a time of rapid change, instability and crisis in European and North American societies, including the physics community. The Weimar Republic in Germany was particularly characterised by political and economic instabilities even as it hosted the world's indisputable leading community of theoretical physicists and maintained its position as a country with leading experimental communities as well.[1] Crisis phenomena in physics were experienced by the economic shortage in the public sector as well as the repeated announcement of an apparent breakdown of the physics worldview.[2] The atmosphere of crisis stood in contrast to the rapid development in theories, experimental methods and instrumentation in physics at the same time. Historical studies have long paid attention to these developments in physics research in the interwar years in Germany and elsewhere, illuminating aspects of the development of theory, experimentation and instrumentation, as well as the economic constraints on research.[3] In contrast to this strong interest in research, the interaction of these developments with aspects of academic teaching, and its evolution in the same period has scarcely been looked at.[4] From an institutional perspective, the interaction between academic teaching and research was omnipresent. In the interwar period, government, philanthropic and industrial research laboratories were of growing importance, but the vast majority of research was nonetheless carried out at universities and institutes of technology (Technische Hochschulen), which were essentially devoted to teaching. The Humboldtian concept of *Wissenschaft* – being the unity of teaching and research – was predominant at these institutions and among science practitioners.[5]

In this paper, I deal with one aspect of academic teaching: the transformation of lecture demonstration experiments. I follow the development of demonstration experiments within the German speaking community from just before the First World War to the Weimar Republic. The development of the *System Pohl* as a rigorous modular and standardized system of lecture demonstrations by Robert Wichard Pohl and its characteristics are central to this story. Pohl's system of demonstrations was based on his epistemology of simplicity, *Anschaulichkeit* (clearness and vividness) and experimentalism. In a final part, I will look at the propagation and adaptation of the System Pohl in the German community, as well as in other countries through his students, his textbooks and his close collaboration with the instrument company *Spindler & Hoyer*.

1 See Kragh (1999), pp. 139–154, and Gregory Breit in Cornell (1986), p. 328.

2 University budgets in the immediate postwar years were far behind prewar levels. Lack of funding for instruments and personnel stood against a drastic increase in the number of students. The shortage of foreign currency and hyperinflation aided to the economic crisis. The crisis of the physics worldview manifested itself in the deficiency of the old quantum mechanics and the widespread conviction that hitherto fundamental concepts of physics, such as causality and energy conservation had to be reconstructed from ground up. See especially Kragh (1999), pp. 139 and 155–161, von Meyenn (1994), and Forman (1971).

3 See, for example, Kragh (1999), Richter (1972), and Galison (1997).

4 An exception is Michael Aaron Dennis' study of Charles Stark Draper's instrumentation laboratory at MIT. See Dennis (1991). Rammer (2004) has argued for the importance of Robert Pohl's teaching in the formation of the Pohl School and its specific style of thought and thought collective. More about this later. Eckert and Schubert (1990) have also discussed Pohl's teaching in their chapter on the Pohl School.

5 The important schools of quantum mechanics of the 1920s, for example, emerged from university departments and were based on the close interaction of senior professors with young and talented students. See Kragh (1999), pp. 159–160.

LECTURE DEMONSTRATIONS AND THE STRUGGLE FOR *ANSCHAULICHKEIT*

A central and characteristic element of academic physics teaching in early twentieth century Germany were the large experimental lectures, the *Experimentalvorlesungen* delivered by experienced chair professors. The smaller classrooms with specialized lectures were delivered by extraordinary professors and *Privatdozenten*.[6] The demonstration experiments of the experimental lecture set a contraposition to the often criticised *Kreidephysik* (the chalk physics) of the theoretical lecture.[7] Despite many different opinions about the contents and the methods of lecturing experimental physics, physicists still agreed upon that there was no way around the experimental lecture based on demonstrations.[8] Inherent to the concept of the experimental lecture was the teaching through experiments to *veranschaulichen* (visualise) the physical concepts to be understood and memorized.

The concept of *Anschaulichkeit* played a central role in the intellectual discourse about physics during the Weimar Republic. *Anschaulichkeit* refers to *becoming clear by looking at something* and has no literal translation into English. It can be, however, understood as a combination of clarity, distinctiveness and vividness. Critics from within the physics community as well as sections of the cultural milieu of the Weimar republic attacked the new abstract theoretical physics, especially quantum mechanics and relativity, for being *unanschaulich* (non-vivid).[9] Representatives of the new theories, such as Werner Heisenberg and the physicist-philosopher Philipp Frank responded to the criticism by arguing that the new theories were actually not less *anschaulich* than the older ones, by trying to redefine *Anschaulichkeit* and to depict the classical mechanical view as at least as abstract as the quantum mechanical view.[10] Heisenberg linked the question of whether a theory was *anschaulich* directly to the question of whether an experiment could be conceptualised to investigate the empirical consequences of the theory.[11] Thought-experiments were thus frequently employed to visualise the theories of relativity and quantum mechanics and their consequences in place of physical experiments that were carried out in laboratories or classrooms. Thus thought-experiments gained their epistemological value and their persuasive power from that scientists agreed upon that they could be carried out in principle,

6 A *Privatdozent* was a lecturer who held a *Habilitation* as the formal qualification to become a university professor, but who had not been appointed on a professor chair yet.

7 For a criticism of Kreidephysik in the context of school teaching, see for example Reuter (1925). The criticism of Kreidephysik has to be understood in the context of the raising status of theoretical physics compared to experimental physics in the interwar period, and the perception of many experimental physicists that theoretical physics became more and more abstract.

8 See explicitly Mecke and Lambertz in Geiger (et al.) 1926, Vol. 1, Vorlesungstechnik, p. 209.

9 von Meyenn (1994). See also Poske (1921).

10 Heisenberg (1927), Frank (1928), and Forman in von Meyenn (1994), pp. 181–200.

11 Heisenberg (1927), pp. 172 and 174.

and that their outcome was unambiguous. Scientists had to be able to imagine the experiment and to accept it as a demonstration.[12] Thereby physicists actively established a close relationship between the discourse about theoretical physics and *Anschaulichkeit* through thought-experiments on one hand, and the *Anschaulichkeit* that was provided by demonstration experiments in the lecture on the other.

Lecture demonstrations were, however, not limited to the visualisation of physical concepts and principles alone. Methods and instruments of measurement, as well as technical concepts and machinery, were demonstrated. In order to fulfil their function of being *anschaulich* and to aid the understanding and the remembering of physical concepts, experimental lectures were carefully staged, aided by demonstration assistants and instrument makers. Lecturing and demonstrating the experiments was considered an art, and the grand masters of this art were frequently used as role models by their students.[13] Demonstration experiments were very different from experiments in research in that unlike a research experiment, they were not open-ended but had a fixed outcome. Compared to research experiments, which are generally long-term processes, demonstration experiments would not usually take more than a few minutes.

DEMONSTRATION EXPERIMENTS BEFORE THE GREAT WAR: TANTALUSBECHER, DONNERHÄUSCHEN UND DIE GESAMTE ZAUBERPHYSIK

The first years of the twentieth century up to the First World War were generally marked by a wealth of public institutions dedicated to scientific research and an expansion of the scientific community, expressed, among other things, through the foundation of many new universities and other scientific institutions. With educational reforms like the ones in Prussia in 1882 and 1892, and in other German states, mandatory experimental exercises were introduced in secondary school physics. Other countries in Europe and America experienced a similar growth in the market for didactical instruments in secondary schools and higher education.[14] An industry emerged to develop and produce ‘teaching technology’, often in close co-operation with university professors. After the Franco-Prussian War, German instrument makers overtook French instrument makers as leaders in the growing instrument market.[15] Companies like Max Kohl of Chemnitz, and E. Leybold’s Nachfolger of Cologne were grazing together on these rich pastures, and we can see this in their huge catalogues that were almost identical in their

12 About the function of thought experiments, see Kuhn (1977).

13 See, for example, about Philipp Lenard in Ramsauer (1949), pp. 107–108; about Pohl’s view on Georg Quincke, see Teichmann (1988), p. 21, about Ferdinand Braun, see Zenneck (1961), pp. 82–83.

14 See Brenni, Cuenca and Simon, Hoffmann, Kremer, and Turner in this volume.

15 See Brenni, this volume, Williams (1994), and Cahan (1989).

offerings of instruments. They even used the same engravings! Max Kohl and E. Leybold's Nachfolger offered almost every instrument that was explained in the *Müller-Pouillets Lehrbuch der Physik und Meteorologie,* a standard textbook of the time, supplemented by even more demonstration apparatus described in other textbooks and manuals for demonstration experiments.[16] The result was a variety of different apparatus proposed for demonstrating the same experiment or phenomena. The textbooks and manuals had been published in several editions by the turn of the century, and with every new edition they had grown bigger by adding new knowledge, new experiments and new instruments to the old. The instrument companies in turn also added the new instruments to their product range and continued to produce the old ones. Together they offered a vast number of instruments that had by then acquired a historical meaning rather than representing a body of knowledge or experimental practice that was up to date.[17]

The claim of the large experimental lecture was to represent the whole of physics, not on the blackboard, but through experiments. Many of the demonstrated experiments were seen as crucial experiments in physics, for example – Charles Augustin de Coulomb's determination of the inverse square law, James Prescott Joule's measurement of the mechanical equivalent of heat, or Heinrich Hertz's experiments on electromagnetic waves. The original experiments could not be performed in a standard lecture theatre and the experiments performed here were didactically reduced versions instead. The instruments sold were modelled after the original instruments but often not as precise as the research instruments. Sometimes the demonstration instruments even lacked vital parts of the research instrument that they were modelled after.[18] The experimental lecture claimed that every piece of knowledge in physics could be set into question and then be proven experimentally, at least in principle. One of its functions was to underline the authority of physics as an empirical body of knowledge, and this proof was established not through open-ended experimentation but through demonstration.

16 Johann Heinrich Jacob Müller published his first edition of the textbook in two volumes between 1842 and 1844. The 10th edition, published by Leopold Pfaundler between 1905 and 1914, had four volumes. Among the other books that the instrument catalogues frequently referred to were Weinhold (1913), Lehmann (1904–1909), and Grimsehl (1909). See also Brenni in this volume.

17 See Brenni, this volume.

18 In the collection of the Department of Physics of the Norwegian Institute of Technology, for example, there is a Coulomb torsion balance, which was delivered by Max Kohl in 1910. The torsion micrometer is a dummy and cannot be turned and the instrument cannot be used as a precision measurement apparatus. The instrument is therefore not a measurement instrument but an icon for the inverse square law. About the problems to determine the inverse square law with the Coulomb balance, see Heering (1998).

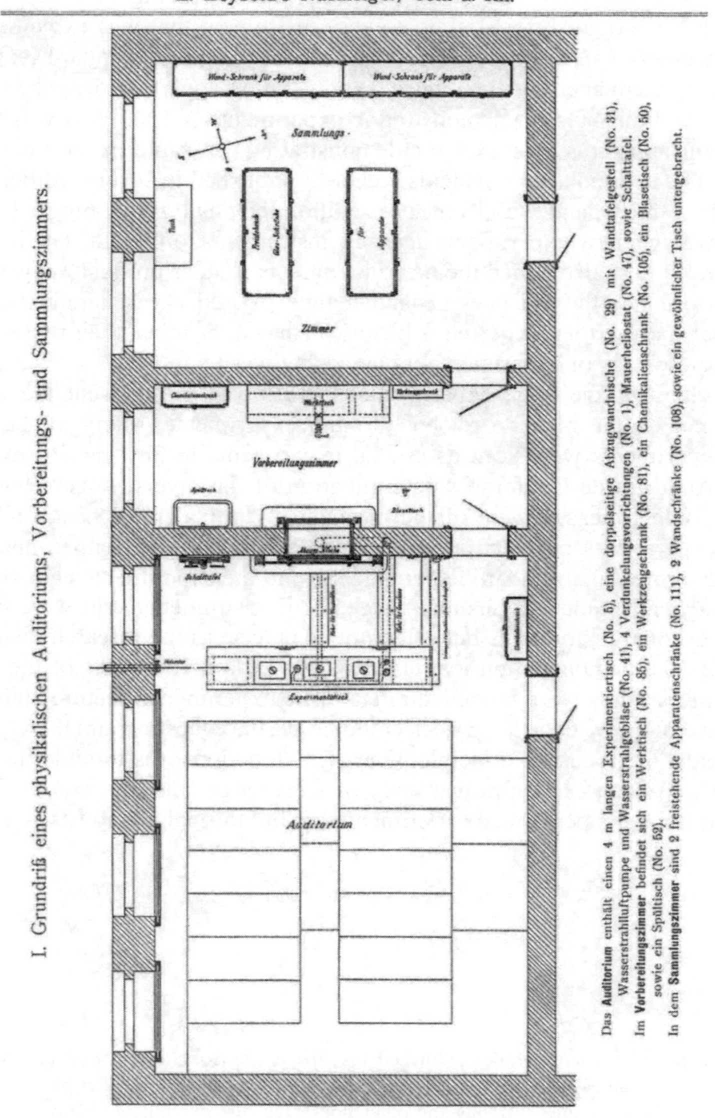

Fig. 1: Outline of the lecture room, preparation room and store-room for scientific instruction. E. Leybold's Nachfolger (ed. R. Bernoulli): Einrichtungen und Apparate für den physikalischen Unterricht sowie für Übungen im Praktikum nebst Literaturangaben, 1913. Both Max Kohl's and E. Leybold's Nachfolger's catalogues before and after the First World War gave such detailed proposals for the furnishing of teaching chambers for science teaching in higher education alongside with explicit references to university science teaching and several pages of further advice.

The instrument catalogues as well as the treatises on demonstration experiments by the physics teachers Joseph Frick and Adolf Ferdinand Weinhold gave detailed descriptions not only of the instruments and demonstrations, but also of the ideal arrangement of the teaching premises, the seating arrangement of the lecture theatre, furniture, natural and artificial lightning, electricity and gas installations, projection arrangements and equipment, etc.[19] The ideal premises were supposed to be divided into the lecture hall, the preparation room, and the collection room where the instruments were stored in cabinets. Central to the lecture theatre was the large experimental table where the experiments were set up and demonstrated by the professor or lecturer and his assistants (see fig. 1 and fig. 2).

The large suppliers of teaching apparatus delivered to universities, technical institutes and secondary schools.[20] There was certainly a difference in scope and complexity between physics teaching at primary and secondary schools and teaching at universities and technical colleges or institutes, but the boundary is not always easily identifiable in the demonstration apparatus on offer. Many secondary schools had excellent teaching collections, which could measure themselves with the collections of universities or technical schools. They also used the same kinds of instruments. There was though one main aspect of secondary school teaching that made it distinct from teaching both at universities and technical schools, and that was frequently discussed: science education in secondary schools was not aimed at serving the needs of vocational training or as technical training for the study of the sciences at universities. The main goal of science education was the ideal of a general humanist *Bildung*. Scientific education was mainly seen as moral training. Science teaching should teach modesty and bring up noble and cultured characters, and train its pupils to formulate precise concepts. The view of *Science as Culture* was placed within a broader vision of cultural history. This idea of science education as *Bildung* therefore gave the historical approach a vital role within science teaching at schools.[21]

19 See Brenni in this volume, the Max Kohl and Leybold catalogues, and Lehmann (1904), vol. 1.

20 See especially the long list of letters of recommendation from 'satisfied customers' around the world in the introduction part of the Max Kohl and Leybold catalogues.

21 See, for example, Poske (1921), Gebhardt (1922), and Rosenberg (1923). See Olesko (1995), and Hoffman in this volume about North America. Pohl saw, as late as in 1952, *Bildung* as the main objective of physics teaching in schools and the education towards 'inner modesty' (Erziehung zur inneren Bescheidenheit) as one its main tasks. See Pohl (1952), especially p. 537. See also Ramsauer (1949), pp. 48–52 and 61–68.

Fig. 2: Lecture demonstrations at the Norges Tekniske Høgskole (Norwegian Institute of Technology) in Trondheim, 20 September 1915, dealing with the formation of the earth according to Laplace. A centrifugal machine with accessories was set up on the large experimental table. The person on the photo was the instrument maker and lecture assistant Thorvald Reed. The setup of the Trondheim experimental lecture was representative for physics lectures of its time. Professor Sem Sæland had been studying with Philipp Lenard in Heidelberg and used his lectures as an example.[22]

In this context arrived the Great War and severe disruption of social and political life for the European, but especially for the German physics community. It was arguably the first war where total scientific and industrial mobilisation made a key impact on warfare, both on the battlefield and at home.[23] Students and professors alike had left universities, either to fight on the battlefields or to work on war-related research and development projects. When they returned to the universities in 1919 the world was not the same. Enthusiasm about German victory had been replaced by the disappointment and effects of defeat. Following the involvement of many scientists in the infamous *Aufruf an die Kulturwelt* of 1914, where German intellectuals defended German warfare, German scientists were boycotted in international collaboration in the postwar years.[24] At home, the relationship between science and industry had also been altered during the war. Many companies continued wartime practices, extended their laboratories and employed an even

22 Wittje (2003), pp. 48-49, and Ramsauer (1949), p. 109.
23 Trischler (1996).
24 See von Meyenn (1994), pp. 29–34, and Kragh (1999), chapter 9.

higher number of scientists in industrial research.[25] The economic situation and hyperinflation in the immediate postwar years made it impossible for educational and scientific institutions to buy new equipment. When the economic situation began to improve towards the end of the 1920s, Max Kohl and E. Leybold's Nachfolger published almost unchanged pre-war editions of their catalogues. The times, however, had changed, and pre-war teaching technology seemed outdated, as Manfred Dunkel the son-in-law of Alfred Schmidt, the owner of E. Leybold's Nachfolger had to confront in 1929:

> "In 1929 the new main catalogue for physical apparatuses on which Dr. Vieth had worked for four years was also completed; again it appeared in German, English and French.
>
> Its' size was smaller than that of the pre-war catalogue. But it still covered 460 pages and specified 5600 apparatus. My father-in-law Schmidt had received the addresses of some physicist friends of mine in Ludwigshafen, and he also sent the catalogues to them. If I would have seen the catalogue before, I would have discouraged him strongly, because the Tantalus cup, the thunder house, and all the magic physics [Tantalusbecher, Donnerhäuschen und die gesamte Zauberphysik] caused derisive laughter among the young physicists working on x-ray structural analysis, electron diffraction, atomic physics etc. They came to me, just to show me the most ridiculous things. I was so embarrassed that I sank in shame and forgot to thank my father-in-law for sending [the catalogue], which he took amiss." [M. Dunkel (1973), pp. 47–48, translated from the German by author].

ZEITGEIST AND RECESSION: THE APPEARANCE OF THE 'SYSTEM POHL' AFTER THE GREAT WAR

Rapid changes in ideas of physics and its practice made an overhaul of physics teaching after the Great War seem unavoidable. Most of the pre-war textbooks reappeared in the 1920's in revised editions, adding on the *new physics*. But the younger generation of Germany physicists were not impressed. They saw these textbooks as out-dated and too voluminous.[26] The books still carried much of the old accumulated knowledge and practice, which now appeared dispensable to many. The idea of including all of the new physics in teaching was used as an argument to get rid of some of the old in order to create space. The same, in turn, accounted for many of the old demonstration experiments, which had now be-

25 Georg Gehlhoff, an industrial physicist, working for the German glass industry and founder of the *Deutsche Gesellschaft für technische Physik,* estimated in 1919 that 2/3rds of Germany's physicists did not work in the public academic sector but in the industry. Helga Schultrich estimated in 1985 the total amount of physicists in Germany in 1930, excluding school teachers, to be about 1,600, of which 500 worked in universities and other institutions of higher learning, just over 100 in public research institutions, 100 in the patent system and about 800 in the industry. See Hoffmann and Swinne (1994), p. 47, n. 37.

26 An 11th and last edition of the *Müller-Pouillet* textbook appeared between 1925 and 1934 and had grown to five volumes. See Eucken et al (1925–1934). Pohl later tried to limit and modernize new editions of his textbooks in order to 'avoid them growing fat as a deceiving sign that they were outdated'. See Mollow (1977).

come mere curiosities. The old catalogues offered dozens of different elaborate apparatus for showing the same phenomena, which Paolo Brenni describes, in his contribution to this volume, as a form of elephantasis. But the problem was not only one of space and money. The pomp of the pre-war demonstration lecture was now perceived as misplaced and out of fashion, its historicism was rejected and had to make way for a New Soberness. The community demanded that entirely new textbooks be written for academic physics teaching based on concepts appropriated to the new times, rather than mere revisions of the old. In the same vein, the demand for developing modernised, rationalised, standardised and simplified concepts of lecture demonstrations and demonstration instruments was also articulated.

Fig. 3: 92 624. House with Tower and Lightning Conductor. A small vessel, containing spirit, is placed in the tower and the sheet iron vessel placed in the house is filled with a mixture of house gas and air, and well corked. A charged Leyden jar is placed in the neighbourhood of the sphere of the lightning conductor and discharged. The spirit, and the gas mixture are then ignited; the former commences to burn and the latter forces the stopper out with a bang, the roof of the house is forced out and the walls collapse. [Max Kohl A. G.: Preisliste Nr. 100, Band III Physikalische Apparate, 1928, p. 849, translation from the pre-war English catalogue].

In response to these demands, Robert Wichard Pohl, Professor of Experimental Physics, head of the *I. Physikalisches Institut* at the University of Göttingen, and one of the founders of what became known as solid state physics, became a predominant figure in the renovation of lecture demonstrations and physics textbooks in Germany.[27] Pohl was only 31 when he was appointed extraordinary professor in 1916. He did not take up his professorship before 1919, because of his military service during the First World War. When Pohl arrived in Göttingen, it was in the process of being established as a distinguished place to tackle innovations in physics teaching. In 1920, Max Born and James Franck were appointed to chairs in theoretical and experimental physics, respectively, and Pohl was promoted to ordinary professor. With Pohl, Franck and Born, Göttingen had appointed three young but leading figures of the German physics community, each of them governing over their own independent *Physikalisches Institut*. Göttingen physics graduates were now jokingly divided into the *Po(h)lierten* (the polished), the *Fran(c)kierten* (the stamped) and the *Bornierten* (the narrow minded), depending on which of the three departments and their respective research schools they belonged to. The Born School was to become, next to the Sommerfeld School in Munich and the Bohr School in Copenhagen, one of the most important centres for the development of quantum mechanics in the 1920s.[28] With Ludwig Prandtl's school of aerodynamics and Max Reich on the chair of Applied Electricity, Göttingen was at the same time a centre of applied and technical physics.[29] Finally, Göttingen was also stronghold of scientific instrument making and had a history of collaboration between scientists at the university designing instruments, and instrument makers building and marketing them.[30]

The Pohl School of experimental physics did not gain the same kind of international repute and connections as the Born School. But it did become a dominant school within the German experimental physics community. While Born and Franck were forced to leave Göttingen after the Nazis came to power in 1933, Pohl continued his school through the Third Reich, and up until his retirement as Professor Emeritus in 1952.[31] The Pohl School was organised around the studies of the optical and electrical properties of films and crystals, and especially colour centres (*Farbenzentrenforschung*) as an experimental system.[32] All of Pohl's students worked with this experimental system. Visitors were surprised by the similarity of experimental arrangements in most laboratory rooms.[33] The so-

27 On Pohl, see Eckert and Schubert (1990), pp. 95–109, Teichmann (1988), and Rammer (2004). Robert Pohl started to use his middle name Wichard between 1925 and 1927.

28 Kragh (1999), pp. 159–160. See Eckert and Schubert (1990), pp. 70–73, about the Sommerfeld School.

29 See Eckert (2006) on Prandtl and the *Kaiser-Wilhelm-Institut für Strömungsforschung* in Göttingen.

30 See Meinhardt (1973).

31 Born was forced to leave his post in 1933 due to the anti-Semitic *Gesetz zur Wiederherstellung des Berufsbeamtentums* (Law for the Restoration of the Professional Civil Service). As a veteran of the First World War, Franck could have stayed on, but left in protest. Latest in 1935, Franck would have been sacked as well, due to the *Nuremberg Laws*. See Lemmerich (2007) for a biography of Franck.

32 For the concept of experimental systems, see Rheinberger (1997).

33 Gerlach (1978), p. 219. Rammer (2004, p. 319) found that in 1945, all members of the *I. Physikalisches Institut* under Pohl worked on the same research topic, whereas the other

cial structure of the Pohl School differed significantly from the Born School or the Bohr School in the sense that Pohl was the undisputed authoritarian patriarch of his school, who did not give his disciples much of a voice, whereas Born and Bohr acted more as patriarchal mentors, who developed or perhaps co-produced the new quantum mechanics in a dialogue with their young students.[34] The Bohr, Born, and Sommerfeld schools were also open to input from outside, and the students moved rather freely between the different schools and nations. The Pohl School, in contrast, remained relatively closed. Its members cited mainly each other's work and collaborated little with researchers from outside the school.[35] Pohl's assistants usually did not do their *Wanderjahre*, but stayed with Pohl from their time as undergraduate students until they were appointed as professors at other universities.

More than for his research in the electrical and optical properties of crystals, Pohl became famous for his experimental lectures based on a fundamentally new system of lecture demonstrations and his novel series of textbooks in experimental physics. In the division of labour among the Göttingen experimental physicists, Pohl taught the large experimental lectures, and only the lectures, year after year, whereas Franck organised the laboratory courses.[36] In the lectures, Pohl was assisted by his *First Assistant* as well as his master mechanic Wilhelm Sperber. These assistants were members of the faculty who completed their *Habilitation* thesis to qualify for the appointment as professor. Pohl's First Assistants were, as a rule, subsequently appointed on physics chairs at other universities and carried Pohl's tradition further. Gerhard Rammer has described the Pohl School as a thought collective after Ludwik Fleck. In Rammer's view, Pohl and his collaborators and students constituted the esoteric circle of the collective, whereas guests of the department and researchers at other institutions working in the same field formed its exoteric circle.[37] Rammer argues that the style of thought of the Pohl School was largely identical with the style of thought of the physicists at the time in general, but displayed a distinct peculiarity in its resistance against contemporary theoretical physics. Theoretical interpretations were secondary in the publications profile of Pohl's institute.[38] This resistance against theory was expressed

physics departments in Göttingen showed more diversity in the assignment and variety of research topics.

34 For the Pohl's style of leadership, see Mollow (1977), p. 232, Eckert and Schubert (1990), pp. 96–109, Gerlach (1978), p. 216, and Minnigerode (1976), p. 141.

35 Rammer (2004), pp. 326–357. The Pohl Schools had its closest international relationship with the institute of Abraham Fedorovich Ioffe in Leningrad. See Teichmann (1988), pp. 130–138.

36 In autumn 1933, however, after Franck had left Göttingen, Pohl advised Franck's former assistant Günther Cario to organise the laboratory course according to Pohl's methodology. The laboratory courses were then taken over by Franck's successor Georg Joos. See Rammer (2004), pp. 396–397.

37 Rammer (2004), pp. 358–372, and Fleck (1979). I have not adopted Rammer's analysis of the Pohl School as a thought collective, because I do not find it suitable to explain the spread of Pohl's teaching system.

38 Pohl's former student Erich Mollow explained Pohl's rejection of mathematical theory with the circumstance that Pohl's mathematical training was weak, which Pohl himself regretted.

in Pohl's dictum *Theorien kommen und gehen, aber die Tatsachen bleiben* (theories [may] come and go, but the facts remain).[39] By facts he meant experimental facts, which in contrast to theories, he claimed were secure. His dictum was also directed against certain fashions in physics, which he rejected.[40] Another characteristic of Pohl's ethos of science was his call for simplicity. Pohl famously placed his motto *Simplex sigillum veri* (Truth lies in simplicity) in the lecture theatre (see fig. 4).[41] He applied this principle claim of simplicity to his lecture demonstrations, his textbooks, his own research practice, and even to his popular writings, which he constructed in short main clauses.[42]

Rammer has argued that the large experimental lecture had an important function in the establishment of the thought collective, because it introduced the students to a distinctive way in which Pohl conducted experiments, and his way of separating between what he called experimental facts and what he identified as theoretical interpretations. Dispersion and absorption, for example, were experimental facts whereas electromagnetic waves propagating in vacuum where not a fact, but a theoretical and mathematical abstraction.[43] Contemporaries of Pohl claimed that one could always spot a student who had studied with Pohl and heard his experimental lectures. Even students of Pohl's former Assistants, such as Pohl's successor Rudolf Hilsch, could be distinctly identified as belonging to the Pohl School.[44] Pohl's lecture certainly made a strong impression on the students who attended it. But these students cannot be identified with belonging to the thought collective of the Pohl School. Whereas Pohl's research school was limited to his students, and Assistants, the influence of Pohl's lectures was much bigger. Also students who later became part of the other Göttingen research schools, as the Born School and the Franck School, attended Pohl's lecture, as well as other science and medicine students at Göttingen. The lecture theatre was thus extended in 1926 with funds from the Rockefeller Foundation. Its capacity was more than doubled and offered space for more than 400 Students.[45] Bernhard Gudden, a former First Assistant of Pohl, estimated in 1944 that more than 10 000 students must have attended the lectures.[46] Pohl was also known to be a great performer

See Mollow (1977), p. 230. Walther Gerlach diagnosed Pohl as renouncing from mathematical thinking. See Gerlach (1978), p. 219.

39 Eckert and Schubert (1990), p. 103. Mollow (1978, p. 17) stated that Pohl himself had mentioned that he had adopted this dictum from his teacher Georg Hermann Quincke.

40 He also rejected the slogan of *Modern Physics*. See Gerlach (1978), p. 219.

41 Pohl is said to have inherited his motto from the famous Göttingen Professor Georg Christoph Lichtenberg. See Eckert and Schubert (1990), pp. 98–99.

42 Mollow (1977), p. 233.

43 Rammer (2004), pp. 360–364.

44 Achilles (1977), p. 159, and Gerlach (1978), p. 216.

45 Mollow (1978), p. 13, Eckert and Schubert (1990), p. 98, Minnigerode (1976), p. 140, and Pohl (1933). The number of regular seats was limited to 400 due to the width of the building, instead of the 800 seats that Pohl estimated as necessary.

46 Achilles (1977), p. 158, and Gudden (1944), p. 168.

and visitors of the university as well as the Göttingen bourgeoisie went to see him lecture.[47]

Fig. 4: Physics lecture theatre of the University of Göttingen after the modification according to R. W. Pohl in 1926. Credit: Deutsches Museum.

Göttingen had a long tradition of experimental physics lectures and lecture demonstrations, dating back to the eighteenth century and the famous Georg Christoph Lichtenberg. Pohl's direct predecessor however was the Dutch theoretical physicist, Peter Debye. Debye had been appointed Professor of Theoretical Physics at Göttingen in 1914 but after Eduard Riecke the professor of experimental physics died in 1915, Debye also took over his duties. When Pohl took over in 1919, both laboratory teaching and research equipment seem to have been in a rather poor state.[48] Instead of continuing existing practices of lecture

47 Mollow (1977), p. 233.

48 An account of the poor quality of Debye's experimental physics teaching regarding student experiments is found in the joint interview with James Franck and Herta Sponer-Franck July 12 1962, conducted by Thomas S. Kuhn and Maria Goeppert-Mayer, tape 23, side A, pp 20 – 21 in the transcriptions, Niels Bohr Library, American Institute of Physics. Herta Sponer described the student experiments as very poor, not being interesting and in need for repair. Debye did not seem to take her complains very serious and did, instead, convince her to do some theoretical work. See tape 23, side A, pp 20 – 21 in the transcriptions, Niels Bohr Library. See Gerlach (1978), p. 217, about the state of the research equipment that Pohl inherited from Debye. Nothing is known about Debye's experimental lectures in Göttingen.

demonstration at Göttingen, and using the existing collection of demonstration apparatus, Pohl decided to make a radical change and to build up from scratch both the conceptual as well as the material basis of his lecture.[49] Within a few years, he brought his scheme of teaching to perfection and made it to a system that could be merchandised as a modular demonstration technology accompanied by a series of textbooks. Pohl's style of teaching had thus broken away from the pre-war lecture and became distinctive.

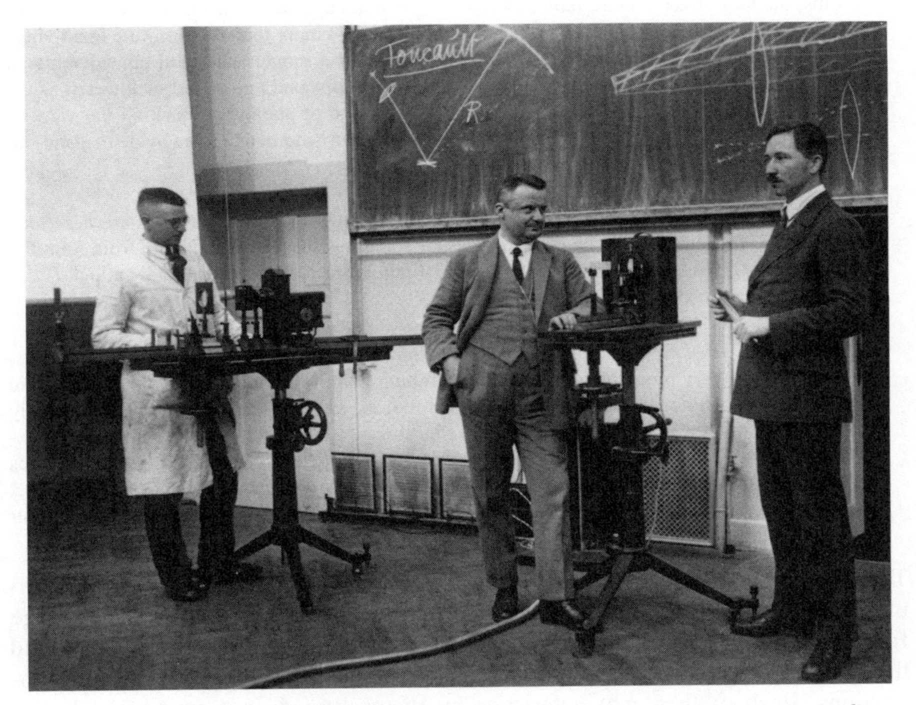

Fig. 5: Robert Pohl in the centre, leaning on one of his tables for shadow projection in the Göttingen lecture hall. At left: chief mechanic Wilhelm Sperber, at right: Pohl's assistant Bernhard Gudden who was appointed professor at the University of Erlangen in 1926. The photo was most probably taken before the extension of the lecture theatre in 1926.
Credit: AIP Emilio Segre Visual Archives, Physics Today Collection.

49 For the historical collection of scientific instruments at the Göttingen physics department, see Beuermann et. al. (1986 and 1988).

What were, then, the main features of Pohl's teaching system, and what made it so different? Pohl himself explained his way of teaching in the preface to the first edition of his *Einführung in die Mechanik und Akustik* of 1930:

> "Fundamental experiments are brought into prominence throughout; they are intended to make the ideas concerned as clear as possible to the student and to indicate the orders of magnitude involved, quantitative details being left in the background.
>
> Some of the experiments require a good deal of space. In the Göttingen lecture-room a smooth parquet floor 12 metres by 5 is available. That troublesome obstacle in old lecture-rooms, the large fixed lecture-table, was got rid of years ago. Instead, small tables are set up as required. These, however, are not fixed to the floor, any more than the furniture in a living-room is. Owing to these handy tables the experimental arrangements gain considerably in clarity and convenience. Most of the tables can be turned round and raised or lowered as required. Troublesome perspective effects due to one piece of apparatus blocking the view of another are thus avoided. The apparatus in actual use at any moment can be made to stand out so that it is easily seen by each member of the audience.
>
> The pieces of apparatus required are simple and not numerous. Many of them are described here for the first time. Like the ordinary lecture apparatus they are obtainable from Spindler and Hoyer, G.m.b.H., Göttingen. *[In Great Britain from Messrs. Baird and Tatlock (London), Ltd., Hatton Garden, E.C.1.]
>
> Most of the figures are based on photographs, almost all by my chief mechanic, Mr. Sperber, to whom I am very much obliged for his continuing help. Many of these figures have again been made into silhouettes. This form of illustration is well adapted for printing and usually gives an idea of the size of the apparatus used. Finally, a silhouette indicates whether an experiment is suited to a large lecture-room, for it is then particularly important that the outlines should be clear and uninterrupted by subsidiary material, such as clamp-stands and the like." [Authorized translation of R. W. Pohl, *Einführung in die Mechanik und Akustik,* Vorwort zur ersten Auflage, 1930, English ed. 1932].

These aspects, the qualitative rather than quantitative character of the experiments, the handy small tables that could be moved around (instead of the large fixed table in the centre), the low number and simplicity of the instruments, and the dominance of the shadow projection, were central to Pohl's lecture demonstrations. Each of these characteristics by themselves do not sound novel. The shadow projection was one of the oldest demonstration techniques then available, used since the eighteenth century.[50] What Pohl accomplished was the development of the shadow projection into an overarching technique, projecting every kind of experiment on the white walls of his lecture theatre. He designed demonstration experiments from all fields of physics, such as mechanics, electromagnetism, optics, acoustics, and hydromechanics to be mounted on optical benches based on triangular bars. These experiments were then projected onto the wall by means of a carbon arc lamp and a lens system (fig. 7). Pohl's projection system was much simpler, more modular and more flexible than the

50 See Hackman and Brenni, both this volume.

pre-war projection apparatus. At many places in his textbooks, Pohl preferred to reproduce the silhouettes of the instruments rather than the instruments themselves (fig. 8). As Pohl argued, the clearness of the uninterrupted shadow would bring to the forefront the simple facts that were to be observed. The shadows only enhanced the simplicity – his chosen virtue for physics and its demonstration.

One prominent element of Pohl's experimental lecture on mechanics, which also featured as silhouettes in Pohl's *Mechanik und Akustik,* was his rotating chair (fig. 6). In the rotating chair, the lecturer or his assistants became part of the demonstration experiment in the lecture. Using the rotating chair, basic concepts of physics, like angular momentum, inertia and centrifugal forces could be visualized by positioning the human experimenter and observer into the rotating frame of reference.

Fig. 6: Silhouettes of different experiments with the rotating chair (Spindler & Hoyer. Der Drehstuhl nach Prof. R. W. Pohl (1930), p. 3).

The movable tables were the most striking elements of Pohl's demonstrations. Since the large experimental table had been removed from the lecture hall, the small tables could be placed in scene and spotlighted during the lecture like in a play in theatre. Pohl was, again, not the first physics lecturer to remove the large experimental table from the lecture hall. After Otto Lehman was appointed Professor of Physics at the *Technische Hochschule* of Karlsruhe in 1889, he removed the large experimental table, and substituted it with a number of small tables, on which the experiments could be arranged.[51] Small tables on wheels had been used for demonstration experiments long before Pohl as well. Walther Gerlach claimed that Pohl had copied the design of his table from Friedrich Paschen.[52]

51 See Lehmann (1904), pp. 10–12, and Zenneck (1961), pp. 94–95.
52 Gerlach (1978), p. 216.

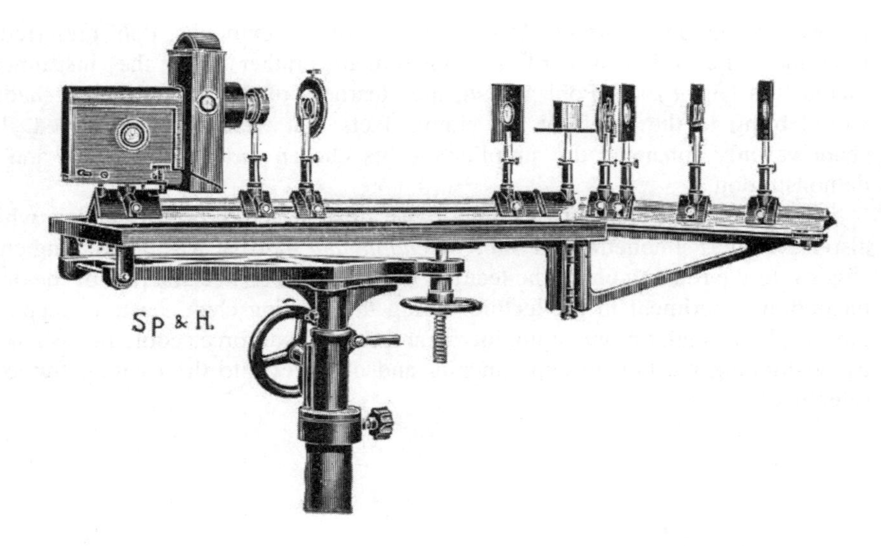

Fig. 7: Projection apparatus after Pohl, mounted on table with extension arm (Spindler & Hoyer. Vorführungsversuche aus der Optik (1933), p. 13).

However, what is undisputed, and therefore the focus of this paper, is that no other physics lecturer tried to transform these alternations into a uniform and universal system of teaching. For Lehmann, for example, removing the large experimental table was an answer to the odd architecture of the Karlsruhe lecture hall and he did not argue for it as an element of a larger philosophy of demonstration lectures. As with everything else in the immediate interwar years, characterised by inflation and the absence of funding to buy equipment, Pohl's rather simple and plain apparatus were made at the institute's shop. Pohl's student Erich Mollow tells us that Pohl read about everything he described in his textbooks or his lecture from their original works, and he replicated most of the experiments. In the process, he kept involved his students in the design and test of the experiments, as well as in the re-working of new editions of his textbook.[53] What was completely novel with Pohl was the rigour with which he applied these devices and made them central to his philosophy of lecturing.

Everything in Pohl's lecture, I argue, was connected, also the demonstration technology and the textbooks – well beyond the walls of the lecture theatre. Pohl always referred to Spindler & Hoyer as the producer of the apparatus that he displayed in his textbooks. Spindler & Hoyer, in turn, organised their pamphlets, in which they listed the Pohl apparatus, according to Pohl's textbooks and explained his philosophy of teaching and his system.[54] The cooperation between Pohl and

53 Mollow (1977), p. 235.

54 Spindler & Hoyer (1929), (1930), (1932a), (1932b), (1932c).

Spindler & Hoyer began in 1924, more or less after Pohl gave a lecture at a conference organised by the *Gesellschaft für Feinmechanik und Optik* (Society for Precision Mechanics and Optics) in Göttingen. Adolf Hoyer, one of the two founders and business director of Spindler & Hoyer proposed and promoted the idea of developing Pohl's demonstration instruments as part of the company's production line. This co-operation was to become the company's biggest economic success in times when other more traditional teaching instrument making companies faced serious economic struggles.[55] Spindler & Hoyer advertised Pohl's system of teaching not only through their pamphlets, but also in their business letters to customers, as in a letter to Marko W. Georgevitsch & Bruder in Belgrade of May 1935:

> "...For the method [of lecture demonstration] by Professor Pohl, the following points are characteristic:
>
> Brilliant experimental technique and many new apparatus and arrangements of greatest clearness and vividness (Anschaulichkeit);
>
> Widely following the requirements of the applied physicist, and therefore also the use of the measure- and unit systems common in technology. Tutorial for the students to measure and calculate in all fields, but without losing the connection to theory anywhere;
>
> Eliminating everything that has to be seen as out-dated. Instead always considering the results of the newest developments;
>
> Structure of the teaching of electricity from the field concept by the very beginning, that means in contrast to the detour through the historical development;
>
> Self-contained interrelated presentation of the basic teachings of the different fields of physics – the teaching of vibrations, waves and radiation."
>
> [From a sales offer letter of Spindler & Hoyer to the company Marko W. Georgevitsch & Bruder, Belgrade, 2 May 1935, Archive 'Spindler & Hoyer', Dep. 104, Nr. 37, Stadtarchiv Göttingen, translated from the German by author].

The letter highlights important features of Pohl's teaching system as we discussed before: the centrality of the claim to *Anschaulichkeit*, the rejection of an historical approach (particularly in the teaching of electricity) as dead weight and detour, and the interrelatedness of the teaching different fields of physics.

THE DIFFUSION AND ADAPTATION OF THE 'SYSTEM POHL'

Pohl's textbooks and demonstration experiments spread quickly and widely all over Germany, but also to many other countries. Pohl's three textbooks (Mechanics and Acoustics, Optics, and Electricity) were published in 21, 17 and 13 editions until the 1970s and can be found in many copies in the libraries of German universities and other institutions of higher education. Equally, Pohl's teaching apparatus is to be found in more or less every teaching collection and sometimes still in use! While the textbooks were translated into English, Italian,

55 See Meinhardt (1977), p. 42.

Marathi, Russian, and Spanish,[56] even more common than the translations were the German originals. These are still found in libraries all over the world, witnessing the importance of German as an international language of science in the interwar years. Many smaller European countries used mainly German textbooks and with the textbooks, Pohl's teaching system and his apparatus also went all over the world.

Fig. 8: Parallel plate capacitor mounted on universal rail for shadow projection (Pohl Einführung in die Elektrizitätslehre, 1927, p. 15).

The textbooks, however, were not the only means of diffusion for Pohl's teaching system. As discussed earlier, Göttingen was one of the most important national as well as international centres for theoretical, experimental and applied physics in the interwar period. Many of the 10 000 students that, according to Bernhard Gudden's estimate, had attended Pohl's lecture by 1944, came into teaching physics themselves and appear to have adopted Pohl's system as an obvious choice as seen through the prevalence of the textbooks and apparatus collections. This was especially true for Pohl's Assistants, like Gudden, who scrupulously followed Pohl's teaching methods when appointed as professors elsewhere, and whose students could be identified as belonging to the Pohl School.[57] Göttingen also had many students from other countries. In 1933, for example, three of the 25

56 See Pohl (1928), (1929), (1930a), (1932a), (1932b), (1932c), (1947), (1964), and (1971).

57 See Achilles (1977), p. 159, and Rammer (2004), pp. 358–372. Among Pohl's students and assistants who were appointed professor were Bernhard Gudden (Erlangen and Prague), Rudolf Hilsch (Erlangen and Göttingen), Werner Martienssen (Stuttgart and Frankfurt), Erich Mollwo (Erlangen), Heinz Pick (Stuttgart), Rudolf Fleischmann (Hamburg and Erlangen), Fritz Stöckmann (Darmstadt and Karlsruhe) and Gerhard Heiland (Aachen).

doctoral candidates in Physics at Göttingen were foreigners.[58] In addition, many scientists from all over the world visited Göttingen. Apart from participating in the Göttingen research communities, they may have well studied the teaching methods, premises, and instruments and implemented these, sometimes modified, in their home institutions. Spindler & Hoyer certainly noted this in their company report of 1931:

> "... The division of scientific apparatus, especially the one for demonstration apparatus in physics, has developed well. Several German university teachers, who got appointments at universities overseas, furnished their institutes with our apparatus. In addition, a large number of foreign university teachers were again in Göttingen, who inspected our apparatus with great interest. The most important retail companies in the United States and England, for example, are eager to reach agreements, in order to especially sell our apparatus. ..." [Company report of Spindler & Hoyer from October 1 1930 to September 30 1931, Archive Spindler & Hoyer, Dep. 104, Nr. 24, Stadtarchiv Göttingen, translated from the German by author].

The company's claim that "the division of scientific apparatus, especially the one for demonstration apparatus in Physics, has developed well" in 1931 is quite noteworthy, when we take into account the effects of the Great Depression on the German economy in general, but especially relevant for us the sorry state of precision mechanics and scientific instrument industry. Alfred Schmidt, the sole owner of the company E. Leybold's Nachfolger, committed suicide in March 1931 because of the company's hopeless financial situation.[59] Beyond the economic difficulties of Great Depression, perhaps one reason for E. Leybold's Nachfolger's financial difficulties was that Schmidt felt he had missed out on modernising the company's product range and tried to sell the same demonstration instruments as before the First World War. After Schmidt's death, his son-in-law Manfred Dunkel continued the company; he not only reduced and modernised the product range but also began retailing Pohl's demonstration apparatus.[60]

E. Leybold's Nachfolger and Max Kohl were not the only competing firms for Spindler & Hoyer. In order to promote their instruments more efficiently, a number of scientific instrument makers from Göttingen had joined forces and founded in 1921 the trade association *Verkaufsvereinigung Göttinger Werkstätten für Feinmechanik, Optik und Elektrotechnik GmbH* (Sales Association of the Göttingen Manufacturers for Precision Mechanics, Optics, and Electro Technology), which was shortened to *Vereinigung Göttinger Werke* in 1925. Apart from Spindler & Hoyer, the *Vereinigung Göttinger Werke* also represented other instrument makers of international importance, as the Satorius Werke and Gebr. Ruhstrat. Spindler & Hoyer had already begun marketing Pohl's instruments in 1925, two years before his first textbook appeared. The company used their ex-

58 Rammer (2004), p. 258. Among Pohl's foreign students were the Hungarian physicist Gyulai Zoltán and the Ukrainian Alexander Smakula, who was appointed as professor at the Massachusetts Institute of Technology in 1951.

59 Dunkel (1973), p. 49. See Meinhardt (1973), pp. 47–48 about Spindler & Hoyer's general situation during the Great Depression.

60 Examples of Pohl's instruments can be found in the sales catalogue of E. Leybold's Nachfolger of 1938.

isting worldwide network of business contacts to market the instruments, by advertising them in their correspondence and distributing pamphlets.[61] The association also produced joint pamphlets, promoting the products of the Göttingen instrument industry and organised the representation on trade fairs. These pamphlets advertised Pohl's demonstration apparatus among others and referred to Spindler & Hoyer's own pamphlets.[62] The executive officer of the *Vereinigung Göttinger Werke,* Erich Löwenstein, represented these Göttingen companies on business trips to European Countries and in the USA.[63] Pohl himself travelled to Bulgaria, England, Holland, Poland, the Soviet Union, Hungary and Switzerland, and promoted not only his research, but also his demonstration apparatus.[64]

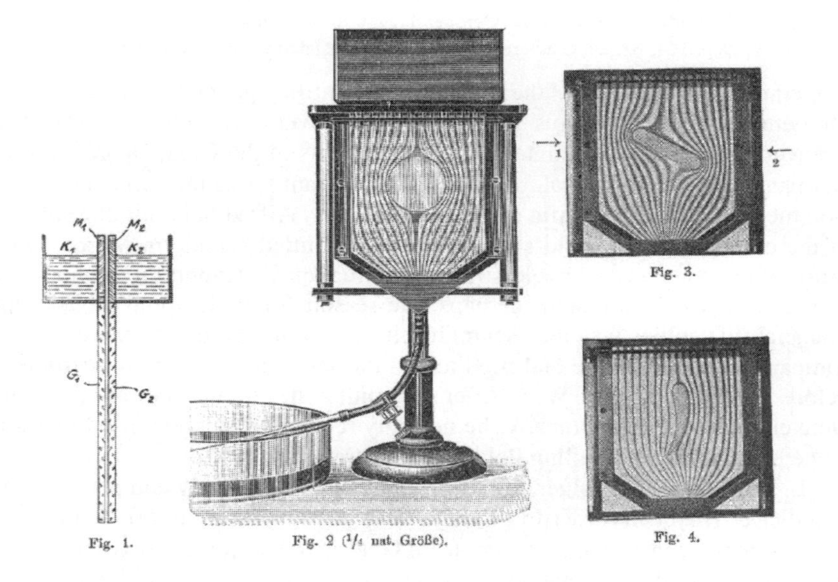

Fig. 9: Flow line apparatus for the shadow projection of demonstration from hydromechanics (Pohl, "Zwei Vorführungsapparate zur Hydrodynamik," 1925, p. 120).

61 These Letters can be found in the Archive Spindler & Hoyer, Dep. 104, Nr. 37, for example letter to *Waltherr Y Hermano*, Habana, 17 April 1926.

62 See *Vereinigung Göttinger Werke* (1935 and 'not dated').

63 See Meinhardt (1973), pp. 68 and 77–79. On October 22 1930, for example, Löwenstein gave a lecture on and a demonstration of the Pohl's apparatus at the Danish Physical Society (*Dansk Fysisk Forening*) in Copenhagen. See Fysisk Forening – Records and Manuscripts, Niels Bohr Archive, Copenhagen. Erich Löwenstein was forced to immigrate to the USA in 1935, where he continued to work as a sales representative for the Göttingen companies.

64 Minnigerode (1976), p. 142.

Pohl's demonstration apparatus were used independently from his textbooks as well, and not only in higher education. In Prussia, where profound reforms affected the education sector in the 1920s, teacher training colleges replaced the old teacher seminars. Secondary schools experienced a fundamental transformation. The importance of ancient and linguistic subjects, in particular Ancient Greek language, was downplayed, and scientific subjects increased in prominence, not only by the number of teaching hours. The sciences were now taught up to the German A-level exams, the *Abitur*. The schools had to be equipped substantially better with demonstration apparatus than before, although the financial situation required strictest thriftiness. Spindler & Hoyer began to exhibit Pohl's apparatus' at meetings of the *Verein zur Förderung des mathematisch naturwissenschaftlichen Unterrichts* (the Association for the Advancement of Mathematics and Science Teaching), as well as the *Deutscher Philologenverband* (the German Association of Secondary School Teachers). Spindler & Hoyer's instruments apparently satisfied the needs of the physics teachers for efficiency combined with low costs, and the business with educational instruments at schools finally contributed also to the larger economic success that Spindler & Hoyer had not expected.[65]

One main reason, I have argued, why Pohl's teaching system spread so successfully, was that it squared well with contemporary discussions placing value upon vividness, simplicity, soberness and the rejection of historicism in academic science teaching, shared or came to be shared by many of Pohl's fellow physicists. Johan Peter Holtsmark, Professor of Physics at the *Norges Tekniske Høgskole* (N.T.H. – the Norwegian Institute of Technology) in Trondheim serves as an example.[66] Holtsmark, aged 29, was appointed Professor of Physics at N.T.H in 1923. He was a representative of the new atomic physics and had strong ties with Göttingen. He had been Peter Debye's assistant in Göttingen for tow years in 1918 and 1919. Holtsmark's own assistant Bjørn Trumpy was to follow his advisor's footsteps in Göttingen and was awarded a Rockefeller scholarship this time to study with Max Born in 1928.[67] Holtsmark's teaching differed significantly from predecessor Sem Sæland. Whereas Sæland's teaching followed the classical pre-First World War canon, Holtsmark introduced modern, especially atomic physics.

65 Meinhardt (1973), pp. 43–44.

66 N.T.H. had been founded in 1910, following the German model of a *Technische Hochschule*. Three of the first professors were Germans; others had studied in Germany and continued to be oriented towards the German scientific and engineering community. Sem Sæland, N.T.H.'s first Professor of Physics, had studied with Philipp Lenard in Heidelberg. Sæland had been impressed by Lenard's lecture and followed his example. Johan Holtsmark succeded Sæland when the latter left for Kristiania (today Oslo) in 1923. See: Wittje (2003), pp. 48–49, and Ramsauer (1949), p. 109.

67 Wittje (2003), pp. 50 and 100.

Holtsmark's lecture script, published in three volumes in 1926, bears resemblance to Pohl's approach. In the introduction to the first volume, Holtsmark states:

> "It is common to begin a series of lectures with an historical overview on that branch of science in question. I shall not engage in such a venture. Such an authentic historical overview demands a very deep study and requires that the earlier [historical] treatises be seen with the same perspective as the time when they were carried out.
>
> This is generally impossible and the historical approach usually leads to a totally wrong judgment of what was done earlier. Therefore I assume it is a lot better to start with the viewpoint of physics today, it is difficult enough to make clear what we mean by physics now...." [J. P. Holtsmark, *Forelesninger i fysikk 1926, NTH, bind I*, Introduction, p.1, translation from the Norwegian by author].

Like Pohl, Holtsmark rejected an historical approach to academic physics teaching. Interestingly, he argued that such an approach would not do justice to the history of physics either since it would give us a wrong judgement of what was done earlier. Like in Spindler & Hoyer's promotion letter for Pohl's demonstration apparatus, Holtsmark rejected the historical approach even more explicitly in the teaching of electrodynamics. Again, the concept of *Anschaulichkeit* (in Norwegian *Anskuelighet*) became central:

> "When electrostatics is taught in schools and high schools, a lot of importance is usually given to the vast number of experiments that can be done. To a large extent these experiments date back to the middle of the 18th century, or, in other words, to the period when electrostatics first was developed. As a result they are in many cases not very useful to visualise (anskueligjjøre) the modern conception of electricity, because at the time other conceptions of electricity were in use than (those in use) today. We should therefore limit ourselves to draw up experiments to the extent, where they can contribute to the understanding of modern teaching of electricity, as we owe it to Faraday and Maxwell...." [J. P. Holtsmark, *Forelesninger i fysikk 1926, NTH, bind III*, Electrostatics, p.272, translated from the Norwegian by author].

It is not clear when or how exactly Holtsmark has been informed about Pohl's teaching system, but it is not surprising that he was drawn towards the system. Holtsmark introduced the Pohl textbooks for his physics lectures in 1932.[68] With the textbooks, Holtsmark also introduced Pohl's lecture demonstrations. The historical collection of the physics department at the Norwegian University of Science and Technology – the successor of N.T.H. – contains a large collection of Pohl's demonstration apparatus, including his rotating chair, a number of rotating tables, carbon arc lamps, and accessories.[69] However, only some of the apparatus was bought from Spindler & Hoyer. Many accessories were actually made at the local workshop of the physics department. Holtsmark also had other sources in Göttingen. In September 1931, he ordered a Pohl projection table from the Göt-

68 J. P. Holtsmark, 1932, note in Archive *NTH-Fysisk institutt*, box Eb:1, korrespondanse, akt1: Rektor, sekretariatet, Statsarkivet i Trondheim.

69 I even found one of Pohl's flow line apparatus (fig. 9) in the collection of abandoned teaching apparatus in the *Katedralskolen*, Trondheim's oldest and most prestigious secondary school.

tingen *Universitätsmechanikermeister* (University Master Mechanic) Albrecht and inquired for the price of a rotating chair.[70] Spindler & Hoyer were the largest, but as the example of Master Mechanic Albrecht shows, clearly not the sole supplier of Pohl apparatus in Göttingen.

Pohl's influence on science teaching can be traced to North America as well. Richard Manliffe Sutton's *Demonstration Experiments in Physics,* of 1938, refers to the English translations of Pohl's textbooks.[71] Before the Second World War, Pohl had never travelled to the USA, but in the postwar period he did and apparently even grew fond of the more informal hierarchies of the American academic system.[72] In 1960, Pohl received the Oersted Medal of the American Association of Physics Teachers for his demonstration experiments in physics lectures. His address in response to the award is remarkable: First, Pohl made clear that "Certainly, the most important task of a university professor is scientific research, either alone, or with a group of students. ... teaching is a rewarding task only for those who consider the fact that every human activity has a small net efficiency."[73] This is surely a surprising statement, considering the immense effort Pohl had invested in teaching throughout his career and the large impact he had on academic physics teaching for several generations, clearly surpassing his impact as a researcher.[74] However, it perhaps reflects the low status that teaching had, and still has, for German university professors. Second, Pohl chose a historical topic for his address, the *Discovery of Interference by Thomas Young*, stating that "... in order to achieve a clear understanding of concepts and in order to recognize among the overwhelming quantity of material that which is most important in each particular case. .... it often is quite useful to occupy oneself a little with historical questions."[75] Pohl, who had cleansed his textbooks as well as his lecture demonstrations from every trace of historicism, acknowledged the usefulness of history of science for science teaching.

CONCLUSIONS

Robert Wichard Pohl, without doubt, has been the most influential figure in the transformation and renewal of experimental physics teaching in interwar and postwar Germany. Given that he was not politically compromised during the the Third Reich, Pohl managed to build a powerful position within the German physics community especially in the immediate post-war years and could recom-

70 Holtsmark to Albrecht, Sept. 21 1931, Archive NTH-Fysisk institutt, Statasarkivet, Ea:2.

71 Sutton (1938), p. 510. It also refers to a volume of "Physical Principles of Optics and Heat" (to be published), which, according to my knowledge, has never appeared.

72 Rammer (2004), pp. 170–171, and Minnigerode (1976), p. 142.

73 Pohl (1960), p. 530.

74 Gerlach (1978, p. 216) stated that Pohl saw teaching as the main task of the university professor throughout his life. As much as Gerlach's statement might have been true for Pohl's practice, it is in conflict with Pohl's own statement.

75 Overbeck (1960), and Pohl (1960), p. 530.

mend his students for vacant physics chairs all over Germany. By the 1970s, 11 of the 55 German doctoral candidates of Pohl had been appointed to professorships in physics in Germany, and of his seven foreign students, six were appointed as professors in other countries.[76] As for his system of lecture demonstrations and textbooks they also reached far beyond the German community.

What, could we say, were the reasons for Pohl's success, and why was his teaching system so attractive? One aspect is certainly that his teaching philosophy squared well with contemporary discourses regarding *Anschaulichkeit* and simplicity in physics. His rejection of historicism in academic physics teaching was also shared by many of his colleagues. His instruments brought a New Sobriety into the demonstration lecture precisely when many were tired of the excessive pomp of pre-war demonstration culture. Most important, Pohl also delivered a complete system for the experimental physics lecture, which included textbooks, lecture methods and modular devices. Instead of developing their own teaching program and means, physics lecturers could simply adopt Pohl's system. Pohl's close collaboration with Spindler & Hoyer made the instruments widely promoted and available through their large sales and marketing network. Finally, Göttingen's place as an international centre of modern physics and Pohl's success in having his students appointed as professors at other universities secured the spread of the teaching system.

Pohl's lectures, demonstrations and textbooks were also criticized. Critics pointed out often to his extreme experimentalism, his renouncement of theory, especially quantum mechanics, and his uncompromising authoritarianism. They often denounced Pohl's lecture as 'Theatre Pohl' or 'Circus Pohl'. Born called Pohl an extreme experimentalist; also others thought that he went too far in his program to reduce the whole of physics to simple vivid experiments.[77] The criticism, however, did not stop the success or the diffusion of Pohl's system of teaching. Pohl edited revised versions of his textbooks until the 1970s. His very first book, the *Elektrizitätslehre* appeared in its 21st edition in 1975, one year before Pohl's death.

The 'Circus Pohl' was probably the last grand school of lecture demonstrations in academic physics teaching. In the course of the second half of the twentieth century, the number of lecture demonstrations in experimental physics courses has decreased dramatically. Some of the old universities keep up their tradition of demonstration experiments. The *I. Physikalisches Institut* of the University of Göttingen maintains the original collection of demonstration experiments developed by Pohl and uses it in its lectures. However, almost everywhere, demonstrations have lost their central place to chalk and blackboard physics, and to new media. In the new millennium a new generation of physics

76 Minnigerode (1976), p. 142. Pohl's student Heinz Pick went, in 1981, so far to say that there was the 'Munich connection' and the 'Pohl connection', and nobody else who supplied German universities with candidates for professors in physics. See Rammer (2004), p. 34.

77 Eckert and Schubert (1990), p. 99, and Achilles (1977), p. 157. For criticism see also Gerlach (1978), p. 217.

students as well as researchers are growing up with PowerPoint and computer simulations.

Why did demonstration experiments lose their importance? A part of the answer lies in the shifting priorities, and in the removal of resources from the demonstration lectures. The experiments for each lecture required elaborate preparations, often over several days, keeping the lecture theatre, one teaching assistant and one mechanic busy.[78] In today's universities, lecture theatres are used for different lectures every other hour, leaving only a few minutes for the setup of experiments. Few universities allocate an assistant and a mechanic to the lecture demonstration alone, which was the norm but appear now as a luxury. There are also other, and we might think better, reasons for the decline of the demonstration culture. In some cases, where resources are available, the authoritative style of the lecture demonstration has given more space to laboratory courses and students perform their own experiments. And then again, physics has changed. Mathematical analysis has become more important and requires more time, also in the experimental lecture. Computer simulations have gained much importance in today's experimental physics. That computer simulations are also used in today's teaching of experimental physics is therefore only consistent.

In 2004 and 2006, Klaus Lüders, Professor of Physics at the Freie Universität Berlin, and Robert Wichard Pohl's son, Robert Otto Pohl published a new edition of Pohl's textbooks with Springer. *Pohl's Einführung in die Physik. Band 1: Mechanik, Akustik und Wärmelehre* has now been published in its 20th revised edition by Springer in 2009. The *Einführung in die Physik. Band 2: Elektrizitätslehre und Optik* has been published in its 23rd revised edition in 2010. Both volumes contain DVDs with altogether 110 videos of demonstration experiments, produced with the original apparatus from the Göttingen collection. The Pohl lecture is not dead and for all we know, we may even see its revival in newer avatars.

ACKNOWLEDGEMENTS

I would like to thank Gustav Beuermann, Robert Otto Pohl, Jürgen Teichmann, the Stadtarchiv Göttingen, and the companies Linos Photonics and Leybold Vacuum, (successors of Spindler & Hoyer and E. Leybold's Nachfolger, respectively), for their invaluable help in tracing the story. I would also like to thank Berthold Wittje for translations from Russian, and Jahnavi Phalkey for translations from Marathi. I am also grateful to Peter Heering and Gerhard Rammer for their suggestions and especially to Jahnavi Phalkey for her skilful editing.

78 See, for example, Zenneck (1961), pp. 156–157 and pp. 242–243.

REFERENCES

Achilles, Manfred (1977). "R.W. Pohl – Zum einjährigen Todestag." *Praxis der Naturwissenschaften – Physik* 6: 155–159.

Beuermann, Gustav, and Thomas Werner (1986). *Die historische Sammlung des I. Physikalischen Instituts der Georg-August-Universität Göttingen. Ausstellungskatalog anläßlich der 250-Jahrfeier der Georg-August-Universität im Jahre 1987*. Göttingen: Akademie der Wissenschaften.

Beuermann, Gustav, M. Czaske, and U. Schulz (1988). *Die historische Sammlung des I. Physikalischen Instituts der Georg-August-Universität Göttingen. Teil II des Katalogs*. Göttingen: Akademie der Wissenschaften.

Cahan, David (1989). *An Institute for an Empire: The Physikalisch-Technische Reichsanstalt, 1871–1918*. Cambridge: Cambridge University Press.

Cornell, Thomas D. (1986). *Merle A. Tuve and his program of nuclear studies at the Department of Terrestrial Magnetism: The early career of a modern American physicist*. Ph.D. dissertation, Johns Hopkins University. Ann Arbor, Michigan: University Microfilm International.

Dennis, Michael Aaron (1991). *A Change of State: The Political Cultures of Technical Practice at the MIT Instrumentation Laboratory and the John Hopkins University Applied Physics Laboratory, 1930 –1945*. PhD dissertation, Johns Hopkins University. Ann Arbor, Michigan: University Microfilm International.

Dunkel, Manfred (1973). *Geschichte der Firma E. Leybold's Nachfolger: 1850 – 1966 mit besonderer Berücksichtigung der Ereignisse in und unmittelbar nach dem 2. Weltkrieg*. Cologne: Grevet & Bechtold.

Eckert, Michael, and Helmut Schubert (1990). *Crystals, electrons, transistors – From scholar's study to industrial research*. New York: American Institute of Physics.

Eckert, Michael (2006). *The dawn of fluid dynamics: A discipline between science and technology*. Weinheim, Wiley-VCH-Verl.

Eucken, Arnold, Otto Lummer, and Erich Waetzmann, eds. (1925–1934). *Müller-Pouillets Lehrbuch der Physik*. 5 vols. Braunschweig: F. Vieweg & Sohn.

Fleck, Ludwik (1979). *Genesis and development of a scientific fact*. Chicago: University of Chicago Press.

Forman, Paul (1971). "Weimar culture, causality and quantum theory: Adoption by German physicists and mathematicians to a hostile intellectual environment." *Historical Studies in the Physical Sciences* 3: 1–115.

Frank, Philipp (1928). "Über die "Anschaulichkeit" physikalischer Theorien." *Die Naturwissenschaften* 16, no. 8: 121–128.

Galison, Peter (1997). *Image and Logic – A Material Culture of Microphysics*. Chicago and London: University of Chicago Press.

Gebhardt, Martin (1922). "Zur Farbenlehre. Ein Beitrag zur Methodik des physikalischen Unterrichts." *Zeitschrift für den physikalischen und chemischen Unterricht* 35, no. 1: 1–9.

Geiger, Hans, and Scheel, Karl, eds. (1926). *Handbuch der Physik, Band 1: Geschichte der Physik – Vorlesungstechnik*. Berlin: Julius Springer.

Gerlach, Walther (1978). "Robert Wichard Pohl, 10.8.1884–5.6.1976." *Jahrbuch der Bayerischen Akademie der Wissenschaften 1978*: 214–219.

Grimsehl, Ernst (1909). *Lehrbuch der Physik – zum Gebrauche beim Unterricht, bei akad. Vorlesungen und zum Selbststudium*. Leipzig: Teubner.

Gudden, Bernhard (1944). "R. W. Pohl zum 60. Geburtstag." *Die Naturwissenschaften* 32: 166–169.

Heering, Peter (1998). *Das Grundgesetz der Elektrostatik – Experimentelle Replikation, wissenschaftshistorische Analyse und didaktische Konsequenzen*. Wiesbaden: DUV.

Heisenberg, Werner (1927). "Über den anschaulichen Inhalt der quantentheoretischen Kinematik und Mechanik." *Zeitschrift für Physik* 43: 172–198.

Hilsch, Rudolf (1964). "Zum 80. Geburtstag von Robert Wichard Pohl am 10. August 1964." *Die Naturwissenschaften* 51, no. 15: 349–350.

Hoffmann, Dieter, and Edgar Swinne (1994). *Über die Geschichte der "technischen Physik" in Deutschland und den Begründer ihrer wissenschaftlichen Gesellschaft Georg Gehlhoff.* Berlin: ERS-Verl.

Holtsmark, Johan Peter (1926). *Forelesninger i fysikk 1926, NTH, bind I – III.* Trondheim: Tapirs forlag.

Joos, Georg (1954). "Robert Pohl 70 Jahre." *Zeitschrift für angewandte Physik* 6: 339.

Max Kohl A. G. (1905) *Preisliste Nr. 50, Band I – Einrichtungsgegenstände für physikalische und chemische Lehrräume.* Chemnitz: Max Kohl A. G.

Max Kohl A. G. (after 1911). *Preisliste Nr. 50, Band II und III – Physikalische Apparate,* Chemnitz: Max Kohl A. G.

Max Kohl A. G. (1925) *Preisliste Nr. 50, Band IV – Unterrichts- und Laboratorienmöbel. Physikalische Apparate.* Chemnitz: Max Kohl A. G.

Max Kohl A. G. (ca. 1926–1928). *Preisliste Nr. 100, Band I–III Unterrichts- und Laboratorienmöbel. Physikalische Apparate.* Chemnitz: Max Kohl A. G.

Kragh, Helge (1999). *Quantum generations – a history of physics in the twentieth century.* New York: Princeton University Press.

Kuhn, Thomas S. (1977). "A function of thought experiments" in Thomas S. Kuhn, *The Essential Tension: Selected studies in scientific tradition and change.* Chicago: University of Chicago Press: 240–265.

Lehmann, Otto (1904 – 1909). *Dr. J. Fricks physikalische Technik oder Anleitung zu Experimentalvorträgen sowie zur Selbstherstellung einfacher Demonstrationsapparate.* 2 vols. Braunschweig: F. Vieweg & Sohn.

Lemmerich, Jost (2007). *Aufrecht im Sturm der Zeit: Der Physiker James Franck, 1882–1964.* Diepholtz, Stuttgart, Berlin: GNT-Verlag.

E. Leybold's Nachfolger, ed. R. Bernoulli (1913): *Einrichtungen und Apparate für den physikalischen Unterricht sowie für Übungen im Praktikum nebst Literaturangaben.* Cologne: Paul Gehly.

E. Leybold's Nachfolger, ed. Georg Vieth (1929): *Einrichtungen und Apparate für den physikalischen Unterricht sowie für Übungen im Praktikum nebst Literaturangaben.* Cologne: Paul Gehly.

E. Leybold's Nachfolger (1938). *Physical apparatus.* Köln Bayental (Cologne).

Lüders, Klaus, and Robert O. Pohl (2009). *Pohl's Einführung in die Physik. Band 1: Mechanik, Akustik und Wärmelehre.* 20th revised ed. Berlin et al: Springer.

Lüders, Klaus, and Robert O. Pohl (2010). *Pohl's Einführung in die Physik. Band 2: Elektrizitätslehre und Optik.* 23rd revised ed. Berlin et al: Springer.

Martienssen, Werner (1974). "Robert Wichard Pohl zum 90. Geburtstag." *Physikalische Blätter* 30: 313–314.

Meinhardt, Günther (1973). *75 Jahre Spindler & Hoyer.* Göttingen: Spindler & Hoyer.

von Meyenn, Karl, ed. (1994). *Quantenmechanik und Weimarer Republik.* Braunschweig: F. Vieweg & Sohn.

Minnigerode, Gunther von (1976). "Robert Pohl. 10. August 1884 – 5. Juni 1976." *Jahrbuch der Akademie der Wissenschaften zu Göttingen 1976*: 137–146.

Minnigerode, Gunther von (1978). "Nachruf auf Robert Wichard Pohl." in *R. W. Pohl Gedächtnis-Kolloquium am 29. November 1976.* Göttingen u.a.: Musterschmidt: 53–58.

Mollwo, Erich (1977). "Erinnerungen an R. W. Pohl." *Physik und Didaktik* 3: 230 – 236.

Mollwo, Erich (1978). "Gedenkrede auf Robert Wichard Pohl." in *R. W. Pohl Gedächtnis-Kolloquium am 29. November 1976.* Göttingen u.a.: Musterschmidt: 13–19.

Müller, Johan H. J. (1842–1843). *Pouillet's Lehrbuch der Physik und Meteorologie für deutsche Verhältnisse frei bearbeitet.* 2 vols. Braunschweig: F. Vieweg & Sohn.

Olesko, Kathryn (1995). "German Models, American Ways: The "New Movement" among American Physics Teacher, 1905–1909." in Henry Geitz, Jürgen Heideking and Jurgen Herbst, *German Influences on Education in the United States to 1917*. Cambridge: Cambridge University Press.

Overbeck, Clarence J. (1960). "Robert Wichard Pohl: Oersted Medalist for 1959." *American Journal of Physics* 28: 528–529.

Pfaundler, Leopold, ed. (1905–1914). *Müller-Pouillets Lehrbuch der Physik und Meteorologie. In vier Bänden* 10. umgearb. und verm. Aufl. Braunschweig: F. Vieweg & Sohn.

Pohl, Robert (1925). "Zwei Vorführungsapparate zur Hydrodynamik." *Zeitschrift für den physikalischen und chemischen Unterricht* 38, no. 3: 119–122.

Pohl, Robert Wichard (1927). *Einführung in die Elektrizitätslehre.* Berlin: Julius Springer.

Pohl, Robert Wichard (1928). *Elementi teorico-practici di elettrofisica moderna. Prima traduzione italiana del' Ing. Carlo Rossi.* Milan: Ulrico Hoepli.

Pohl, Robert Wichard (1929). *Einführung in die Elektrizitätslehre.* Translated into Russian by L. A. Tumermana and edited by Professor Ja. N. Schpilrein. Moscow, Leningrad: State Publishing House.

Pohl, Robert Wichard (1930a). *Physical principles of electricity and magnetism.* London: Blackie & Son.

Pohl, Robert Wichard (1930b). *Einführung in die Mechanik und Akustik. Einführung in die Physik Band 1.* Berlin: Julius Springer.

Pohl, Robert Wichard (1932a). *Einführung in die Mechanik und Akustik.* Translated into Russian by Professor K. A. Leontef. Moscow, Leningrad: State Publishing House for Science and Technology.

Pohl, Robert Wichard (1932b). *Physical principles of mechanics and acoustics.* London: Blackie & Son.

Pohl, Robert Wichard (1932c). *Tratado de física. II, Electricidad por el Dr. R. W. Pohl; traducción de la 3a. edición alemana por el Dr. J. Baltá Elías.* Barcelona: Gustavo Gili.

Pohl, Robert Wichard (1933). "Der physikalische Hörsaal der Universität Göttingen." *Physikalische Zeitschrift* 34: 408–410.

Pohl, Robert Wichard (1940a). *Einführung in die Optik, Einführung in die Physik Band 3.* Berlin: Julius Springer.

Pohl, Robert Wichard (1940b). "Ein neuer Versuch zur Interferenz divergierender Lichtbündel." *Physikalische Zeitschrift* 41: 498–499.

Pohl, Robert Wichard (1940c). "Ein einfacher Interferenzversuch mit divergierenden Lichtbündeln." *Die Naturwissenschaften* 28: 585–586.

Pohl, Robert Wichard (1947). *Einführung in die Optik.* Translated into Russian and edited by N. P. Suworov. Moscow, Leningrad: State Publishing House for Technical and Theoretical Literature.

Pohl, Robert Wichard (1952). "Physik und Schule." *Die Naturwissenschaften* 39: 535–537.

Pohl, Robert Wichard (1960). "Discovery of Interference by Thomas Young." *American Journal of Physics* 28: 530–532.

Pohl, Robert Wichard (1964). *Yantrashastra ani Dhwanishastra yanchi Mulatatve* (Einführung in die Mechanik und Akustik). Translated by R. D. Godbole (Marathi). Pune: Pune University Press.

Pohl, Robert Wichard (1971). *Meccanica, acustica e termologia. 17.ed migliorata ed ampliata. Traduzione di Arturo Loria.* Padova: Piccin.

Poske, Friedrich (1921a). "Anschauliche und abstrakte Begriffsdefinitionen im physikalischen Unterricht." *Zeitschrift für den physikalischen und chemischen Unterricht* 34, no. 3: 97–103.

Poske, Friedrich (1921b). "Die Formen des physikalischen Unterrichts [review of P. Johanneson]." *Zeitschrift für den physikalischen und chemischen Unterricht* 34, no. 5: 233–234.

Rammer, Gerhard (2004). *Die Nazifizierung und Entnazifizierung der Physik an der Universität Göttingen.* Göttingen, Univ., Philos. Fak., Diss.

Ramsauer, Carl (1949). *Physik, Technik, Pädagogik: Erfahrungen und Erinnerungen.* Wissenschaftliche Bücherei [1], Karlsruhe: Braun.

Reuter, Hans (1925). *Fort mit der Kreidephysik! ein Führer beim Gebrauch physikalischer Schulapparate.* Volume 1. Göttingen: PhyWe.

Rheinberger, Hans-Jörg (1997). *Towards a History of Epistemic Things – Synthesizing Proteins in the Test Tube.* Stanford: Stanford University Press.

Richter, Steffen (1972). *Forschungsförderung in Deutschland 1920 – 1936, dargestellt an Beispielen der Notgemeinschaft der Deutschen Wissenschaft und ihrem Wirken für das Fach Physik.* Technikgeschichte in Einzeldarstellungen, no. 23. Düsseldorf: VDI-Verlag.

Rosenberg, Karl (1923). "Zum Unterricht in der Elektrizitätslehre." *Zeitschrift für den physikalischen und chemischen Unterricht* 36, no. 3: 145–150.

Schmidt, Alfred (1935). *Geschichte der Firma E. Leybold's Nachfolger 1850–1925.* Cologne: Paul Gehly.

Spindler & Hoyer (1925a). *Der drehbare Experimentiertisch nach Prof. R. Pohl.* Liste 50. Göttingen: Spindler & Hoyer.

Spindler & Hoyer (1925b). *Projektionseinrichtung.* Liste 51. Göttingen: Spindler & Hoyer.

Spindler & Hoyer (1925d). *Stativreiter und Hilfsapparate für den drehbaren Experimentiertisch.* Liste 53. Göttingen: Spindler & Hoyer.

Spindler & Hoyer (1925e). *Demonstrationsapparate zur Schwingungslehre nach Prof. R. Pohl.* Liste 51. Göttingen: Spindler & Hoyer.

Spindler & Hoyer (1929). *Apparate zur Elektrizitätslehre nach Prof. R. W. Pohl.* Liste 60. Göttingen: Spindler & Hoyer.

Spindler & Hoyer (1930). *Der Drehstuhl nach Prof. R. W. Pohl.* Liste 61. Göttingen: Spindler & Hoyer.

Spindler & Hoyer (1932a). *Der drehbare Experimentiertisch nach Prof. Dr. R. W. Pohl mit Projektionseinrichtung, Stativreitern und Hilfsapparaten.* Liste 50/32. Göttingen: Spindler & Hoyer.

Spindler & Hoyer (1932b). *Apparate zur Elektrizitätslehre nach Prof. Dr. R. W. Pohl.* Liste 60/32. Göttingen: Spindler & Hoyer.

Spindler & Hoyer (1932c). *Apparate zur Mechanik und Akustik nach Prof. Dr. R. W. Pohl.* Liste 64. Göttingen: Spindler & Hoyer.

Spindler & Hoyer (1933). *Vorführungsversuche aus der Optik.* Liste 65/33. Göttingen: Spindler & Hoyer.

Spindler & Hoyer (1935). *Vorführungsapparate zur Mechanik der Bewegungen in Flüssigkeiten und Gasen nach Prof. Dr. R. W. Pohl.* Liste 66. Göttingen: Spindler & Hoyer.

Sutton, Richard Manliffe (1938). *Demonstration Experiments in Physics.* New York and London: McGraw-Hill.

Teichmann, Jürgen (1988). *Zur Geschichte der Festkörperphysik: Farbzentrenforschung bis 1940.* Stuttgart: Franz Steiner.

Trischler, Helmuth (1996). "Die neue Räumlichkeit des Krieges: Wissenschaft und Technik im 1. Weltkrieg." *Berichte zur Wissenschaftsgeschichte* 19: 95–103.

Vereinigung Göttinger Werke. Prospekt M 7. Not dated but stamped 'Sept. 1935' by the Department of Physics, Norwegian Institute of Technology, Trondheim.

Die Vereinigung Göttinger Werke für Feinmechanik, Optik und Elektrotechnik – G. m. b. H. Satorius-Werke, Gebr. Ruhstrat, Spindler & Hoyer, Elektroschaltwerk und Gen. Not dated, after December 1935.

Weinhold, Adolf F. (1913). *Physikalische Demonstrationen – Anleitung zum Experimentieren im Unterricht an Gymnasien, Realgymnasien, Realschulen und Gewerbeschulen,* 5th revised ed. Leipzig: Johann Ambrosius Barth.

Williams, Mari E. W. (1994). *The Precision Makers: A History of the Instrument Industry in Britain and France, 1870–1939.* London: Routledge.

Wittje, Roland (2003). *Acoustics, Atom Smashing and Amateur Radio: Physics and Instrumentation at the Norwegian Institute of Technology in the Interwar Period.* Dr.philos dissertation. Trondheim: Norwegian University of Science and Technology.
Zenneck, Jonathan (1961). *Erinnerungen eines Physikers*. Munich: Deutsches Museum.
Zoller, Paul (2009). "Physics Experiments for Everyone: German Makers." *Bulletin of the Scientific Instrument Society* 102: 21–28.

Archives

Firma Spindler & Hoyer, Stadtarchiv Göttingen, Germany
Fysisk Forening – Records and Manuscripts, Niels Bohr Archive, Niels Bohr Institute, Copenhagen, Denmark
NTH-Fysisk institutt, Staatsarkivet i Trondheim, Norway
Oral history interviews, Niels Bohr Library and Archives, American Institute of Physics, College Park, MD, USA

THE ROLE OF INSTRUMENTS IN TEACHING SCIENCE: A MACHIAN VIEW

Hayo Siemsen

INTRODUCTION

Even though Ernst Mach is more known today as a physicist, he also played an influential role in science education (see Blüh, 1967; Euler, 2006; Hoffmann and Manthei, 1991; Hohenester, 1988; Matthews, 1990; Thiele 1963),[1] psychology, and the philosophy of science (Erkenntnistheorie). This philosophy of science provides a specific world view which tries to reconstruct science from the perspective at the beginning of culture and mind, like the perspective of a young child. By following the genesis process and the psychical development of humans, concepts are adapted as directly as possible from elementary sensations. Thus, scientific concepts become intuitively connected to their empirical meanings. This process is primarily unconscious and phenomenal, instead of conscious and logical.

This paper tries to take this Machian world view and show its implications for the understanding of scientific instruments, especially their meanings and uses in science teaching.[2] It will show that with this perspective, the concept of 'instru-

1 For instance, many of the physics textbooks used in Germany in different types of schools and for different age groups from around 1890 into the late 1920s (titled *Grundriss der Naturlehre* and *Lehrbuch/Grundriss der Physik* with about 20 editions) were actually written or heavily influenced by Mach (see Hohenester [1988]; see also the later didactical works of Mach's successors Poske and Höfler). Mach co-founded one of the first journals for 'physics and chemistry education' (*Zeitschrift für den chemischen und physikalischen Unterricht*, published by Springer from 1887 to 1943). He also delivered a lecture in 1886 on the educational value of mathematics and natural sciences in school (Mach, 1893b), which became very influential for school reforms, for instance, those by Klein or the American maths reforms in the 1960s (see Hohenester, 1988; Memorandum, 1962). Even the physics curriculum for junior high schools (Realschulen) for 1879 in Austria was actually written by Mach (Hohenester, 1988). The books were introduced thus: "'Apparatus' are not described in detail. These systems of finely polished hooks and screws are unfitting for a compendium. Such things only have a value in the hands of a scientist if they are worked with and science is produced" (quoted in Hohenester, 1988, p. 144). Mach's laboratory in Prague was known for its elaborate apparatus mostly designed by himself; some of the apparatuses soon found their way into many university laboratories.

2 The Machian world view requires a shift in the meanings of many basic concepts, as well as the relations between the concepts: the concept of 'concept' itself, for instance, is understood as an entity of thought and not of language; phenomena are holistic gestalts, etc. Some of

ments' acquires a different meaning. Thereby, the use of instruments as part of the scientific process changes as well as the type of instruments, which are meaningful in the sense of teaching science. I will argue that this change is better adapted to how children learn science in general, regardless of age and cultural background.

THE PHENOMENAL AND PHYSICAL LANGUAGES OF SCIENTIFIC INSTRUMENTS

Scientific instruments[3] have a phenomenology. Educated scientists already learnt a physical 'language', which is constructed from this phenomenology in order to make it a 'scientific instrument' (the term itself is already highly constructed to start with). For scientists, the concept of the instrument has empirical meanings related to natural phenomena. For students, this construction initially does not exist. For them, natural phenomena exist unrelated to any instrument. In experiments, the instrument can only help in "forcing nature to answer in a specific way" (see Kurki-Suonio, 2010) for instance, by allowing the quantification of perceptions related to a specific phenomenon.

For the student, what scientists call instrument is itself a phenomenon. This 'instrument as phenomenon' might be completely unrelated to the scientific or educational function of the instrument. For instance, the student could perceive shiny brass ('the baroque of Gelsenkirchen'[4]) as an interesting phenomenon on which to focus attention. Thus, attention is distracted from the meaning in science. Or the student might observe certain circumstances of the use of the instrument, such as "the instrument works when the teacher operates it.[5] This phenomenology is by and large not conscious and, therefore, also not accessible to conscious introspection by the student. By extension, it is also not accessible to linguistic inquiry by the teacher or by researchers in science teaching.

These examples show that the phenomenology of scientific instruments is different for children from the view and purpose they are constructed for. This

these uses might be familiar to the reader, some not. For reasons of brevity, I will not try to redefine all the concepts in detail in a Machian way, but hope that the differences in meanings become clear through the context.

3 I use the concept of 'instrument' as it is used by most physicists and physics educators. In this meaning, the instrument is supposed to be detached and independent of the observer (and its inventor). It is the empirical meaning from a materialist point of view of a fully trained scientist.

4 Gelsenkirchen is a city in Germany well known for steel making. It became synonymous for a furniture style which was very opulent and detailed, but was regarded in general as kitch for its lack of style (inner cultural consistency) and outdatedness. Mach's view on the use of apparatuses in books for student instruction implies a seemingly similar critique.

5 This is a correct observation in many classrooms. If this observation is repeated often, the intuitively natural assumption is that this relation must be causal.

phenomenology might be completely unrelated to the "physical language"[6] used by the teacher. The teacher could then be talking Chinese in a German classroom with the same effect. With Chinese, the students might actually understand more, because their attention would focus less on the words and more on the actions of the teacher. As the haptic/enactive 'doing' of the teacher is, for the students, closer to the phenomenology of the instrument than his physical language, it might thereby make more sense of the phenomenon for the students.[7]

Nowadays, the choice and use of scientific instruments in science teaching is mainly based on the 'physical language' used in science (description, precision, etc.).[8] But is this language adequate for the understanding of school children?

6 The differentiation between phenomenal construction and physical language is shown by Eino Kaila (1962, 1979), a Finnish psychologist and philosopher of science, who was strongly influenced by Mach and, in turn, was central for current Finnish science education (see Siemsen and Siemsen, 2009). The phenomenal/physical differentiation is not the same as Mach uses in his terminology, but for the purposes of this paper, it is sufficiently similar to highlight the main problem of the current use of instruments in science education from a Machian perspective.

7 This genetic problem in principle is important in all areas, which are taught 'new', that is, for which no analogies to areas of knowledge with previously laid empirical meanings exist (see Siemsen, 2010a). Children cannot be assumed to already be small scientists in the sense that they can be expected to have full experience in any field of science. They do not live in that scientific culture. As a result, the abstraction from phenomenon to usage does not work. They intuitively use the method which is basic for them, for instance, the method of variation. Their use of it though is still unlike scientific use, such as experimentation, in order to jump over the difference. The bridging of this difference requires an increasing understanding of the whole scientific process, on the one hand, and a minimum of experiences in any scientific field, on the other. Only then can the method be related to an intuitivized meaningful empirical basis. Mach describes this in terms of a general principle of teaching (1866, pp. 2–3):

"Once a part of science belongs to the literature, a second task remains, which is to popularize it, if possible. This second task also has its importance, but it is a difficult one. It has its importance, because – regardless of the distribution of knowledge that increases its value – it is not unimportant either for the further development of science itself how much knowledge has been disseminated into the public. The difficulty is to know the soil very well in which one wants to plant the knowledge. It is a prevalent but wrong opinion that children are not able to form precise concepts and come to the right conclusions. The child is often more sensible than the teacher. The child is very well able to comprehend, if one does not offer too much new at a time, but properly connects the new to the old. The adult is a child when facing the completely new. Even the scholar is a child when confronted with a foreign subject. The child is a child everywhere, as everything is new to him. The art of popularization lies in avoiding too much of the new at one time."

8 It should be noted here that the term 'physical language' encompasses much more than one might intuitively assume (see Siemsen, 2010a). As John Bradley, a chemist strongly influenced by Mach and a critic of the Nuffield experimental approach in the United Kingdom, recognised in his retirement speech (1975, p. 9): "Why have teachers … including myself, failed so miserably? … We have answered the question: Where does theory begin?: wrongly. … So with good intentions, we have said to Robert: What matters is the atom, or the molecule or the equation. Poor Robert has been stranded; he resembles a child aged six given logarithms to multiply three by two or like David he is too small and weak to carry the

Even many so called ‘basic concepts’ are highly constructed. The reason for this is an often found confusion of *logical* construction of scientists (and science teachers) with *psychical* construction of students in the epistemological assumptions today. The main methodology of science is currently based on the axiomatic method, that is, everything is constructed from a few basic axioms and definitions (such as force, straight lines, numbers, chemical elements, etc.). This method is transferred into the classroom already at a very early age (in most countries after primary school). It is assumed that the definitions and axioms are so intuitive, that every student will understand them immediately. But if the definitions and axioms are already constructed and the students lack this prior construction, they will not be able to bridge this cognitive gap. Only the students who have already learnt this construction, for example, from their parents, will be able to overcome the gap.

Now one might argue that forces, straight lines, numbers, chemical elements, etc., are basic and therefore must be understandable. How could one otherwise start teaching science with the ‘instrument’ of logic? In order to see the problem in the first place, one needs to shift the view from the *logical/axiomatic perspective* to the *genetic/phenomenal perspective*. The concept of force, for instance, is not understandable without the concept of interaction (see, for instance, Kurki-Suonio and Kurki-Suonio, 1994, or Stein, 1990). According to the conventions of cultures other than the western scientific culture, and even according to modern physics (see Wilczek, 2004), this concept might not exist (it is at best superfluous or even inconsistent with the more basic concept of energy). Nevertheless, it is currently taught in all science lessons at some time. Similarly, the concept of ‘straight lines’ is not understandable without the volume comparison of thin bodies, such as tightened strings (see Mach, 1905). A light ray–often used as the example for a straight line–is already relatively abstract as a concept. Even physically, the light ray is not a straight line, at least according to the theory of relativity (if there is some mass nearby, a problem from which scientists or teachers tend to abstract without explicitly noticing). Also, the arithmetic concept of generally transferrable numbers is the result of several syntheses of concepts, such as equality versus the identity of a number (see Mach, 1905, or Wertheimer, 1912). For instance, even for adults, it is not intuitive to understand that

armour of Mendeleeff and Cannizzaro [both eminent chemists whose concepts Bradley had earlier suggested be taught to students]. I am convinced that almost all of us have answered the question wrongly. Where has been our mistake? We have forgotten that all thought is theory, and that classification is thought and therefore also theory.” Who has started to teach science without a priori classification? For instance, any Socratic or axiomatic approach tacitly takes for granted an existing world view. The step from phenomena to axioms looks seemingly small for a scientist, but opens a yawning chasm for many students. As Mach (1890, p. 4) observed, “… This view seems to be shared little in the circles of teachers, and even those, who agree with them theoretically, in practice abdicate from them again and again, which manifests itself in the overestimation of the *logical* and a disregard of the *psychological* moment in education. [In education,] criticism cannot begin where empirical meanings [konkrete Vorstellungen] are still lacking.”

100+1equals 1+100. If we (as persons in general, not as mathematicians) have a hundred of something and we get one more, it does not make much of a difference in most cases. If we have one of something and we gain a hundred extra, it clearly does make a difference. Mathematically (and metrically) there is of course no such difference.

Children or persons without much experience of scientific concepts do not understand these constructs so easily and they cannot be assumed to think logically in order to construct them. They might adapt their thoughts so that after years of training in the science classroom, the result might look like a *logical* process, but it fundamentally remains a *psychical* process. Einstein (as an "educated scientist", in Hadamard, 1945, p. 142) described the connection between the "psychical entities which seem to serve as elements in thought" and "relevant logical concepts" as no more than a scientific "desire":

> It is … clear that the desire to arrive finally at logically connected concepts is the emotional basis of this rather vague play with the above mentioned elements. But taken from a psychological viewpoint, this combinatory play seems to be the essential feature in productive thought– before there is any connection with logical construction in words or any other kinds of signs which can be communicated to others.

As Einstein meant this for the educated scientist, it must apply even more so for children not yet educated in western scientific thinking. Can we, for instance, assume schoolchildren to have already developed a "desire" for science? And can we assume that they have already understood the role of instruments in learning science? The question more likely is whether scientists understand the role of instruments in the first place. And even if they understood it, can it be communicated in a 'physical language' in a way which will enable students to understand it?

AN ANTHROPOLOGICAL EXAMPLE OF THE ROLE OF AN INSTRUMENT IN A GESTALT SWITCH

Mach gives an anthropological example of the use of an instrument for deciding among alternative descriptions. This type of example is, of course, equally applicable for deciding between alternative descriptions by students. Chuar is an American Indian chief who, together with a cowboy, tries to throw stones over a chasm. Both of them fail, although the chasm does not seem too wide. Chuar interprets the phenomenon as a physical 'pull' of the depth of the chasm. He likens it to the 'attraction' of depths one feels when, for instance, one climbs a tree. Seen from physical language, Chuar thus interprets a *psychical* phenomenon as a *physical* phenomenon (Mach, 1905, p. 120). Mach suggested that an experiment should resolve the issue. For instance, Chuar would probably agree to an experiment where a scale containing two stones equal in weight on each side is moved halfway over a chasm. If the scale would not tilt as a result of the experiment, Chuar's interpretation would not be suitable for describing this fact. The resulting intellectual tension should lead to an adaptation of his thoughts in order to find a more general description (which could but does not have to be the western scientific one).[9]

Mach adds a similar example from his youth: he was carrying an empty flower pot when his father asked him if there was anything in it. Mach said he thought not, so his father asked him to take the flower pot to a sink with water and put it upside-down into the water. When he turned the pot under the water, bubbles would appear instead of "nothing".[10] As we can see from this example, the Chuar case also applies for conceptual gestalt changes for children.

Interestingly, Piaget criticised Mach for taking chief Chuar seriously. What for Mach were two concepts, from which the one closer to the description of phenomena could be found by appropriate experiments (that is, by putting a scale in equilibrium half-way over the chasm), Piaget saw as an absurd example. "One can ask oneself whether the answers just analyzed are really primary and the first stage of the childlike animism. We have found 5 to 6-year-old children, who had already reached later stages" (1978, p. 150). The question is if we should take the concepts of children seriously, even if they are already older than the age up to which they are supposed to have "overcome" the "earlier stages" according to Piaget. Chuar is an adult, but from a completely different culture with different conceptual frames based on different cultural conventions and constructions. Wertheimer (1912) had warned about cultural anthropomorphisms in anthropological genetic research of established scientific concepts, such as number: "It is heuristically necessary to approach those things not with the prejudice that only less perfect, vague, imprecise pre-forms of our categorial entities exist; it can happen that categorial entities are formally conceptualized differently, have different goals."

The example shows that Piaget's "stages" are not primarily age, but culture-dependent. Piaget carried out his experiments around 1920 with mostly middle class Swiss and French children coming from a similar cultural background.[11] He acknowledged this experimental bias only at a late age (Piaget, 1972). But the problem was picked up earlier by Jerome Bruner's critique on Piaget (see Bruner, et al., 1966): the omission of culture on the development (cultural genesis) of children (for details, see Siemsen 2010b; Siemsen and Siemsen 2008, 2009). If

9 One should note here that the 'resolution' does not imply the new gestalt in terms of absolute 'stages' of knowledge. Ignoring facts is in principle also a psychological way of at least temporarily resolving mental tension. But this does not relieve the teacher from the psychological question of creating more irresistible mental tension.

10 See, for instance, Mach's autobiography in Hoffmann and Laitko (1913; 1991). Mach later reflected that his father (who was a private teacher and also taught his son in his earlier years) had adapted many examples from Philo of Byzantium.

11 One can find a similar empirically induced insight from other great educators at the end of their lives, such as Alfred Binet (see Siemsen, 2010b) or Henry Edward Armstrong (ibid., 2010a). Unfortunately, they all died shortly after this insight and their successors did not follow their changed mind or, as in the case of Piaget and Binet, their later works were not (or only much later) translated to English.

one does not take historically earlier or culturally different explanatory gestalt concepts seriously, one will never be able to help people see their inconsistencies, using them to switch to broader and more consistent (scientific) concepts. If not addressed and connected to the old gestalt, newly taught concepts will remain superficial. Gestalts tend to be psychically very stable (see also Lorenz, 1959).

THE MACHIAN USE OF INSTRUMENTS IN SCIENCE TEACHING

How can this gap in science teaching be avoided in the first place? I will here provide a suggestion which is based on the epistemology of science (Erkenntnistheorie) of Ernst Mach. For Mach, scientific instruments do not exist apart from the empirical meanings and the ideas embodied in them. We adapt our thoughts to the facts (empiricism) and our thoughts to each other (theory, or metaphysics).[12] What is specific for instruments is the possibility to combine haptic/enactive empirical meanings with theoretical ideas.[13] This dual aspect defines strengths and weaknesses in their use for teaching science.

Instruments are tools. They convey no other scientific meaning than in the way they are used. But these uses are connected to the physical language of western science. For a person not trained in this language (such as children or people from cultures not yet trained in schools oriented on western culture), an instrument will be meaningless[14] unless it can be connected to already constructed meanings, that is, meanings from everyday experience. The concepts resulting from such everyday experiences might not seem very 'scientific', that is, sophisticated from our perspective, but they–also historically–formed the genetic basis of our scientific concepts (for extensive examples, see Mach, 1893a).[15] In this sense,

12 In this sense, Mach criticises metaphysics, but in the sense of reducing it to a minimum, not with the idea of abolishing it altogether. What is needed instead is its genetic analysis in order to distinguish between empirical and conventional elements in metaphysical construction/ adaptation. The conventional and speculative elements can then be submitted to regular critique instead of becoming successively unconscious and intuitive. Through this process, one can avoid what Einstein (in his obituary to Mach [1916], p. 102–03) called the "excessive authority" of such conventions.

13 George Sarton (1918) had in his early works elaborated the interrelation between teaching, science history, experiments, and instruments.

14 It is meaningless in the sense of a western trained scientist. It can have other meanings such as a theological ("magical") one, though.

15 In his last years, Mach wrote a little known book on culture and mechanics (*Kultur und Mechanik*, published 1915, not translated into English), in which he asks the genetic question of the origins of mechanics in the experiences of handcrafts. He especially draws on anthropological, biological as well as early childhood observations. Like in reconstructing evolutionary developments from fossils, genetic reconstruction needs a continuous research into the details, especially when most details are missing. The first intuition might not necessarily be final, but the question itself of the "genetic origin of our instruments" as a starting point for the foundation of a "general genetic technology" could be, at least according to Mach (1915, p. 5), a fruitful one.

sophistication is a matter of degree and not of principle.[16] For Mach, science is only a specific form of continuation of what goes on in daily life: "The adaptation of thoughts to facts [and the thoughts to each other] is the aim of all scientific research. In this, science only deliberately and consciously pursues what in daily life goes on unnoticed and of its own accord" (1896, p. 156). Both scientific thought and the thought of daily life are already constructed. "As soon as we become capable of self-observation, we find our thoughts, in large measure, already adjusted to the facts. [But] almost every new fact necessitates a new adaptation, which finds its expression in the operation known as *judgement*." For Mach, thus, judgement is an unconscious adaptive construct, not a conscious logical one (see Mach 1890; Siemsen 2010a).

As a result, instruments should intuitively connect to the empirical meanings of the every day life of the students. They should be haptically usable for all of them.[17] Thus, self-made instruments from nature or household materials are preferable not only because they are cheaper, but also because of their phenomenal value. Experiments not only work in the classroom, but also at home. Different instruments or materials can be shown to reproduce similar phenomena. In such a way, ideas can be deduced from the concrete instrument[18] and the concrete circumstances of its use in the science lesson.

Finally, the use of scientific instruments, especially for teaching purposes, is also an epistemological question regarding the scientific process. There is, for instance, a fundamental difference of *the use of instruments for demonstrating phenomena* and their use for conducting experiments. Both uses should not be mixed up. In showing phenomena, for example, one should use different instruments and materials in order to demonstrate the generality of a phenomenon and the similarities of the circumstances under which it can be observed. Phenomena pose questions and should therefore not be used for premature "framing" in the sense of narrowing the question to a pre-defined "problem".[19]

The use of instruments for experiments, on the contrary, normally varies in only one specific aspect. This aspect should be discussed in a thought experiment with the students beforehand, so that they are aware of the question posed to nature and the expected answer(s) to be gotten from the experiment (for a more detailed discussion of this, see Mach, 1905 or Kurki-Suonio, 1998 and 2006). For

16 It is also anthropomorphic in the sense that it is dependent on our cultural perspective (culturally teleological).

17 In order to be 'haptically usable' the haptical/enactive concepts (gestalts), which are genetically prior to the concept of the instrument in question, need to have been taught (see Mach, 1915).

18 The Machian idea of reducing the experience of an instrument towards its "whole" gestalt, away from its baroque hooks and screws, seems central for the idea of the "shadow play" projections of instruments by Pohl (see the paper by Roland Wittje in this volume).

19 This is close to the question posed by Wagenschein (1983) in his plea to "save the phenomena" in science education. How bad is the "state of the phenomena" that it needs saving?

instance, there can never be an experimental ‘proof’ of a hypothesis or theory. Experiments should not be misused for this purpose in the science lesson.

As one can see from this brief overview of the role of phenomena and experiments, it is central to see them as part of a psychical process of the students (see Mach, 1890). Thus, for a meaningful process, a psychical theory is required as well. Finnish science educators achieved this by a unique synthesis of Gestalt psychology with a Machian philosophy of science introduced by Kaila (see Siemsen and Siemsen, 2009). In the Czech Republic on the contrary—where Mach has also been very influential in science teaching—this aspect is missing. The science teachers there conduct many experiments, but the manner in which these are carried out has no strong epistemological consistency.[20] As a result, the teachers do not construct the necessary ‘gestalt frame’ for a successful use of instruments that would show empirically in the OECD PISA study (see Siemsen, 2009).

A Machian epistemology provides a different world view, which in turn requires applying different empirical meanings also for concepts, such as ‘scientific process’ or ‘scientific instruments’. As this epistemology tries to start its construction process from the point of view similar to that of a very young child (see Mach, 1920), it is also extremely well suited for teaching science. This has been elaborated and applied in detail by Finnish science teachers in their education system, especially by Kaila, Rolf Nevanlinna, Kurki-Suonio and their students (see Lavonen et al., 2004; Siemsen, 2010c; Siemsen and Siemsen, 2009).[21] The ‘success’ of this approach was shown empirically in the OECD PISA study.[22] With a similar epistemology, it can in principle be reproduced in other cultures (see also Siemsen, 1981).

ACKNOWLEDGEMENTS

I would like to thank Peter Heering and Roland Wittje for providing the idea for this paper, as well as my father Karl Hayo Siemsen for his invaluable help in proofreading.

20 The Machian use of physical instruments, for instance, was continued by one of Mach’s students and later physics professor Čeněk Strouhal in Prague.

21 At least part of this epistemology has also been applied to science museums by Frank Oppenheimer in his ‘Exploratorium’ and his museum culture.

22 This is by no means a claim for the absoluteness of this interpretation of the Finnish and Czech PISA results. There can be other interpretations. Nevertheless, up to now, no alternative consistent and general interpretation has been provided, which focuses on factors specific only to Finnish (or Czech) science education and its history and philosophy (see Siemsen, 2010c). Also, this interpretation is consistent regarding the PISA results and the history and philosophy of science education in a number of other countries. But this paper is not concerned with that point. The final question is the one from Wagenschein (1970) of “What remains?” (Was bleibt?) of science education after school. Unfortunately, often seemingly not much (also in the view of others asking a similar question; see, for instance, Arons, 1997). It therefore seems time for a change in the point of view.

REFERENCES

Arons, A.B. 1997. *Teaching Introductory Physics*. New York: John Wiley & Sons.
Bradley, J. 1975. 'Where Does Theory Begin?', *Education in Chemistry,* March, pp. 8–11.
Bruner, J.S., R.R. Olver, P.M. Greenfield, Hornsby, J., Kenney, H.J., Maccoby, M., Modiano, N., Mosher, F.A., Olson, D.R., Potter, M.C., Reich, L.C., McKinnon Sonstroem, A. 1966; 1967. *Studies in Cognitive Growth: A Collaboration at the Center for Cognitive Studies*. New York: John Wiley & Sons.
Blüh, O. 1967. 'Ernst Mach as Teacher and Thinker', *Physics Today*, 20 (6).
Einstein, A. 1916. 'Ernst Mach', *Physikalische Zeitschrift*, 17 (7), 1 April, pp. 101–04.
Euler, M. 2007. 'Revitalizing Ernst Mach's Popular Scientific Lectures', *Science and Education* 16 (6), pp. 603–611.
Hadamard, J. 1945; 1954. *The Psychology of Invention in the Mathematical Field*. New York: Dover.
Hoffmann, D. and W. Manthei. 1991. 'Ernst Mach als Schulphysiker, Didaktiker und Bildungstheoretiker', i D. Hoffmann and H. Laitko (eds), *Ernst Mach – Studien und Dokumente zu Leben und Werk*. Berlin: Deutscher Verlag der Wissenschaften.
Hohenester, A. 1988. 'Ernst Mach als Didaktiker, Lehrbuch- und Lehrplanverfasser', in R. Haller and F. Stadler (eds), *Ernst Mach Werk und Wirkung*. Wien: Hölder, Pichler, Tempski.
Kaila, E. 1962. *Die perzeptuellen und konzeptuellen Komponenten der Alltagserfahrung*. Helsinki: Acta Philosophica Fennica, Fasc. XIII.
———.1979. *Reality and Experience. Four Philosophical Essays*, in R.S. Cohen and G.H. von Wright (eds), Dordrecht: Reidel Publishing Company.
Kurki-Suonio, K. and R. Kurki Suonio. 1994. 'The Concept of Force in the Perceptual Approach', in H. Silfverberg and K. Seinelä (eds), *Ainedidaktiikan teorian ja käytännön kohtaaminen. Matematiikan ja luonnontieteiden opetuksen tutkimuspäivät 24*. Reports from the Department of Teacher Education in Tampere, University of Tampere A18/1994, pp. 321–34, downloaded from http://per.physics.helsinki.fi/~kurkisuo/, accessed 23 November 2009.
Kurki-Suonio, K. 1998. 'Products and Processes', farewell speech presented at Helsinki University, 29 January 1998, downloaded from http://per.physics.helsinki.fi/~kurkisuo/, accessed 23 November 2009.
———. 2006. 'Kolme luentoa käsitteenmuodostuksesta. Three lectures on Concept Formation', in Fysiikan täydennyskoulutuskurssi: *Fysiikan historia ja filosofia,* 5–9 June 2006. (Physics teachers' complementary education course: *History and Philosophy of Physics,* 5–9 June 2006, Department of Physical Sciences, University of Helsinki. The author kindly provided an English translation of the lectures.)
———. 2010. 'Principles Supporting the Perceptional Teaching of Physics: A "Practical Teaching Philosophy"', *Science and Education*, in print.
Lavonen, J., J. Jauhiainen, I. T. Koponen, and K. Kurki-Suonio. 2004. 'Effect of a Long-term In-service Training Program on Teacher's Beliefs about the Role of Experiments in Physics Education', *International Journal of Science Education*, 26 (3), pp. 309–28.
Lorenz, K. 1959; 1968. 'Gestaltwahrnehmung als Quelle wissenschaftlicher Erkenntnis', in: *Vom Weltbild des Verhaltensforschers*, München: dtv.
Mach, E. 1866. *Einleitung in die Helmholtz'sche Musiktheorie – Populär für Musiker dargestellt*. Graz: Leuschner & Lubensky.
———. 1893a; 1960. *The Science of Mechanics: A Critical and Historical Account of Its Development*. La Salle, IL: Open Court.
———.1893b; 1986. *Popular Scientific Lectures*. La Salle, IL: Open Court.
———. 1886; 1919. *Die Analyse der Empfindungen und das Verhältnis vom Physischen zum Psychischen*. Jena: Gustav Fischer.
———. 1890. 'Über das psychologische und logische Moment im Naturwissenschaftlichen Unterricht', *Zeitschrift für den physikalischen und chemischen Unterricht*, 4 (1), October, pp. 1–5.

———. 1885; 1896. *Contributions to the Analysis of the Sensations*, reprint. La Salle, IL: Open Court.
———. 1905;1926; 2002. *Erkenntnis und Irrtum: Skizzen zur Psychologie der Forschung*, Fifth Edition, Leipzig. Reprint, Düsseldorf: rePRINT Berlin.
———. 1913;1991. 'Selbstbiographie', in D. Hoffmann and H. Laitko (eds), *Ernst Mach – Studien und Dokumente zu Leben und Werk*. Berlin: Deutscher Verlag der Wissenschaften.
———. 1915. *Kultur und Mechanik*. Stuttgart: Spemann.
———. 1920. Letters to Gabriele Rabel, in G. Rabel, 'Mach und die Realität der Außenwelt', *Physikalische Zeitschrift*, XXI, pp. 433–37.
Matthews, M.R. 1990. 'Ernst Mach and Contemporary Science Education Reforms', *International Journal of Science Education*, 12 (3), pp. 317–25.
Memorandum. 1962. 'On The Mathematics Curriculum of The High School', *American Mathematical Monthly*, March.
Piaget, J. 1972; 2008. 'Intellectual Evolution from Adolescence to Adulthood', *Human Development*, 15, pp. 1–12, reprint 2008, 51, pp. 40–47.
Piaget, J. 1978. *Das Weltbild des Kindes*. Stuttgart: Klett-Cotta.
Sarton, G. 1918. 'The Teaching of the History of Science', *The Scientific Monthly*, 7 (3), September, pp. 193–211.
Siemsen, K.H. 1981. *Genetisch-adaptativ aufgebauter rechnergestützter Kleingruppenunterricht: Begründungen für einen genetischen Unterricht*. Frankfurt am Main: Peter D. Lang.
Siemsen, K.H. and H. Siemsen. 2008. 'Ideas of Ernst Mach Teaching Science', in Gh. Asachi University, Cetex, Iasi, 5th Int. Seminar on Quality Management in Higher Education *(QM 2006)*, Tulcea, Romania, June.
Siemsen, H. and Siemsen, K.H. 2009. 'Resettling the Thoughts of Ernst Mach and the Vienna Circle to Europe—The Cases of Finland and Germany', *Science and Education*, 18 (3), pp. 299–323.
Siemsen, H. 2010a. 'Intuition in the Scientific Process and the Intuitive "Error" of Science', in A.M. Columbus (ed.), *Advances in Psychology Research* 72, Hauppauge: Nova Science.
Siemsen, H. 2010b. 'Alfred Binet – Ernst Mach: Similarities, Differences and Influences', *Revue Recherches & Éducations*, 3, pp. 351–403.
Siemsen, H. 2010c. 'Ernst Mach and Genetic Epistemology in Finnish Science Education', *Science and Education*, in print.
Siemsen, H. 2011. 'Mach's Science Education, the PISA Study and Czech Science Education', in A. Mizerova (ed.), *Ernst Mach: Fyzika – Filosofie – Vzdělávání*. Brno: Masaryk University Press, in print.
Stein, H. 1990. 'From the Phenomena of Motions to the Forces of Nature: Hypothesis of Deduction?', *Philosophy of Science Association*, 2, pp. 209–22.
Thiele, J. 1963. 'Ernst Mach als Pädagoge', *Schweizerische Hochschulzeitung*, 36, pp. 219–22.
Wagenschein, M. 1970. 'Was bleibt?', in J. Fluegge (ed.), *Zur Pathologie des Unterrichts*. Bad Heilbrunn: Klinkhardt, pp. 74–91.
Wagenschein, M. 1983. 'Rettet die Phänomene', in Wagenschein: *Erinnerungen für morgen*. Weinheim: Beltz,.
Wertheimer, M. 1912; 1925. 'Über das Denken der Naturvölker, Zahlen und Zahlgebilde', in *Drei Abhandlungen zur Gestalttheorie*. Erlangen: Verlag der Philosophischen Akademie, pp. 106–63.
Wilczek, F. 2004. 'Whence the Force of F = ma?', Culture Shock, *Physics today.org*.

NOTES ON CONTRIBUTORS

Paolo Brenni is researcher of the National Research Council in Florence, Italy. He is currently President of the Scientific Instrument Commission of the IHUPS.

Mar Cuenca-Lorente is "JAE-CSIC" predoctoral funded research student at the Instituto de Historia de la Medicina y de la Ciencia "López Piñero" (Consejo Superior de Investigaciones Científicas and University of Valencia), Spain.

Pere Grapí is associate professor in the Centre d'Estudis d'Història de la Ciència, Universitat Autònoma de Barcelona, Spain.

Willem Hackmann is a former curator at the Museum for the History of Science in Oxford, and an Emeritus Reader and Emeritus Fellow of the University of Oxford, UK.

Peter Heering is Professor of Physics and Physics Didactics at the University of Flensburg, Germany.

Michelle Hoffman is a PhD candidate at the Institute for the History and Philosophy of Science and Technology at the University of Toronto, Canada.

Gianna Katsiampoura is senior researcher in History of Medieval Science at the National Hellenic Research Foundation, Athens, Greece.

Richard Kremer is Associate Professor of History at Dartmouth College, New Hampshire, USA. He also curates Dartmouth's collection of historic scientific instruments, which contains apparatus from the 1770s through the present.

Pete Langman is an independent academic and writer who has lectured at Sussex and Brunel Universities, Goldsmiths College, Queen Mary, and the Central School of Speech and Drama, all UK.

Efthymios Nicolaidis is Director of Research in the Program of History of Science at the National Hellenic Research Foundation, Athens, Greece.

Lissa Roberts is Professor of Long-Term Development of Science and Technology in the Department for the Study of Science, Technology and Society (STePS), University of Twente, the Netherlands.

Dawn Sanders is Scientific Associate of the botany department at the Natural History Museum, London, UK, and a consultant for The Charles Darwin Trust.

Hayo Siemsen is a researcher at the Ernst Mach Institute for Philosophy of Science and INK, Hochschule Emden/Leer (University of Applied Sciences), Germany.

Josep Simon is 'Marie Curie' postdoctoral fellow at the Institut de Recherches Philosophiques, Université Paris Ouest, France.

Constantine Skordoulis is Professor of Physics and Epistemology at the Department of Education, National and Kapodistrian University of Athens, Greece.

Steven Turner is curator of physical sciences at the Smithsonian Institution's National Museum of American History, Washington DC, USA.

Roland Wittje is lecturer in history of science at the University of Regensburg, Germany.